EXTREMELY LOW FREQUENCY ELECTROMAGNETIC FIELDS: THE QUESTION OF CANCER

Edited by
Bary W. Wilson,
Richard G. Stevens, and
Larry E. Anderson

 **BATTELLE PRESS**

Columbus • Richland

RC268.7
E43
E97
1990

A portion of the funding for this research was provided by the U.S. Department of Energy. However, the views and findings of the various papers are solely those of the authors and do not necessarily represent the policy of the agency. The agency encourages wide dissemination of the technical information contained herein, with due respect for the Publisher's rights regarding the complete volume.

Printed in the United States of America.

Battelle Press
505 King Avenue
Columbus Ohio 43201-2693
614-424-6393; 1-800-451-3543
FAX: 614-424-5263

Contents

PART IV. Possible Mechanisms

PART V. Possible Consequences

Contributors

W. Ross Adey (211),* Research Service, Pettis Memorial Veterans Administration Center and Loma Linda University School of Medicine, Loma Linda, California 92357

Larry E. Anderson (17, 139, 159, 361), Life Sciences Center, Pacific Northwest Laboratory, P.O. Box 999, Richland, Washington 99352

Carl L. Blackman (187), Health Effects Research Laboratory, U.S. Environmental Protection Agency, Research Triangle Park, North Carolina 27711

David E. Blask (319), Department of Anatomy, College of Medicine, University of Arizona, Tucson, Arizona 85724

Charles F. Ehret (47), Biological, Environmental, and Medical Research Division, Argonne National Laboratory, 9700 S. Cass Ave., Argonne, Illinois 60439

Kenneth R. Groh (47), Biological, Environmental, and Medical Research Division, Argonne National Laboratory, 9700 S. Cass Ave., Argonne, Illinois 60439

William P. Hammond, IV (109), Division of Hematology, Department of Medicine, University of Washington School of Medicine, Seattle, Washington 98195

---

*Numbers in parentheses indicate the pages on which the authors' contributions begin.

William T. Kaune (17), Enertech Consultants, 300 Orchard City Drive, Suite 132, Campbell, California 95008

Abraham R. Liboff (251), Department of Physics, Oakland University, Rochester, Michigan 48063

Bruce R. McLeod (251), Electrical Engineering Department, Montana State University, Bozeman, Montana 59717

Gary A. Pascoe (337), Tetra Tech, Inc., 11820 Northup Way, Suite 100, Bellevue, Washington 98005

Marijo A. Readey (47), Biological, Environmental, and Medical Research Division, Argonne National Laboratory, 9700 S. Cass Ave., Argonne, Illinois 60439

Russel J. Reiter (87), Cellular and Structural Biology, University of Texas Health Sciences Center, 7703 Floyd Curl Drive, San Antonio, Texas 78284

Stephen D. Smith (251), Department of Anatomy, University of Kentucky, Lexington, Kentucky 40506

Richard G. Stevens (9, 361), Epidemiology and Biometry, Pacific Northwest Laboratory, P.O. Box 999, Richland, Washington 99352

Thomas S. Tenforde (291), Chief Scientist, Life Sciences Center, Pacific Northwest Laboratory, P.O. Box 999, Richland, Washington 99352

Bary W. Wilson (3, 159, 361), Chemical Sciences Department, Pacific Northwest Laboratory, P.O. Box 999, Richland, Washington 99352

Preface

Electric and magnetic fields can affect biological systems in ways that are not yet fully appreciated. Of particular interest during the past decade have been the possible effects of extremely low frequency (ELF) electromagnetic fields on humans. The debate concerning possible health consequences from the increasing worldwide exposure to ELF electromagnetic fields has waxed and waned, and has often been driven by epidemiological studies in an area in which few relevant laboratory data have been available. Beneficial effects of ELF field exposure, such as enhanced bone fracture healing, have been reported and are now well documented in the literature. Public attention, however, has focused on the less well documented detrimental effects and associated uncertainties. As one result of the recent public awareness of the ELF question, the quality of epidemiological studies and laboratory research has improved. In the past few years, much insight into ELF bioeffects and their possible mechanisms has been gained.

This book is not intended, however, as a general treatment of the subject of ELF bioeffects. In particular, it does not address questions related to risk assessment or possible hazards of what we term ELF exposure. We have elected, rather, to focus on that which seems to be the central scientific issue emerging from current ELF research in epidemiology and in the laboratory; namely, can ELF electromagnetic fields interact with biological systems in such a way as to increase cancer risk? We examine how cancer risk might be related to two reproducible biological effects of ELF exposure: effects on the pineal gland and circadian biology, and effects on calcium homeostasis in cells. Because we are concerned with the possible biological mechanisms of carcinogenesis, epidemiological studies are only briefly reviewed.

To accommodate a broad readership, introductory chapters on the various disciplines relevant to later sections follow the background material given in Part I. Part II contains introductions to the physics and measurement of ELF fields, chronobiology, pineal gland and neuroendocrine function, and the mechanisms of carcinogenesis. In Part III are presented observed ELF effects, and in Part IV, the possible mechanisms involved. Part V discusses possible consequences of the ELF fields, including carcinogenesis. This book is intended to be provocative and useful to the professional investigator as well as to nonprofessionals interested in ELF bioeffects.

The editors of this volume wish to acknowledge the generous support of the U.S. Department of Energy, Office of Energy Storage and Distribution, and, in part, the Office of Health and Environmental Research (to R.G.S.), during the preparation of the manuscripts for book production. We also wish to acknowledge the uncompromising technical assistance of Susan A. Kreml in the production of this book.

PART I.
Introduction

1 The Moscow Signal

BARY W. WILSON
Pacific Northwest Laboratory
Richland, Washington

CONTENTS

During the past century, dramatic changes have occurred in the human habitat. One of the most significant is the development and use of electric power. Electricity has formed a base for modern technology, and its benefits as a keystone of industrial societies can scarcely be quantified. Its increased generation and use, however, have caused radical changes in human exposure to virtually the entire electromagnetic spectrum, including non-ionizing radiation such as microwaves, radio-frequency signals (RF), and the extremely low frequency (ELF) electric and magnetic fields associated with electric-power transmission.

THE MOSCOW SIGNAL

In 1966 two Russian scientists, T. P. Asanova and A. I. Rakov, reported neurological symptoms including headache, excitability, fatigue, and diminished libido in workers at a high-voltage power switchyard in the U.S.S.R. Western scientists criticized the lack of proper controls for the study, and the work of Asanova and Rakov received little attention in the

West until nearly a decade later. Serious U.S. scientific interest in possible health effects from ELF electromagnetic fields resulted instead, to a large extent, from work on two national security-related issues in the 1970s.

The first effort derived from congressional hearings on a large naval communications facility known first as Sanguine, later as Seafarer, and most recently as Project ELF. This facility was intended for use as an extensive antenna system, transmitting at low frequencies and high power, to provide communications with submarines at sea. The U.S. Department of the Navy established a committee to study possible health effects of the signal emanating from the Seafarer system. A panel of seven scientists who drafted recommendations for further Seafarer study recognized that high-tension power-distribution lines could create electrical fields orders of magnitude greater than those that would be associated with the Navy project. This opinion was of interest because results of preliminary studies that were available to the panel appeared to show adverse health effects of nonionizing electromagnetic radiation (NIEMR) exposure on small animals. The panel's deliberations recorded that NIEMR exposure appeared to "lower the body's natural resistance" and give rise to "induced stress."

The second issue was considerably more bizarre. At about the time that Asanova and Rakov were completing their studies, the U.S. State Department became aware of inordinately high microwave flux densities in the U.S. Embassy building in Moscow. A secret project, "Pandora," was set up to determine what, if anything, the Russians were trying to accomplish by beaming this "Moscow Signal" at the embassy. At that time, it was believed by many Pandora members that the Russian objective was to adversely affect the performance levels of embassy staff. Some evidence existing at that time suggested, for example, that nonthermal levels of microwave radiation could contribute to formation of eye cataracts. Part of Pandora's objective was to determine if there might be other nonthermal health effects of microwave radiation. In 1987, *Time* magazine reported that the microwaves may have been used to gather intelligence from inside the embassy.

From these early efforts to examine NIEMR exposures for possible adverse health consequences have come a number of important research efforts in several countries. These programs, spanning the frequency range from radio frequencies, including microwaves, to extremely low frequencies (ELF), have been expanded in scope and depth to pose questions beyond the initial claims that NIEMR was hazardous to human health. Thus, although it has long been recognized that microwave power densities sufficient to cause tissue warming could cause biological damage, only more recently has it been acknowledged that nonthermal power densities may also contribute to biological effects. Much of the recent scientific effort has been directed toward understanding biological effects of ELF electric and magnetic fields. Significant progress has been achieved in both defining the ways in which living organisms interact with ELF fields and in describing biological effects, both real and potential, from such fields.

Although it is now clear that ELF fields do cause biological effects, the basis for those effects, and the underlying mechanisms of interaction between the fields and living organisms, are less well understood. There is some disagreement among scientists as to their relevance to human health concerns. In the ELF range, vigorous research is now being sponsored by the U.S. Department of Energy and the Electric Power Research Institute as well as a few states in the United States. Active research programs are also to be found in Canada, Sweden, England, Italy, and the U.S.S.R., and smaller programs are being developed in various other countries.

THE QUESTION OF ELF ELECTRIC- AND MAGNETIC-FIELD EFFECTS ON BIOLOGICAL SYSTEMS

Perception of the earth's magnetic field is important for some animals in receiving temporal and spatial information. Pigeons and rats, for example, can detect changes in the angle of incidence for artificial earth-strength magnetic fields. Changes in magnetic-field strength can suppress the normal nocturnal increase in rat pineal melatonin levels. That the earth's magnetic field, part of the environment of biological systems throughout their evolution, is used by certain species in their adaptive strategies is not surprising. Sharks and rays, for example, use magnetic-field information for navigation. Environmental ELF electrical fields, on the other hand, are a relatively new phenomenon, yet interactions also occur between these fields and biological systems. Many of these do not depend on the ability to consciously sense the field. People live and work in ambient fields that normally cannot be consciously perceived.

The ways in which exposure to ELF electric and magnetic fields may affect biological systems are not obvious. Ionizing radiation can interact with neutral molecules to form chemically reactive radical or ionic species; however, ELF radiation transfers energy to tissues at a level lower than is already present in the form of thermal energy. ELF electromagnetic fields, nonetheless, appear to interact with tissue, and in particular with neural tissue in some whole-animal and cellular systems.

In reviewing self-reported human symptomatology related to ELF exposure, it is clear that complaints are most often associated with subjective-state effects and often involve feelings of fatigue and general malaise, including headache. Data from laboratory and controlled field studies also suggest that strong ELF electromagnetic fields may affect normal physiological function in humans. For example, transient changes in evoked response measures, in the ability to distinguish flickering lights in low-light conditions, and in cardiac interbeat interval have been reported as a result of the ELF electromagnetic field. Subtle endocrine changes in humans have also been reported. Many of these effects may arise from interaction of ELF fields with the neuronal system. Much of the information presented in this book suggests that the nervous system may be an important locus of effects of ELF fields.

OVERVIEW OF ELF-INDUCED BIOLOGICAL EFFECTS

Review of the literature on biological effects of ELF electric and magnetic fields shows that now, in contrast to the 1970s, there is a substantial body of data on the subject. For the purpose of this introduction, published work has been separated into three categories:

1. Areas that have been investigated wherein no effects have been found;
2. Areas wherein preliminary work has indicated that effects may exist. However, these effects are not yet considered (by this author) to be established consequences of ELF field exposure;
3. Areas wherein investigators in the field generally agree that reproducible effects have been found and confirmed.

The areas showing no effects may be summarized as follows. There is no evidence that ELF electric and magnetic fields can be directly or acutely lethal. Studies on several species of animals have not shown any reproducible gross changes in the structure of organs, muscle tissue, or skeletal characteristics. Well-executed developmental studies on animals have also failed to show any exceptional effects of ELF electric- or magnetic-field exposure. There have been no reproducible demonstrated negative effects on reproduction in animal studies using ELF electric fields. However, magnetic-field studies on reproduction are ongoing.

Some laboratory evidence from preliminary work suggests that genetic replication in rapidly dividing cells may be affected by ELF electric fields having specific waveforms. Work in other laboratories using standard in vitro assays, however, has failed to show any effects on gene structure or function as a result of exposure to ELF electric and magnetic fields. Standard in vivo tests to assess immune function have shown no compromise of the animal's ability to respond to several types of antigens. Results from in vitro tests are more difficult to assess, and appear contradictory on the issue of effects on immune function. Thus, immune function and genetic effects are two areas that require additional work to determine whether effects can be demonstrated.

There is general agreement that certain types of ELF fields can affect calcium ion transport across cell membranes in vitro, and that these fields may affect the structure and function of the membranes themselves. There is general agreement that reproducible effects can be demonstrated in certain aspects of behavior and nervous system function. Effects on normal circadian rhythms, as determined by oxygen consumption, motor activity, core body temperature, neurotransmitter synthesis, and pineal gland function, have been reported by several laboratories. Changes in corticosterone, testosterone, and melatonin secretion have also been shown to result from ELF electric-field exposure.

ELF electric- and magnetic-field effects in the areas of nervous and endocrine system function and of circadian rhythms, and ways in which these changes might be linked to cancer, are important topics of the chapters that follow. We focus on these effects because they can be demonstrated in whole-animal mammalian systems, including, to some extent, human beings. Whether or not ELF exposure increases cancer risk remains to be determined. It is the purpose of this book to bring together the work of researchers in fields related to the question. In so doing, we suggest possible mechanisms that may be tested experimentally.

THE BIOELECTROMAGNETIC DISCIPLINE

In the 1990s, public attention will again be focused, as it was in the 1970s, on the question of ELF electric- and magnetic-field health effects. Researchers in the field welcome the renewed interest of the public, the electric power industry, and the government. It must be recognized, however, that there is much fundamental understanding that research in this field has yet to acquire. It is now clear that ELF fields can interact with biological systems to produce certain effects, and that there are statistically significant and reproducible results which warrant continued investigations into the basic mechanisms of ELF effects interactions with living organisms.

Some support for bioelectromagnetics research will necessarily be tied to the regulatory aspects of health and safety, but it is basic scientific research that will eventually provide the foundation for understanding effects arising from ELF field exposure. Whether or not these physiological changes are clinically significant will only be determined after substantially more investigation has been completed.

Bioelectromagnetics research has undergone significant changes since the 1970s. There are now two or more active scientific societies that support journals devoted to bioelectromagnetics, and investigators from disciplines outside biology have entered the field. This influx of multidisciplinary expertise has led to development of several testable hypotheses regarding the mechanisms of action and possible consequences of organismic response to ELF exposure. Some of these hypotheses are presented and discussed in this book.

It would be an oversimplification to state that events in the U.S.S.R. in the 1960s were responsible for subsequent U.S. research into ELF field health effects. Nonetheless, the importance of the Russian reports on ELF exposure and the presence of "The Moscow Signal" in bringing this issue to the attention of the U.S. government and government funding agencies is now a matter of historical record.

2 Overview: ELF and Carcinogenesis

RICHARD G. STEVENS
Pacific Northwest Laboratory
Richland, Washington

CONTENTS

In this overview, background material and the motivation for interest in extremely low frequency (ELF) electric and magnetic fields as they relate to increased cancer risk are presented.

Our human environment has changed greatly during the past several hundred years. Many of these changes have been caused by our own human activities, and some have been detrimental. Infectious diseases, the scourges of earlier human populations, have been overshadowed by the greater problems of chronic diseases in our current industrialized societies. Industrialization has changed our transportation, our shelter, our work, our leisure, and the quantity and quality of our food – and our electromagnetic environment. It is the purpose of this book to examine one possible consequence of the introduction and increasing use of electric power as societies become industrialized: cancer.

We focus on two biological effects of ELF fields that have been identified and replicated in the laboratory: (1) ELF electromagnetic impact on pineal gland production of the hormone melatonin in whole animals; and (2) calcium homeostasis in cellular systems. Very few reliable data for human systems bear directly on the question: "Can ELF field exposure increase cancer risk?" The available evidence is reviewed by Tenforde in Chapter 12. For the most part, the book concentrates on characterization and engineering of ELF fields, experiments on pineal melatonin production and calcium homeostasis, and membrane effects as well as investigations outside the bioelectromagnetics community on the relationship between these biological phenomena and cell injury, DNA damage, and cancer.

PINEAL MELATONIN

Wilson et al. (1981, 1983) showed that exposure to a 60-Hz electric field could reduce the normal nocturnal rise in pineal melatonin content in the rat. Welker et al. (1983) showed that magnetic-field exposure can have the same effect. Melatonin has been implicated in the etiology of breast cancer induction (Cohen et al., 1978; Tamarkin et al., 1981; Blask and Hill, 1986). On this basis, Stevens (1987, 1988) has suggested that electric power may account, in part, for the large geographic variation in human breast cancer risk.

CALCIUM

There is evidence that nonionizing electromagnetic radiation can affect calcium homeostasis at certain combinations of frequency and intensity (Bawin and Adey, 1976; Blackman et al., 1985; Smith et al., 1987). Calcium homeostasis is important in maintenance of cellular reducing agents, in particular, glutathione (Reed and Fariss, 1984). Smith et al. (1981) found that incubation of freshly isolated hepatocytes in calcium-free media greatly increased the toxicity of carbon tetrachloride, bromobenzene, and ethyl methane sulfonate. Reed and Fariss (1984) speculated that calcium efflux leads to depletion of glutathione and consequently to greater susceptibility to oxidative damage by the hepatotoxins. This suggests the intriguing possibility that nonionizing radiation may, under certain circumstances, act as a radiosensitizer, and that exposure to nonionizing radiation might increase toxicity of carbon tetrachloride to hepatocytes in vitro. So-called window effects of nonionizing radiation on fibroblast protein synthesis (McLeod et al., 1987) and photochemical reactions (Okazaki and Shiga, 1986) have been reported. Mouse embryo fibroblast transformation by x rays and phorbol ester has been affected by a 2.45-GHz microwave (Balcer-Kubiczek and Harrison, 1985).

One of the important late effects of radiation is on the bone marrow (Fliedner et al., 1986). Leukemia, in particular acute nonlymphocytic leukemia, is strongly related to ionizing radiation exposure (Finch, 1984; Miller and Beebe, 1986). Because nonlymphocytes such as neutrophils and macrophages generate oxygen radicals during normal functioning (Johnston et al., 1975; McCord and Fridovich, 1978), nonlymphocytes may rely more heavily on cellular reducing agents and may have greater radiation sensitivity than lymphocytes, and thus be more sensitive to changes in calcium balance. The evidence is not yet adequate to determine whether electromagnetic fields influence risk of leukemia, but this has been suggested from occupational studies (Savitz and Calle, 1987), and again nonlymphocytic leukemia has been cited as the cancer most strongly associated with exposure in adults. Age at exposure to either ionizing or non-ionizing radiation would be expected to be important because the baseline risk of lymphocytic and nonlymphocytic leukemias changes greatly with age (Stevens, 1986).

STRUCTURE OF THE BOOK

We have developed this book in five sections. The Introduction is composed of Wilson's chapter on historical perspective and this overview. The Background section includes four chapters important to the interpretation of later chapters. Kaune and Anderson discuss the practical aspects of ELF field levels in various environments and the methods and techniques used to measure these levels. Groh and colleagues discuss observations in their laboratory on the effects of ELF fields on circadian biology. Specific biological mechanisms are not directly addressed, but the importance of circadian biology to the ELF story, a recurring theme of the book, ties directly into following presentations on the pineal gland. Reiter's primer on pineal gland physiology, with particular emphasis on pineal response to acute and chronic stress, is a starting point for understanding the ELF–pineal gland connection. Hammond then discusses the role of oncogenes in leukemia and the significance of calcium to oncogenesis: leukemia can be considered as a model for cancer in general.

The section on Observed ELF Effects includes three chapters describing ELF research that demonstrates the effect of ELF fields on two biological processes. Anderson describes what is known about the interaction of ELF fields with the neural and neuroendocrine systems, and Wilson and Anderson describe experiments specific to the effect of ELF fields on pineal gland function. Blackman then presents his series of experiments on the effect of ELF fields on calcium homeostasis.

The Possible Mechanisms section contains three chapters. Adey discusses his work, and that of others, that is directed at understanding at the most fundamental level how ELF fields could interact with biological systems: the primary thrust is interactions with cell surfaces. Liboff and colleagues discuss a theory to explain the "window" effects seen in many calcium experiments – why certain frequency-intensity combinations demonstrate an effect on calcium homeostasis whereas others do not. Tenforde gives an expanded review of the theories of interaction of ELF fields with biological systems to complete the section.

The final section, Possible Consequences, contains three chapters. Blask presents data on melatonin and cancer from his own and others' work, and Pascoe describes the role of calcium in oxidative stress in cells. In the final chapter, Stevens, Wilson, and Anderson connect the themes of these chapters and suggest avenues for future research.

REFERENCES

Balcer-Kubiczek, E. K., Harrison, G. H. 1985. Evidence for microwave carcinogenesis in vitro. Carcinogenesis 6:859–864.

Bawin, S. M., Adey, W. R. 1976. Sensitivity of calcium binding in cerebral tissue to weak environmental electric fields oscillating at low frequency. Proc. Natl. Acad. Sci. USA 73:1999–2003.

Blackman, D. F., Benane, S. G., House, D. E., et al. 1985. Effects of ELF (1–20 Hz) and modulated (50 Hz) RF fields on the efflux of calcium ions from brain tissue in vitro. Bioelectromagnetics 6:1–11.

Blask, D., Hill, S. 1986. Effects of melatonin on cancer: studies on MCF-7 human breast cancer cells in culture. J. Neural Transm. (Suppl.) 21:443–449.

Cohen, M., Lippman, M., Chabner, B. 1978. Role of pineal gland in aetiology of breast cancer. Lancet 2:814–816.

Finch, S. C. 1984. Leukemia and lymphoma in atomic bomb survivors. In: Radiation Carcinogeneis. Progress in Cancer Research and Therapy, Vol. 26, Boice, J. D., Fraumeni, J. F., eds., pp. 37–44. New York: Raven Press.

Fliedner, T. M., Nothdurft, W., Calvo, W. 1986. The development of radiation late effects to the bone marrow after single and chronic exposure. Int. J. Radiat. Biol. 49:35–46.

Johnston, R. B., Keele, B. B., Misra, H. P., et al. 1975. The role of superoxide anion generation in phagocytic bactericidal activity. J. Clin. Invest. 55:1357–1372.

McCord, J. M., Fridovich, I. 1978. The biology and pathology of oxygen radicals. Ann. Intern. Med. 89:122–127.

McLeod, K. J., Lee, R. C., Ehrlich, H. P. 1987. Frequency dependence of electric field modulation of fibroblast protein synthesis. Science 236:1465–1469.

Miller, R. W., Beebe, G. W. 1986. Leukemia, lymphoma, and multiple myeloma. In: Radiation Carcinogensis, Upton, A. C., Albert, R. E., Burns, F. J., Shore, R. E., eds., pp. 245–260. New York: Elsevier.

Okazaki, M., Shiga, T. 1986. Product yield of magnetic-field-dependent photochemical reaction modulated by electron spin resonance. Nature (London) 323: 240–243.

Reed, D. J., Fariss, M. W. 1984. Glutathione depletion and susceptibility. Pharmacol. Rev. 36:25S–33S.

Savitz, D. A., Calle, E. E. 1987. Leukemia and occupational exposure to electromagnetic fields: review of epidemiologic surveys. J. Occup. Med. 29:47–51.

Smith, M. T., Thor, H., Orrenius, S. 1981. Toxic injury to isolated hepatocytes is not dependent on extracellular calcium. Science 213:1257–1259.

Smith, S. D., McLeod, B. R., Liboff, A. R., et al. 1987. Calcium cyclotron resonance and diatom mobility. Bioelectromagnetics 8:215–227.

Stevens, R. G. 1986. Age and risk of acute leukemia. *J. Natl. Cancer Inst.* 76:845–848.

Stevens, R. G. 1987. Electric power and breast cancer: a hypothesis. *Am. J. Epidemiol.* 125:556–561.

Stevens, R. G. 1988. Electric power, melatonin, and breast cancer. In: *The Pineal Gland and Cancer*, Gupta, D., Attanasio, A., Reiter, R. J., eds. London: Brain Research Promotion.

Tamarkin, L., Cohen, M., Roselle, D., et al. 1981. Melatonin inhibition and pinealectomy enhancement of 7,12-dimethylbenz(a)anthracene-induced mammary tumors in the rat. *Cancer Res.* 41:4432–4436.

Welker, H. A., Semm, P., Willig, R. P., et al. 1983. Effects of an artificial magnetic field on serotonin *N*-acetyltransferase activity and melatonin content of the rat pineal gland. *Exp. Brain Res.* 50:426–432.

Wilson, B. W., Anderson, L. A., Hilton, D. I., et al. 1981. Chronic exposure to 60-Hz electric fields: effects on pineal function in the rat. *Bioelectromagnetics* 2:371–380.

Wilson, B. W., Anderson, L. A., Hilton, D. I., et al. 1983. Addendum. *Bioelectromagnetics* 4:29.

PART II.
Background

3 Physical Aspects of ELF Electric and Magnetic Fields: Measurements and Dosimetry

WILLIAM T. KAUNE
Enertech Consultants
Campbell, California

LARRY E. ANDERSON
Pacific Northwest Laboratory
Richland, Washington

CONTENTS

Levels of electromagnetic fields present in the environment have increased greatly in the twentieth century. The pervasiveness of fields in the extremely low frequency (ELF) range results, in large measure, from the rapid growth in generation, transmission, and distribution of electric power and use of this power in most of the machines and appliances found in modern industrialized society. There is no doubt that the use of electric power is one of the cornerstones of the standard of living enjoyed in developed countries. It is, however, appropriate to inquire as to whether there are adverse side effects from interactions between the electromagnetic environment associated with electric-power use and living organisms, including man. In the past three decades, research programs throughout the world have made significant progress in defining the physical interaction between electric and magnetic fields and living organisms.

The purpose of this chapter is to discuss, from a physical point of view, interactions between biological systems and electric and/or magnetic fields characterized by ELF frequencies. The major sources of electric and magnetic fields in the ELF range are the 50- or 60-Hz fields generated in electric-power systems throughout the world. Other ELF sources that have received significant attention are military communication and navigation systems. Almost all the heretofore-published dosimetric studies have been based on 60-Hz research, and data reviewed in this chapter focus on 60 Hz. Methods are presented, however, to extrapolate these data to other frequencies.

A number of exposure mechanisms result in the application to the body surfaces of humans and animals, and the induction inside these bodies, of electric and/or magnetic fields. The more important of these include:

1. Electric fields in the air act on the surfaces of, and induce electric fields and currents inside of, the bodies of humans or animals.
2. Magnetic fields in the air act directly on moving charged particles present in the bodies of humans and animals, and also induce electric fields and currents inside these bodies.
3. Electric currents induced in a conducting object (e.g., an automobile) exposed to an electric field can pass through a human or animal in contact with the field.
4. Magnetic-field coupling to a fenceline or to other long conductors can, in certain circumstances, cause currents to pass through a human or animal who touches the conductor.
5. An animal or human who is standing on earth that is carrying electric currents may be exposed to a "step potential" that will cause currents to flow in the body.
6. While primarily a higher frequency phenomenon, transient (often called spark) discharges that occur when two bodies exposed to an electric field come into contact can be a significant annoyance near strong field sources.

Of these six coupling mechanisms, currents produced by the latter four can be characterized as usually of short duration and potentially of sufficient magnitude to be perceptible, perhaps annoying, and possibly even dangerous to humans and animals. For these reasons, they have, by and large, been studied for decades by organizations concerned with electrical safety. By comparison, currents resulting from the first two mechanisms listed are almost always too small to be perceptible. However, because these currents are chronically present in the bodies of almost all humans and animals living in developed countries, questions as to possible long-term effects on individuals have been raised. This chapter addresses the physical basis for these concerns and concentrates on the inductive coupling of humans and animals to electric and magnetic fields.

The following sections define electric and magnetic fields, identify the units in which their magnitudes are generally expressed, and discuss physical and biophysical aspects of the coupling between these fields and living organisms.

ELECTRIC AND MAGNETIC FIELDS

The Electric Field

Any system of electric charges produces an electric field, E, at all points in space. E is a vector quantity, which means that it is characterized by both a direction and magnitude. The fundamental significance of the electric field is that any other electric charge placed in it will experience a force, F, which is related to E by the expression

$$F = qE \qquad [1]$$

where q is the size of the charge placed in the field. Note that the force on a positive charge (e.g., a proton) is in the same direction as E while the force on a negative charge (e.g., an electron) is in the opposite direction.

Except for certain areas of microscopic science, electric- and magnetic-field quantities are expressed in the International System of Units (SI). In this system, the unit of electric charge is the coulomb (C). The smallest quantity of electric charge that has been observed in nature is that of an electron (-1.6×10^{-19} C) or a proton ($+1.6 \times 10^{-19}$ C). All larger quantities of charge are thought to be integral multiples of the electronic charge.

In the SI system, the units of electric-field strength are volts per meter (V/m). It is generally much easier to measure the electric potential, V, than the electric field, because the potential (which has units of volts) is much less dependent on the physical geometry of a given system (e.g., locations and sizes of conductors). The electric potential is often defined in such a way that the earth's potential is zero.

In many systems, the relationship between V and E can be easily determined. The simplest example, and one of considerable practical importance, is two parallel, conducting plates that are initially uncharged. Suppose that a charge, q, is removed from one of these plates and is placed on the other. (This transfer of charge could be accomplished using a battery or a transformer connected between the plates.) A potential, V, and an electric field, E, are thus established between the two plates. Except near the edges, the electric field between parallel plates is uniform and is perpendicular to their surfaces. In this region, E and V are related by

$$E = V/h \qquad [2]$$

where h is the separation distance between the plates.

As shown by Eq. 1, electric fields exert forces on charged particles. In an electrically conductive material, such as living tissue, these forces will set charges into motion to form an electric current, which is measured in amperes (A). The distribution of current in a three-dimensional volume is frequently specified using the current-density vector, J, whose direction is the direction of current flow at a particular point and whose magnitude is equal to dI/dA, where dI is the current crossing a very small surface element of area, dA, oriented perpendicular to J. The units of current density are A/m^2. J is directly proportional to E in a wide variety of materials, that is,

$$J = \sigma E \qquad [3]$$

where the constant of proportionality, σ, is called the electrical conductivity of the medium. The units of σ are siemens per meter (S/m). Conductivities of living tissues, as measured by several groups, lie in the approximate range 0.01 to 1.5 S/m (Schwan and Kay, 1956, 1957; Geddes and Baker, 1967; Schwan, 1977a).

Electric-Field Sources

Experiments have shown that a vertical electric field is present in the lower portion of the earth's atmosphere. The source of this field is positive charge carried from the ground to the upper atmosphere by thunderstorm activity. The mean strength of the ground-level atmospheric electric field is about 130 V/m and its direction is vertically downward (Israel, 1973), but over time this field is highly variable. Ground-level field strengths in excess of 100 kV/m have been observed on flat, unobstructed surfaces during thunderstorms (Israel, 1973; Toland and Vonnegut, 1977).

Electric fields with frequencies from about 30 Hz to 100 kHz have predominantly manmade sources. The strongest of these fields to which humans are normally exposed, with the exception of a few occupational settings, are those produced by electric-power generation, transmission, and distribution systems. Considerable data have been published on measurements of the electric fields under high-voltage transmission lines (EPRI, 1975; Bracken, 1976; Bridges and Preache, 1981). Theoretical work has shown that the electric fields produced by these sources can be calculated quite accurately (Schneider et al., 1974; Deuse and Pirotte, 1976; Deno and Zaffanella, 1978).

Table I gives field intensities for practical powerline configurations (Hauf, 1982). These data show that the largest electric fields produced at ground level by transmission power lines now in service are about 15 kV/m. Electric fields under even higher voltage power lines that may be built in the future will probably not significantly exceed this value because of the need to limit shock hazards to personnel in the vicinity of the lines. However, as line voltages are increased, the widths of land on either side of transmission lines that are exposed to fields larger than, for example, 1 kV/m, are increased. The ground-level electric fields found in substations do not significantly exceed the values found in Table I.

Electric-field sources typically consist of more than one charged electrode. For example, alternating-current (AC) transmission lines have three phases wherein each phase consists of a single conductor or, at higher voltages, a bundle of several closely spaced conductors. The voltages of the three phases are approximately equal in magnitude but are separated from each other in electrical phase by a nominal 120°. The electric fields produced by these lines are not in phase, which means that the three spatial components of the electric-field vector produced by such lines will, in general, have different phases.

It can be shown (Deno and Zaffanella, 1978) that the tip of the electric-

TABLE I.
Maximum Electric-Field Strengths at Midspan Under Various Configurations and Voltages of Electric-Power Transmission Lines

Highest system voltage (kV)	Electric-field strength under line at midspan (kV/m)
123	1– 2
245	2– 3
420	5– 6
800	10–12
1200	15–17

field vector produced by a polyphasic sinusoidal source traces, as a function of time, a perfect ellipse. In the plane of this ellipse, the electric field can be described by two orthogonal vector components. Two limiting cases are interesting because they have been used in biological investigations. In one, the field ellipse collapses to a line. This case, referred to as linear polarization, is the one that has been simulated most frequently in laboratory research. The second limiting case occurs when the two independent electric-field components are equal in magnitude but are 90° out of phase. Here, the field ellipse becomes a circle and the field is said to be circularly polarized. (The less precise term "rotating field" is sometimes used to describe this condition.) Electric fields near transmission lines are approximately linearly polarized for distances from the line's phase conductors greater than about 15 m (Deno and Zaffanella, 1978).

Work underway at present in the United States, Canada, the United Kingdom, Sweden, Japan, and possibly other countries has the goal of determining actual exposures to electric fields accumulated by humans working and living in the vicinities of transmission lines and other electric-power facilities (Lovstrand et al., 1979; Looms, 1983; Deno and Silva, 1984; Bracken, 1985; Chartier et al., 1985; Silva et al., 1985). This work has demonstrated that exposure estimates can significantly overstate actual exposure when exposure levels are based on unperturbed field values (i.e., fields measured with no humans present) and on simple estimates of a person's location as a function of time.

The Magnetic Field

Magnetic fields, like electric fields, are produced by electric charge, but only electric charge in physical motion. Magnetic fields exert forces on other charges, but again, only moving charges. Because the most common manifestation of electric charge in motion is an electric current, it is often said that magnetic fields are produced by electric currents and interact with other electric currents.

The vector Lorentz force, F, acting on an electric charge, q, moving with a velocity, v, in a magnetic field, B, is given by the expression

$$F = q\mathbf{v} \times B \qquad [4]$$

where $\mathbf{v} \times B$ is a vector whose magnitude is equal to $vB\sin r$ (r is the acute angle between $\mathbf{v}$ and B) and whose direction is perpendicular to both $\mathbf{v}$ and B in the sense of a right-hand screw (i.e., $\mathbf{v} \times B$ is the vector cross product of $\mathbf{v}$ and B). It follows directly from Eq. 4 that a magnetic field cannot, itself, impart any energy to a charge moving through it because the Lorentz force is always perpendicular to the direction of motion. However, magnetic fields can facilitate the transformation of one form of energy into another, as for example in an electric generator where mechanical energy is transformed into electrical energy. Time-varying magnetic fields are able to induce electric fields and, in conductive media, "eddy" currents. The physical law that governs this phenomenon is called Faraday's law. In integral form, this law is

$$\xi = -\frac{d\phi}{dt} \qquad [5]$$

where ξ is the electromotive force (voltage) induced in a conducting loop and ϕ is the magnetic flux passing through the loop (Reitz and Milford, 1960).

Magnetic fields are specified by two vector quantities, the magnetic flux density, **B**, and the magnetic-field strength, **H**. In the SI system of units, **B** and **H** have units of tesla (T) and amperes/meter (A/m), respectively. In vacuum, air, and to a lesser but still adequate approximation in nonmagnetic materials such as living tissues, **B** and **H** are related by

$$\boldsymbol{B/H} = 4\pi \times 10^{-7} \quad (\text{T} \cdot \text{m})/\text{A} \qquad [6]$$

Thus, only one of the quantities **H** and **B** is independent.

Unfortunately, there is currently no consensus as to the reporting of magnetic-field levels related to biological effects. An additional complication facing individuals reading the literature is that both SI (T) and CGS (gauss, G) units have been and are still being used to express flux-density values.

An argument that supports the use of flux density as the primary specification of magnetic exposure in biologically oriented studies is that it is **B**, not **H**, that enters into the equations (Eqs. 4 and 5) that determine the physical interactions between matter and magnetic fields. Some counter this argument by noting that it is **H**, not **B**; that is, the magnetic analogue of the electric-field vector. The symmetry between these two fields becomes manifest when higher frequencies are considered (i.e., when electromagnetic radiation is important), and for this reason **H** is almost universally used in the radio-frequency and microwave literature on biological effects.

In this chapter, the primary specification of magnetic-field levels will be made using the flux density (**B**) vector because of its role, as just described, in evaluating magnetic effects at the tissue level. However, for those readers with other preferences, each magnetic flux density value quoted will be followed by the equivalent CGS flux-density value in gauss (G).

Magnetic-Field Sources

Natural phenomena, such as thunderstorms and solar activity, produce time-varying magnetic fields in the ELF range (Grandolfo and Vecchia, 1985). Such fields are generally of low strength, approximately 0.01 μT (0.1 mG). However, during intense magnetic storms, these fields can reach intensities of about 0.5 μT (5 mG) (Tenforde, 1985).

Of greater importance, in the context of possible biological effects, are the numerous magnetic fields arising from man-made sources. In the lowest intensity range, generally less than 0.3 μT (3 mG), are fields found in the home (Caola et al., 1983; Male et al., 1987; Kaune et al., 1987b) and in office environments (e.g., near video display equipment) (Stuchly et al., 1983). Magnetic fields from communications and electric-power transmission and distribution systems are somewhat higher and can approach a level as high as about 30 μT (0.3 G) (Haubrich, 1974; Scott-Walton et al., 1979; Guy and Chou, 1982; Kaune, 1985).

Higher flux densities can occur in the immediate proximity of industrial processes using large induction motors or heating devices. Lovsund and co-workers have documented magnetic fields from 8 to 70 mT (80 to 700 G) in the steel industry in Sweden (Lovsund et al., 1982). Recently, significant developments in specific areas of medical care have allowed the use of pulsed magnetic fields for various diagnostic and treatment procedures (Bassett et al., 1982; Margulis et al., 1983; Budinger and Lauterbur, 1984). Flux densities from these new technologies range from 1 to 10 mT (10 in 100 G).

Magnetic fields can have polyphasic sources such as alternating current (AC) transmission and distribution lines, in which case the fields will, in general, be elliptically polarized. Most biological research has used linearly polarized magnetic fields, but in a few recent studies circular polarization has been employed.

ELECTRIC- AND MAGNETIC-FIELD COUPLING TO LIVING ORGANISMS

Exposure of a living organism to electric and/or magnetic fields is normally specified by the unperturbed field strength, that is, the field strength measured or calculated with the subject removed from the system. The use of this field as a description of exposure is convenient because it involves a quantity that is relatively easy to measure or calculate. However, this field must be carefully distinguished from the fields that actually act on an organism, namely, the fields that act on the outer surface of the body and those that are induced inside the body. These latter fields, different from the exposure field because of the perturbations produced by the body of the exposed organism, must be determined to specify exposure at the level of living tissues or to relate exposure levels and conditions from one species and/or exposure geometry to another.

Theory of Electric-Field Coupling to Living Organisms

Because the actual time rate of change of 100-kHz or lower frequency electric fields is quite slow, it is useful to think first of an animal or man exposed to a static electric field, E_0. In this limit, the effect of E_0 is to induce an electric charge on the surface of the exposed body (Reitz and Milford, 1960). This charge produces an electric field, E_0, of its own. The total electric field acting on the exposed organism is, then, $E_0 + E_1$. The magnitude and distribution of the induced surface charge is such that $E_0 + E_1 = 0$ inside the body and $E_0 + E_1$ is usually enhanced relative to E_0 at most points on the exterior surface of the body. Because the electric field inside the body is zero, it is clear that the electrical structure inside the body cannot affect the electric fields outside the body. Thus, the only important properties of the body are its external shape and its location relative to other bodies and the ground.

Now suppose that the electric field oscillates with time. As the field oscillates, the induced surface charge density correspondingly oscillates. As this requires the electric charge to be continually changed on the surface of the subject, there will be currents inside the subject and, in general, currents between it and any other conducting bodies that are in electrical contact with it. Because living tissues have finite conductivities, these currents cannot flow without the existence of corresponding electric fields. In this way, currents and fields are induced inside living organisms.

The variation in surface charge density, however, is so slow that the currents and fields generated inside the human or animal are very small. Estimates show that ELF electric fields induced inside these bodies are generally less than about 10^{-7} of the field outside the body and probably never exceed about 10^{-4} of the external field (Kaune and Gillis, 1981). Because these internal ELF fields are so small, the induced surface charge density at any instant in time is very nearly identical to the charge density that would be induced by a static field that is equal to the instantaneous value of the applied AC field. Given this, it also follows that, as in the static case, the electric field outside the body and the charge induced on the surface of the body are determined only by its shape and location relative to other bodies. Because currents inside the body have as their source induced surface charge, it is also true that the total current passing through any section through a body is independent of its internal structure.

The ELF results can be generalized to frequencies up to 100 kHz by noting that the magnitudes of the electric field acting at the surface of the body and the charge density induced on the surface of the body are independent of frequency and that the magnitudes of the currents and current densities induced inside the body are linearly related to frequency (Kaune and Gillis, 1981). Thus, useful information can be obtained using human and animal models that simulate only the conducting surfaces of the bodies of humans and animals. This technique has been used by

Schneider et al. (1974), by Deno (1977, 1979), and by Kaune and colleagues (Kaune and Phillips, 1980; Kaune, 1981; Kaune and Miller, 1984).

Data on Electric-Field Coupling to Living Organisms

A number of theoretical papers have been published on modeling exposure of animals and humans. Most of these papers (Barnes et al., 1967; Spiegel, 1976; Bayer et al., 1977; Lattarulo and Mastronardi, 1981; Shiau and Valentino, 1981; Hart and Marino, 1982) simulated the body of the exposed organism with a sphere, spheroid, or ellipsoid, but Spiegel (1977, 1981) used more refined models. Because available experimental data currently exceed theoretical data in scope and detail, these theoretical papers are not summarized in this chapter. The interested reader is referred to the reviews of Kaune (1985) and Kaune and Phillips (1985). As discussed in the preceding section, considerable information can be obtained using conducting models that simulate the body shapes of humans and animals to a greater or lesser extent. The following paragraphs summarize data obtained using this technique.

Because exposure to electric fields often occurs when the subject is electrically grounded, a parameter of considerable importance is the short-circuit current that passes between the subject and ground. Published short-circuit current data are summarized in Table II for humans (Bracken, 1976; Deno, 1977), horses and cows (EPRI, 1975), pigs (Kaune et al., 1978), guinea pigs (Kaune and Miller, 1984), and rats (Kaune and Phillips, 1980). Most of these measurements were taken at only one frequency and body weight; they have been extrapolated to other frequencies, f, and body weights, W, by assuming an $fW^{2/3}$ dependence (Kaune and Gillis, 1981). The human data were expressed by Deno (1977) in terms of body height rather than

TABLE II.
Short-Circuit Currents Induced in Grounded Humans and Animals by Vertical Electric Fields[a]

Species	Short-circuit current (mA)
Human	$15. \times 10^{-8} \, fW^{2/3} E_0$
Horse	$8.5 \times 10^{-8} \, fW^{2/3} E_0$
Cow	$8.6 \times 10^{-8} \, fW^{2/3} E_0$
Pig	$7.7 \times 10^{-8} \, fW^{2/3} E_0$
Guinea pig	$4.2 \times 10^{-8} \, fW^{2/3} E_0$
Rat	$4.0 \times 10^{-8} \, fW^{2/3} E_0$

[a] Frequency and strength of the applied field are f (Hz) and E (V/m), respectively; weight of subject is W (g).

body weight. In the preparation of Table II, it was assumed that a body height of 1.7 m is equivalent to a body weight of 70 kg (ICRP, 1975).

The accuracy of the extrapolation procedure described in the preceding paragraph was demonstrated experimentally by Gandhi and Chatterjee (1982), who extrapolated 60-Hz short-circuit current data for jeeps and cars to 765 kHz and found good agreement with measured data at this frequency.

Deno (1977) published the first data on induced currents in anatomically detailed human models; these data were measurements, made with a copper-covered human mannequin, of the currents induced in various parts of the body. Deno also developed a simple technique for measuring the external electric fields acting directly on the surface of the body. As pointed out earlier, the accuracy of surface electric-field measurements made with conducting models like Deno's mannequin is very good if the model accurately simulates the human shape. Kaune and Phillips (1980) used Deno's methods to measure surface electric fields and induced current distributions in grounded rats and pigs exposed to a vertical 60-Hz electric field. Average axial (i.e., along the horizontal axis of the body) current densities were estimated by dividing the total induced current crossing a vertical section through the body by the area of that section.

Surface electric-field and axial current density data, derived from Deno's human measurements and Kaune and Phillips' animal data, are presented in Fig. 1; the magnitude and frequency of the unperturbed electric field were assumed to be 10 kV/m and 60 Hz, respectively. Current-density data may be extrapolated from 60 Hz to any other frequency in the 1-Hz to 100-kHz range by multiplying by $f/60$ (Kaune and Gillis, 1981). The surface electric-field values shown in Fig. 1 are valid for all frequencies in this range.

Figure 1 shows, for a vertical external electric field, that the axial current densities induced inside the body of a human are considerably larger than the corresponding quantities for animals, even though the external electric fields are the same. This conclusion is also true for induced electric fields because of the conditions shown by Eq. 3. These differences mean that external unperturbed fields, which are almost always used to specify exposure, must be scaled to equalize surface electric fields and/or internal current densities and electric fields to extrapolate biological data from one species to another; this is complicated by the fact that the actual value of the scaling factor depends on which quantity is being scaled. For example, a scaling factor for the peak electric-field strength acting on the outer surface of the body would be about 4.9:1 for humans compared to rats while the scaling factor for axial current density in the neck would be about 20:1 for the same species-to-species comparison (see Fig. 1). Evidently, knowledge about the site of action for a particular biological effect is needed before data can be reliably extrapolated across species.

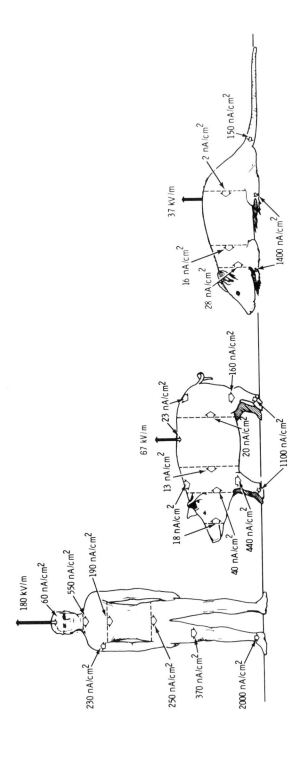

Fig. 1. Grounded man, pig, and rat exposed to vertical, 60-Hz, 10-kV/m electric field. Relative body sizes are not to scale. Surface electric-field measurements for man and pig, surface electric-field estimates for rats, and estimated axial current densities averaged over selected sections through bodies are shown. Calculated current densities perpendicular to surface of body are shown for man and pig. (Reprinted with permission from Kaune, W. T., Phillips, R. D. 1980. Comparison of the coupling of grounded humans, swine, and rats to vertical, 60-Hz electric fields. *Bioelectromagnetics* 1:117–129.)

The surface electric-field data given in Fig. 1 are quite limited in that measurements were made at only a few points on the body. Kaune (1981) described a simple calculation of electric-field strengths averaged over the entire body surface of a grounded subject. This method requires only that the subject's short-circuit current, I_{sc}, and the surface area of its body, A_b, be measured or estimated. In terms of these parameters, the average electric field, E_{avg}, acting on the surface of the body is

$$E_{avg} = \frac{I_{sc}}{\omega \varepsilon_0 A_b} \qquad [7]$$

where $\omega = 2\pi f$ is angular frequency, f is the frequency of the field, and ε_0 is the permittivity of space (8.85×10^{-12} F/m). Table III gives approximate maximum and average electric fields acting on the surfaces of the bodies of grounded humans, swine, rats, guinea pigs, horses, and cows exposed to a vertical 1-kV/m electric field.

The current density data given in Fig. 1 are only one (the axial) of three components of the total current density vector. This is a serious limitation in the animal data because it is certain that significant vertical current is also present. Measurements in three-dimensional models of humans and animals are required to overcome this limitation.

Guy et al. (1982) and Kaune and Forsythe (1985) measured induced electric fields and current densities in grounded homogeneous models of humans exposed to 60-Hz electric fields. Figures 2 and 3 summarize Kaune and Forsythe's data for human models, assuming a frequency of 60 Hz, an exposure field strength of 10 kV/m, and that the models were grounded equally through both feet. Note the enhancements in current density that occur in the axillae (armpits). It is interesting that these researchers also observed a strong enhancement in horizontal current density in the lower

TABLE III.
Peak and Average Electric Fields Acting on the Surfaces of Grounded Humans and Animals Exposed to Vertical, 1-kV/m Electric Fields

Species	Average E (kV/m)	Peak E (kV/m)
Human	2.7	18
Swine	1.4	6.7
Rat (resting)	0.73	3.7
Rat (rearing)	1.5	—[a]
Horse	1.5	—[a]
Cow	1.5	—[a]

[a] Not available.

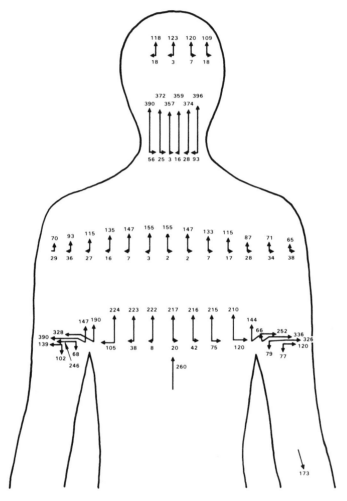

Fig. 2.   Current densities measured in head and upper torso, midfrontal plane, of saline model of standing man exposed to 60-Hz, 10-kV/m electric field. Induced rms current densities are in nanoamperes per centimeter squared (nA/cm²). Model was grounded with equal currents passing through both feet.

(Reprinted with permission from Kaune, W. T., Forsythe, W. C. 1985. Current densities measured in human models exposed to 60-Hz electric fields. *Bioelectromagnetics* 2:1–11.)

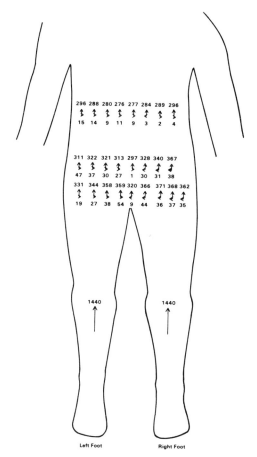

Fig. 3. Current densities measured in lower torso, midfrontal plane, of saline model of standing man exposed to 60-Hz, 10-kV/m electric field. Induced rms current densities are in nanoamperes per centimeter squared (nA/cm²). Model was grounded with equal currents passing through both feet.
(Reprinted with permission from Kaune, W. T., Forsythe, W. C. 1985. Current densities measured in human models exposed to 60-Hz electric fields. *Bioelectromagnetics* 2:1–11.)

pelvic region when only one foot was grounded. One limitation of the current-density data given in Figs. 2 and 3 is that they pertain only to the exposure of grounded humans. It is more usual for humans to be exposed when they are partially insulated from ground by their shoes. Deno (1977) and Kaune et al. (1987a) have made measurements that enable grounded data to be extrapolated to ungrounded exposure situations. Figure 4 summarizes the results of the latter reference.

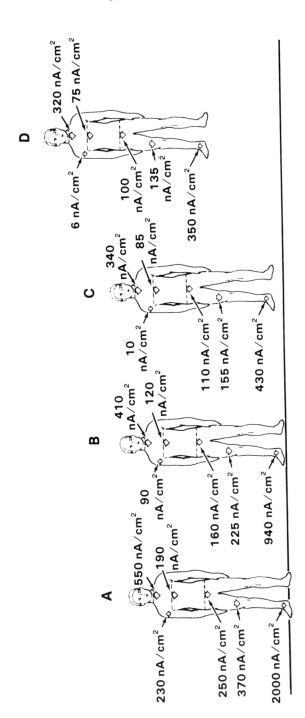

Fig. 4. Average axial current densities induced in 1.7-m-tall human exposed to vertical 10-kV/m, 60-Hz electric field. Four positions of the body relative to ground are shown: (**A**) standing on and in electrical contact with ground; (**B**) feet elevated 1.1 cm above ground to approximately simulate insulating footwear; (**C**) feet elevated 12.8 cm above ground; (**D**) feet elevated 123 cm above ground. (Reprinted with permission from Kaune, W. T., Kistler, L. M., Miller, M. C. 1987. Comparison of the coupling of grounded and ungrounded humans to vertical 60-Hz electric fields. In: *Interaction of Biological Systems with Static and ELF Electric and Magnetic Fields*, Anderson, L. E., Weigel, R. J., Kelman, B. J., eds. Proceedings of the 23rd Annual Hanford Life Sciences Symposium, Richland, Washington. Springfield, Virginia: National Technical Information Service.)

A human or animal may or may not be grounded, depending on the sizes of the contact resistances between those parts of its body in contact with grounded objects and ground. The degree of grounding may be assessed as follows. Let C_g be the capacitance between the body and ground and let R_c be the net contact resistance between the body and ground. In the ungrounded state, R_c will be infinite, or at least large with respect to the reactance of C_g. On the other hand, if $R_c \ll 1/\omega C_g$, the subject can be said to be grounded.

The capacitance between an adult human and ground is about 100 pF (EPRI, 1975). At 60 Hz, therefore, a contact resistance less than about 1 MΩ is required to electrically ground a person, whereas at 100 kHz this value drops to about 500 Ω. One can therefore conclude that humans are probably often grounded when exposed to power-frequency electric fields but are probably seldom grounded when exposed to electric fields with frequencies in the 10- to 100-kHz range.

Laboratory animal data similar to those in Figs. 2 through 4 are needed to determine quantitative factors to scale animal data to human exposure situations; these data should become available in the next few years.

Biophysical Analysis of Electric-Field Coupling

As discussed previously in this chapter, the electric field acting on the surface of the body of a human or animal is enhanced over most of the body surface relative to the unperturbed electric field. It is well known that ELF electric fields can be perceived (Reilly, 1978). One known mechanism of perception is hair stimulation (piloerection), that is, the erection and promotion of oscillatory hair movement by electric forces. The frequency of this vibration can be equal or double the frequency of the applied electric field (Gillis and Kaune, 1978; Cabanes and Gary, 1981), depending on relative humidity and possibly other factors. This mechanism of interaction is important because it may be responsible for biological effects in animals that have no relevance for assessments of risks to humans resulting from exposure to electric fields.

Another well-known mechanism of interaction between electric fields and biological tissues is the direct stimulation of excitable (e.g., neural) cells. It accounts for the ability of humans and animals to perceive electric currents in their bodies and for the possibility that they might be shocked or even electrocuted by such currents. At the tissue level, the mechanism consists of the conduction of depolarizing currents through the membranes of excitable cells and the resultant generation of action potentials.

Most research on this mechanism has been done in the context of electric-shock hazards (Dalziel, 1972): Fixed currents were introduced into the body and the reactions of the subjects were noted. From a mechanistic point of view, a more basic and useful quantity is the current density in the affected part of the body rather than the total current passing through the body. It has been estimated that the threshold ELF current density required to stimulate most excitable cells is about 1 to 2 mA/cm², but that very long nerve cells oriented parallel to the current-density vector may be sensitive to values as small as about 0.1 mA/cm² (Schwan, 1977b; Bernhardt, 1979, 1983, 1985).

Figures 2 through 4, for humans and animals exposed to external 10-kV/m, 60-Hz electric fields, demonstrate that current densities induced by transmission-line electric fields are far smaller than values required for the direct stimulation of excitable cells. However, grounded humans touching large objects such as cars or trucks may be exposed to quite substantial currents (Gandhi and Chatterjee, 1982; Guy and Chou, 1982).

It is difficult to know whether a particular electric-field-induced effect on a human or animal is caused by the field acting at the outer surface of the body or by the fields induced inside the body. No such complication exists for experiments involving cell suspensions, and the existence of effects at low ELF field strengths (e.g., effects on the release of calcium ions from brain tissue) (Bawin and Adey, 1976; Blackman et al., 1979, 1982) shows that there must be mechanisms of interaction between electric fields and biological tissues in addition to the direct stimulation of excitable cells.

Currently, biophysicists are exploring the possibility that communication between brain cells occurs not only through synaptic connections but also through each cell's modulation of, and sensitivity to, the extracellular electrical environment. Because extracellular electric fields are markedly smaller than those in the membrane, it is hoped that this mechanistic picture might prove capable of explaining the sensitivity of certain types of tissues to ELF electric fields. Based on this concept, Bernhardt (1983) argued that extracellular electric fields induced by external fields could not, a priori, be judged safe unless they were less strong than the endogenous extracellular fields present in living tissues. This author used electrocardiographic and electroencephalographic data to estimate endogenous fields in the brain and torso and arrived at a lower limit current density of about 100 nA/cm². Examination of Figs. 2 through 4 shows that current densities above this level are induced in grounded and ungrounded humans exposed to electric fields characteristic of maximum levels produced at ground level by electric-power transmission lines.

Data are available indicating that Bernhardt's basic hypothesis – that induced ELF electric fields at levels below endogenous field levels are too small to cause biological effects – is not valid. These data come from the calcium-efflux experiments mentioned previously in which chick brain tissue was immersed in culture medium and exposed to electric fields through the air dielectric surrounding the preparation. We will estimate the fields induced in the brain tissue and show that these fields are far lower than 100 nA/cm², Bernhardt's estimate of endogenous extracellular levels occurring in living tissues.

Our analysis is based on the geometrical assumptions used by Joines and Blackman (1980, 1981) in analyzing coupling of radio-frequency fields to brain tissue preparations. These authors modeled the preparation as a 0.41-cm-radius sphere of brain tissue surrounded by a 0.71-cm-radius spherical shell of growth medium. The current density, J_b, induced in the brain tissue is

$$J_b = 3D\omega\varepsilon E_0 \qquad [8]$$

where E_0 is the unperturbed field strength in air outside the tissue sample and D is a factor that accounts for the shielding of the brain tissue by the surrounding culture medium.

It is straightforward, though somewhat tedious, to show that D is given by the following expression:

$$D = \left[(R_b/R_m)^3 + \frac{(1 + 2\sigma_m/\sigma_b)}{3}[1 - (R_b/R_m)^3] \right]^{-1} \qquad [9]$$

where R_b and σ_b are the radius and conductivity, respectively, of the brain tissue sphere, and R_m and σ_m are the outer radius and conductivity, respectively, of the spherical shell of culture medium. Values for σ_m and σ_m have not been published. Cerebrospinal fluid, which bathes the brain in vivo, has an ELF conductivity of about 1.5 S/m (Geddes and Baker, 1967). It seems likely that a similar medium was used for the laboratory experiments. Brain tissue has a much lower conductivity of about 0.1 S/m (Bernhardt, 1979); with these values, D ~ 0.08.

Changes in calcium-efflux rates were observed by Bawin and Adey (1976) and by Blackman et al. (1982) when brain tissue preparations were exposed to 3-V/m (rms), 16-Hz electric fields. Using Eq. 8 and the value of D calculated in the preceding paragraph, we estimate the current density induced in the brain tissue to be approximately 0.1 pA/cm². This value, six orders of magnitude lower than Bernhardt's estimate of endogenous levels, indicates that ELF fields far smaller than those that occur normally in living tissues can affect brain tissues.

There have been several attempts to develop models that can explain the interaction of biological tissues with extremely small ELF electric fields. These models have been recently reviewed (Postow and Swicord, 1985; Swicord, 1985; Swicord and Postow, 1986). It appears that at present none of these models has been developed to the point that they can make quantitative predictions. In addition, none appears able to explain, even qualitatively, the intensity windows (i.e., the appearance of biological effects only in a narrow range of field strength) that have been observed in calcium-efflux experiments.

Theory of Magnetic-Field Coupling to Living Organisms

In contrast to electric-field exposure, the bodies of humans, animals, and other living organisms cause almost no perturbation in a magnetic field to which they are exposed. This is true for two major reasons. (1) Excluding a few highly specialized tissues that contain magnetite, living tissues contain no magnetic materials and therefore have magnetic properties almost identical to those of air. (2) The modification in the applied magnetic field, caused by secondary magnetic fields produced by currents induced in the body of the subject, is small (Kaune, 1985, 1986). (An equivalent way of making this latter statement is to say that the electrical skin depth in living tissues is large relative to the sizes of humans and animals.)

Faraday's law of induction states that time-varying magnetic fields generate electric fields through induction. Therefore, a living organism exposed to a magnetic field will also be exposed to an induced electric field that causes eddy currents to flow in its body. These currents circulate in closed loops that tend to lie in planes perpendicular to the direction of the magnetic field.

A fairly useful model for a human or animal exposed to a uniform magnetic field is a homogeneous ellipsoid, because the induced electric field can be calculated exactly (Hart and Marino, 1982). An ellipsoid is defined by three parameters: the semimajor axes, which are the x, y, and z coordinates where the surface of the body intersects the x, y, and z axes, respectively. (Assume that the symmetry axes of the ellipsoid coincide with the coordinate axes.) The electric field, E_m, induced by a magnetic field, B_0, of angular frequency, ω, oriented parallel to, for example, the z axis is

$$E_m = \frac{j\omega B_0}{a^2 + b^2} (a^2 y\hat{x} - b^2 x\hat{y}) \qquad [10]$$

where a and b are the semimajor axes of the ellipsoid in the x and y directions, respectively, x and y are unit vectors in the x and y directions, respectively, and $j = \sqrt{-1}$ indicates that E_m is phase shifted relative to B_0 by 90°.

Published Work on Magnetic-Field Coupling to Living Organisms

Very little theoretical or experimental work has actually investigated the coupling of magnetic fields to living organisms. Spiegel (1976) published a paper describing electric- and magnetic-field coupling to spherical models. The magnetic-field portion of this work essentially involved an application of Eq. 10, where $a = b =$ the radius of the sphere.

Gandhi et al. (1984) described a method for the calculation of currents induced by magnetic fields in complex bodies that simulates the exposed body by a two- or three-dimensional lattice of resistors. Part of such a two-dimensional resistor lattice is shown in Fig. 5; this lattice might repre-

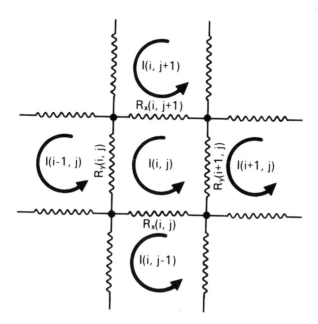

Fig. 5. Section through body of human or animal can be simulated by two-dimensional lattice of resistors. Values of resistors are related to electrical properties of underlying tissue. Currents are induced in lattice by an alternating magnetic field. (Reprinted with permission from Kaune, W. T. 1988. Physical interaction of 1-Hz to 100-kHz electric and magnetic fields with living organisms. In: *Nonionizing Electromagnetic Radiations and Ultrasound, Proceedings of the 22nd Annual Meeting of the National Council on Radiation Protection and Measurements*, Proc. No. 8. Washington, DC: NCRP.)

sent, for example, a cross section through a human torso that is exposed to a vertical magnetic field. (Because eddy currents tend to circulate in planes perpendicular to the applied magnetic field, it seems best to select a planar section through the body that is also perpendicular to the field.)

Electromotive forces (EMFs) will be induced by an applied magnetic field in each of the loops shown in Fig. 5, and loop currents will flow as a result. Let $I(i,j)$ be the current in the (i,j)th loop. The EMF induced in this loop is $-j\omega B_0 A(i,j)$, where $A(i,j)$ is the loop's area. The sum of the voltage drops around this loop must equal the induced EMF:

$$R_x\,(i,j)\,[I(i,j)-I(i,j-1)] + R_y\,(i+1,j)\,[I(i,j)-I(i+1,j)]$$
$$+\,R_x\,(i,j+1)\,[I(i,j)-I(i,j+1)] + R_y\,(i,j)\,[I(i,j)-I(i-1,j)]$$
$$=\,-j\omega B_0 A(i,j) \qquad\qquad [11]$$

The collection of these equations for all the loops in the system constitutes a linear system. This system may be iteratively solved using the successive overrelaxation (SOR) method, in which improved estimates of the loop currents are generated from a previous estimation of these quantities. The basic SOR equation is

$$I^{n+1}(i,j) = \alpha K^{n+1}(i,j) + (1 - \alpha)I^n(i,j) \tag{12}$$

where α is the relaxation factor, $I^{n+1}(i,j)$ is the $n+1$th estimate of the current induced in the (i,j)th loop, and $I^n(i,j)$ is the result of the nth estimate of this current. $K^{n+1}(i,j)$ is an estimate of the (i,j)th loop current obtained from Eq. 11:

$$K^{n+1}(i,j) = [R_x(i,j)I^n(i,j-1) + R_y(i+1,j)I^n(i+1,j) \tag{13}$$
$$+ R_x(i,j+1)I^n(i,j+1) + R_y(i,j)I^n(i-1,j)$$
$$- j\omega B_0 A(i,j)]/[R_x(i,j) + R_y(i+1,j) + R_x(i,j+1) + R_y(i,j)]$$

Values for α of about 1.8 seem to give the most rapid convergence in the types of problems discussed here.

This method appears to be straightforward to use, and it should provide greatly improved estimates of the currents induced in humans and animals exposed to power-frequency magnetic fields. By selecting the resistances R_x and R_y appropriately, it is also possible to simulate the complex electrical structure of living organisms.

Biophysical Analysis of Magnetic-Field Coupling

As discussed, alternating magnetic fields induce electric fields inside the bodies of exposed humans and animals. External alternating electric fields also induce electric fields inside bodies. The distributions of the fields induced by these two types of coupling are different, but at the level of the cell there would appear to be no fundamental difference. Thus, the biophysical analysis provided earlier in this chapter for electric-field induction can also be applied to the electric fields induced by alternating magnetic fields. One interesting question is how large a magnetic field is required to induce current densities sufficient to potentially stimulate excitable cells.

Magnetic induction of currents can be modeled using a simple ellipsoidal approximation of a man. A typical man has a height of 1.7 m, a mass of 70 kg (ICRP, 1975), and a body-width-to-body-thickness ratio of about 2. An ellipsoid with semimajor axes of 0.85 cm, 0.20 cm, and 0.10 cm has the same body height, the same width-to-thickness ratio, and a body volume of 7.1×10^4 cm^3 (assume a specific gravity of about 1). Using Eq. 10, the maximum electric field, E_m, induced in this model can be shown to occur when the magnetic flux-density vector, B, is horizontal and perpendicular to the front of the body. The actual value of E_m is 1.2 fB, where f is the frequency of the field in Hertz. Equation 3 can be used to estimate the induced current density from E_m. Using an average tissue conductivity of 0.2 S/m, the maximum induced current density is

$$J_m \sim 0.24\,fB \qquad\qquad [14]$$

As discussed earlier, the minimum current density that is required to stimulate excitable cells is about 0.1 mA/cm^2 in the ELF range and about 10 mA/cm^2 at 100 kHz. Using Eq. 14, the estimated magnetic flux densities required to induce such current densities are about 0.07 T (700 G) at 60 Hz and about 4 mT (40 G) at 100 kHz. These flux densities are considerably larger than magnetic flux densities produced by electric-power facilities or by communications and navigation systems, but values of this size may be found in certain specialized industrial environments.

In addition to the induction of currents in living bodies, magnetic fields penetrate living organisms without significant perturbation and exert Lorentz forces on charged particles that are in motion within their bodies. Equation 4 describes the dependence of this force on the particle's velocity, its electrical charge, and the magnetic flux density. The most prevalent types of motion in matter are the motion of electrons in atoms, nucleons in nuclei, and the intrinsic spins of these particles. These motions lead to the existence of magnetic dipole moments, which may be either permanent or induced by the applied magnetic field. A magnetic dipole interacts with a magnetic field in such a way that it experiences a torque in a direction to align the dipole parallel to the magnetic field. However, this alignment is resisted by random thermal motion and a statistical distribution of dipole directions is thereby established. Let Θ be the angle between a dipole and the magnetic field, B. For an assembly of similar dipoles in a magnetic field, the average value of $\cos \Theta$, denoted $<\cos \Theta>$, can be calculated (Reitz and Milford, 1960; Kaune, 1985). Complete alignment of the dipoles would be characterized by $<\cos \Theta> = 1$, whereas $<\cos \Theta> = 0$ would indicate no alignment.

At body temperature (310 K) and at the flux density of about 30 μT (0.3 G) that is characteristic of a heavily loaded transmission line, it can be shown that $<\cos \Theta>$ is less than about 10^{-8} for electronic and nuclear magnetic moments that might occur in living tissues. Obviously, the effect on the magnetic dipoles that are part of the body of a subject exposed to such a magnetic field is very small. Of course, every single dipole is subject to this effect and, conceivably, some sort of process might exist that is sensitive to the average response of a large number of dipoles.

Magnetic dipoles placed in a nonuniform magnetic field experience a net force. This force can be estimated (Kaune, 1986), for the conditions described in the preceding paragraph, to be of the order of 10^{-28} N. By comparison, the forces that bind atoms and molecules together are much larger. (For example, the force of attraction between the two atoms that comprise NaCl is about 10^{-9} N.)

Charged particles are also carried by the bulk motion of various parts of the body. For example, charged ions are carried by blood flow. These ions are both positively and negatively charged and will experience Lorentz forces in opposite directions, resulting in a separation of the two polarities of electric charge and, therefore, in the generation of electric potentials. It has been shown that these potentials can produce artifacts in the electrocardiograms of rats (Gaffey and Tenforde, 1981) exposed to static magnetic fields with flux densities above 0.3 T (3 kG).

Another potential mechanism involving the interaction of charged particles in living tissues with magnetic fields has been proposed by Polk (1987). This author noted that living cells are surrounded by counterions, that is, ions in the extracellular fluid that are electrostatically attracted to fixed charges located in the membrane of the cell. This attraction is counterbalanced by the diffusion of these ions away from the cell, and an ionic atmosphere in the neighborhood of the cell is thereby established. Polk's hypothesis is that magnetic fields might distort this atmosphere, possibly leading to some observable effect on the cell.

CONCLUSIONS

Considerable progress has been made during the past 10 years in understanding the physical and biological interactions between living organisms and electric and magnetic fields whose frequencies are in the range of 1 Hz to 100 kHz. Even so, our ability to estimate internal electric fields and current densities resulting from exposure to external electric and/or magnetic fields is still limited. The only exposure situation that can be analyzed with quantitative depth is a homogeneous grounded model of a human or animal exposed to an electric field that is vertical and uniform. Some data are available for ungrounded exposure situations, but they are not sufficient to determine, for example, current densities induced at local points in the body. Questions as to effects on current distribution resulting from the nonuniform electrical properties of tissues are unanswered at this time.

Even less is known about magnetic-field coupling to humans and animals. However, it appears that progress in this area will be simpler than for electric-field exposure because the body of a human or animal does not perturb the applied field. Another limitation of our present knowledge is that it ignores the underlying cellular structure of living tissues.

In conclusion, problems of interest exist in the area of electric- and magnetic-field dosimetry. Measurements need to be made in real animals of electric fields induced by external electric fields. Computer models should be developed that can simulate the exposure of grounded and ungrounded animals and humans to uniform and nonuniform electric or magnetic fields. For these models to be most useful, it should be possible to define quickly and conveniently body shapes and postures and to simulate the major organ systems in the body.

ACKNOWLEDGMENTS

Portions of this chapter were previously published in NCRP publication *Proceedings No. 8 Nonionizing Electromagnetic Radiations and Ultrasound*, Proceedings of the Twenty-Second Annual Meeting of the National Council on Radiation Protection and Measurements. Work was supported by the U.S. Department of Energy, Office of Energy Storage and Distribution, under contract DE–AC06–76RLO 1830.

REFERENCES

Barnes, H. C., McElroy, A. J., Charkow, J. H. 1967. National analysis of electric fields in live line working. *IEEE Trans. Power Appar. Syst.* PAS 86:482–492.

Bassett, C. A. L., Mitchell, S. N., Gaston, S. R. 1982. Pulsing electromagnetic field treatment in ununited fractures and failed arthrodeses. *JAMA* 247:623–628.

Bawin, S. M., Adey, W. R. 1976. Sensitivity of calcium binding in cerebral tissue to weak environmental electric fields oscillating at low frequency. *Proc. Natl. Acad. Sci. USA* 73:1999–2003.

Bayer, A., Brinkmann, J., Wittke, G. 1977. Experimental research on rats for determining the effect of electrical ac fields on living beings. *Elektrizitaetswirtschaft* 76:77–81.

Bernhardt, J. H. 1979. The direct influence of electromagnetic fields on nerve and muscle cells of man within the frequency range of 1 Hz to 30 MHz. *Radiat. Environ. Biophys.* 16:309–323.

Bernhardt, J. H. 1983. *On the Rating of Human Exposition to Electric and Magnetic Fields with Frequencies Below 100 kHz*. Ispra, Italy: Commission of the European Communities, Joint Research Center.

Bernhardt, J. H. 1985. Assessment of experimentally observed bioeffects in view of their clinical relevance and the exposures at work places. In: *Proceedings of the Symposium on Biological Effects of Static and ELF-Magnetic Fields*, Bernhardt, J. H., ed. Munchen, FRG: BGA-Schriftenreihe, MMV Medizin Verlag.

Blackman, C. F., Benane, S. G., Kinney, L. S., Joines, W. T., House, D. E. 1982. Effects of ELF fields on calcium ion efflux from brain tissue in vitro. *Radiat. Res.* 92:510–520.

Blackman, C. F., Elder, J. A., Weil, C. M., Benane, S. G., Eichinger, D. C. 1979. Induction of calcium-ion efflux from brain tissue by radio-frequency radiation: effects of modulation frequency and field strength. *Radio Sci.* 14:93–98.

Bracken, T. D. 1976. Field measurements and calculations of electrostatic effects of overhead transmission lines. *IEEE Power Appar. Syst.* PAS 95:494–504.

Bracken, T. D. 1985. *Comparison of Electric Field Exposure Monitoring Instrumentation*, Final Report, Research Project 799-19. Palo Alto, California: Electric Power Research Institute.

Bridges, J. E., Preache, M. 1981. Biological influences of power frequency electric fields—a tutorial review from a physical and experimental viewpoint. *Proc. IEEE* 69:1092-1120.

Budinger, T. F., Lauterbur, P. C. 1984. Nuclear magnetic resonance technology for medical studies. *Science* 226:288-298.

Cabanes, J., Gary, C. 1981. La perception directe du champ electrique. *CIGRE Report No. 233*-08, International Conference on Large High Tension Electric Systems (Paris). Paris: CIGRE.

Caola, R. J., Jr., Deno, D. W., Dymek, V. S. W. 1983. Measurements of electric and magnetic fields in and around homes near a 500-kV transmission line. *IEEE Trans. Power Appar. Syst.* PAS 102:3338-3347.

Chartier, V. L., Bracken, T. D., Capon, A. S. 1985. BPA study of occupational exposure to 60-Hz electric fields. *IEEE Trans. Power Appar. Syst.* PAS 104:733-744.

Dalziel, C. F. 1972. Electric shock hazard. *IEEE Spectrum*, pp. 41-50.

Deno, D. W. 1977. Currents induced in the human body by high voltage transmission line electric field—measurement and calculation of distribution and dose. *IEEE Trans. Power Appar. Syst.* PAS 96:1517-1527.

Deno, D. W. 1979. Monitoring of personnel exposed to a 60-Hz electric field. In: *Biological Effects of Extremely Low Frequency Electromagnetic Fields*, Phillips, R. D., Gillis, M. F., Kaune, W. T., Mahlum, D. D., eds., pp. 93-108, Proceedings of the 18th Hanford Life Sciences Symposium, Richland, Washington. CONF-781016. Springfield, Virginia: National Technical Information Service.

Deno, D. W., Silva, M. 1984. Method for evaluating human exposure to 60-Hz electric fields. *IEEE Trans. Power Appar. Syst.* PAS 103:1699-1706.

Deno, D. W., Zaffanella, L. E. 1978. *Electrostatic and Electromagnetic Effects of Ultrahigh-Voltage Transmission Lines*, Final Report, Research Project 566-1. Palo Alto, California: Electric Power Research Institute.

Deuse, J., Pirotte, P. 1976. Calculation and measurement of electric field strength near H. V. structures. In: *International Conference on Large High Voltage Systems*, Paper 36-03. Paris: CIGRE.

EPRI. 1975. *Transmission Line Reference Book—345 kV and Above*. Palo Alto, California: Electric Power Research Institute.

Gaffey, C. T., Tenforde, T. S. 1981. Alterations in the rat electrocardiogram induced by stationary magnetic fields. *Bioelectromagnetics* 2:357-370.

Gandhi, O. P., Chatterjee, I. 1982. Radio-frequency hazards in the VLF to MF band. *Proc. IEEE* 70:1462-1464.

Gandhi, O. P., DeFord, J. F., Kanai, H. 1984. Impedance method for calculation of power deposition patterns in magnetically induced hyperthermia. *IEEE Trans. Biomed. Eng.* BME 31:644-651.

Geddes, L. A., Baker, L. E. 1967. The specific resistance of biological material—a compendium of data for the biomedical engineer and physiologist. *Med. Biol. Eng.* 5:271-293.

Gillis, M. F., Kaune, W. T. 1978. Hair vibration in ELF electric fields. In: *Abstracts, Symposium on the Biological Effects of Electromagnetic Waves, 19th General Assembly of the International Union of Radio Sciences*, August 1-8, 1978, Helsinki, Finland, p. 51. Paris: URSI.

Grandolfo, M., Vecchia, P. 1985. Natural and man-made environmental exposures to static and ELF electromagnetic fields. In: *Biological Effects and Dosimetry of Non-Ionizing Radiation; Static and ELF Electromagnetic Fields*, Grandolfo, M., Michaelson, S. M., Rindi, A., eds. New York: Plenum.

Guy, A. W., Chou, C. K. 1982. *Hazard Analysis: Very Low Frequency Through Medium Frequency Range*, Final Report, USAF SAM Contract F33615-78-D-0617, Task 0065. Brooks Air Force Base, Texas: Aerospace Medical Division.

Guy, A. W., Davidow, S., Yang, G. Y., Chou, C. K. 1982. Determination of electric current distributions in animals and humans exposed to a uniform 60-Hz high-intensity electric field. *Bioelectromagnetics* 3:47–71.

Hart, F. X., Marino, A. A. 1982. ELF dosage in ellipsoidal models of man due to high voltage transmission lines. *J. Bioelectr.* 1:129–154.

Haubrich, H. J. 1974. Das magnetfeld im nahbereich von drehstrom- freileitungen [The magnetic field in the proximity of polyphase alternating current overhead transmission lines]. *Elektrizitaetswirtschaft* 73:511–517.

Hauf, R. 1982. Electric and magnetic fields at power frequencies with particular reference to 50 and 60 Hz. In: *Nonionizing Radiation Protection, Vol. VIII*, Seuss, M. J., ed., pp. 175–198. ISBN 92-890-1101-7, Eur. Ser. 10. Copenhagen: World Health Organization.

ICRP. 1975. *Report of the Task Group on Reference Man, ICRP 23*. Oxford: Pergamon Press.

Israel, H. 1973. *Atmospheric Electricity*. U. S. Department of Commerce. Springfield, Virginia: National Technical Information Service.

Joines, W. T., Blackman, C. F. 1980. Power density, field intensity, and carrier frequency determinants of RF-energy-induced calcium-ion efflux from brain tissue. *Bioelectromagnetics* 2:271–275.

Joines, W. T., Blackman, C. F. 1981. Equalizing the electric field intensity within chick brain immersed in buffer solution at different carrier frequencies. *Bioelectromagnetics* 2:411–413.

Kaune, W. T. 1981. Power-frequency electric fields averaged over the body surfaces of grounded humans and animals. *Bioelectromagnetics* 2:403–406.

Kaune, W. T. 1985. Coupling of living organisms to ELF electric and magnetic fields. In: *Biological and Human Health Effects of Extremely Low Frequency Electromagnetic Fields*, pp. 25–60. Arlington, Virginia: American Institute of Biological Sciences.

Kaune, W. T. 1986. *Physical Interaction of Humans and Animals with Power-Frequency Electric and Magnetic Fields*. IEEE Spec. Pub. 86TH0139-6-PWR. Piscataway, New Jersey: IEEE Service Center.

Kaune, W. T., Forsythe, W. C. 1985. Current densities measured in human models exposed to 60-Hz electric fields. *Bioelectromagnetics* 6:13–32.

Kaune, W. T., Gillis, M. F. 1981. General properties of the interaction between animals and ELF electric fields. *Bioelectromagnetics* 2:1–11.

Kaune, W. T., Miller, M. C. 1984. Short-circuit currents, surface electric fields, and axial current densities for guinea pigs exposed to ELF electric fields. *Bioelectromagnetics* 5:361–264.

Kaune, W. T., Phillips, R. D. 1980. Comparison of the coupling of grounded humans, swine and rats to vertical, 60-Hz electric fields. *Bioelectromagnetics* 1:117–129.

Kaune, W. T., Phillips, R. D. 1985. Dosimetry for extremely low-frequency electric fields. In: *Biological Effects and Dosimetry of Non-Ionizing Radiation; Static and ELF Electromagnetic Fields*, Grandolfo, M., Michaelson, S. M., Rindi, A., eds. New York: Plenum.

Kaune, W. T., Kistler, L. M., Miller, M. C. 1987a. Comparison of the coupling of grounded and ungrounded humans to vertical 60-Hz electric fields. In: *Interaction of Biological Systems with Static and ELF Electric and Magnetic Fields*, Anderson, L. E., Kelman, B. J., Weigel, R. J., eds., Proceedings of the 23rd Hanford Life Sciences Symposium, Richland, Washington. CONF-841041. Springfield, Virginia: National Technical Information Service.

Kaune, W. T., Stevens, R. G., Callahan, N. J., Severson, R. K., Thomas, D. B. 1987b. Residential magnetic and electric fields. *Bioelectromagnetics* 8:315–335.

Kaune, W. T., Phillips, R. D., Hjeresen, D. L., Richardson, R. L., Beamer, J. L. 1978. A method for the exposure of miniature swine to vertical 60-Hz electric fields. *IEEE Trans. Biomed. Eng.* BME 25:276–283.

Lattarulo, F., Mastronardi, G. 1981. Equivalence criteria among man and animals in experimental investigations of high voltage power frequency exposure hazards. *Appl. Math. Modeling* 5:92–96.

Looms, J. S. T. 1983. Power frequency electric fields: dosimetry. In: *Biological Effects and Dosimetry of Non-Ionizing Radiation: Radiofrequency and Microwave Energies*. NATO Advanced Study Institute Ser. A, Life Sciences. New York: Plenum Press.

Lovstrand, K. G., Lundquist, S., Bergstrom, S., Birke, E. 1979. Exposure of personnel to electric fields in Swedish extra-high-voltage substations: field strength and dose measurements. In: *Biological Effects of Extremely-Low-Frequency Electromagnetic Fields*, Phillips, R. D., Gillis, M. F., Kaune, W. T., Mahlum, D. D., eds., Proceedings of the 18th Hanford Life Sciences Symposium, Richland, Washington. CONF-781016. Springfield, Virginia: National Technical Information Service.

Lövsund, P., Oberg, P. A., Nilsson, S. E. G. 1982. ELF magnetic fields in electrosteel and welding industries. *Radio Sci.* 17:35S-38S.

Male, J. C., Norris, W. T., Watts, M. W. 1987. Human exposure to power-frequency electric and magnetic fields. In: *Interaction of Biological Systems with Static and ELF Electric and Magnetic Fields*, Anderson, L. E., Weigel, R. J., Kelman, B. J., eds., Proceedings of the 23rd Hanford Life Sciences Symposium, Richland, Washington. Springfield, Virginia: National Technical Information Service.

Margulis, A. R., Higgins, C. B., Kaufman, L., Crooks, L. E., eds. 1983. *Clinical Magnetic Resonance Imaging*. San Francisco: University of California Press.

Polk, C. 1987. Motion of counterions on a cylindrical cell surface: a possible mechanism for the action of low-frequency, low-intensity magnetic fields which displays unsuspected frequency dependence. In: *Interaction of Biological Systems with Static and ELF Electric and Magnetic Fields*, Anderson, L. E., Weigel, R. J., Kelman, B. J., eds., Proceedings of the 23rd Annual Hanford Life Sciences Symposium. Springfield, Virginia: National Technical Information Service.

Postow, E., Swicord, M. L. 1985. Modulated field and "window" effects. In: *Handbook of Biological Effects of Electromagnetic Radiation*, Polk, C., Postow, E., eds. Boca Raton, Florida: CRC Press.

Reilly, J. P. 1978. Electric and magnetic field coupling from high voltage ac power transmission lines – classification of short-term effects on people. *IEEE Trans. Power Appar. System.* PAS 97:2243–2252.

Reitz, J. R., Milford, F. J. 1960. *Foundations of Electromagnetic Theory.* Reading, Pennsylvania: Addison-Wesley.

Schneider, K. H., Studinger, H., Weck, K. H., Steinbigler, H., Utmischi, D., Wiesinger, J. 1974. Displacement currents to the human body caused by the dielectric field under overhead lines. In: *International Conference on Large High Voltage Electric System,* Paper 36-04. Paris: CIGRE.

Schwan, H. P. 1977a. Electrical properties of tissue and cell suspensions. *Adv. Biol. Med. Phys.* 5:147–209.

Schwan, H. P. 1977b. Field interaction with biological matter. *Ann. N.Y. Acad. Sci.* 103:198–213.

Schwan, H. P., Kay, C. F. 1956. Specific resistance of body tissue. *Circ. Res.* 4:664–670.

Schwan, H. P., Kay, C. F. 1957. The conductivity of living tissues. *Ann. N.Y. Acad. Sci.* 65:1007–1013.

Scott-Walton, B., Clark, K. M., Holt, B. R., Jones, D. C., Kaplan, S. D., Krebs, J. S., Polson, P., Shepher, R. A., Young, J. R. 1979. *Potential Environmental Effects of 765 kV Transmission Lines: Views Before the New York State Public Service Commission, Cases 26529 and 26559, 1976–1978.* DOE/EV-0056. Springfield, Virginia: National Technical Information Service.

Shiau, Y., Valentino, A. R. 1981. ELF electric field coupling to dielectric spheroidal models of biological objects. *IEEE Trans. Biomed. Eng.* BME 28:429–437.

Silva, M., Zaffanella, L. E., Hummon, N. 1985. An activity systems model for estimating human exposure to 60 Hz electric fields. *IEEE Trans. Power Appar. System.* 104:1923–1929.

Spiegel, R. J. 1976. ELF coupling to spherical models of man and animals. *IEEE Trans. Biomed. Eng.* BME 23:387–391.

Spiegel, R. J. 1977. High-voltage electric field coupling to humans using moment method techniques. *IEEE Trans. Biomed. Eng.* BME 24:466–472.

Spiegel, R. J. 1981. Numerical determination of induced currents in humans and baboons exposed to 60-Hz electric fields. *IEEE Electromag. Compat.* EC 23:382–390.

Stuchly, M. A., Lecuyer, D. W., Mann, R. D. 1983. Extremely low frequency electromagnetic emissions from video display terminals and other devices. *Health Phys.* 45:713–722.

Swicord, M. L. 1985. Possible biophysical mechanisms of electromagnetic interactions with biological systems. In: *Biological and Human Health Effects of Extremely Low Frequency Electromagnetic Fields.* Arlington, Virginia: American Institute of Biological Sciences.

Swicord, M. L., Postow, E. 1986. *Hypothesis of Weak Electric and Magnetic Interactions with Biological Systems.* IEEE Spec. Pub. 86TH0139-6–PWR. Piscataway, New Jersey: IEEE Service Center.

Tenforde, T. S. 1985. Biological effects of ELF magnetic fields. In: *Biological and Human Health Effects of Extremely Low Frequency Electromagnetic Fields,* pp. 79–128. Arlington, Virginia: American Institute of Biological Sciences.

Toland, R. B., Vonnegut, B. 1977. Measurement of maximum electric field intensities over water during thunderstorms. *J. Geophys. Res.* 82:438–440.

4 Chronobiological Effects of Electric Fields

KENNETH R. GROH
MARIJO A. READEY
CHARLES F. EHRET
Biological, Environmental, and Medical Research Division
Argonne National Laboratory
Argonne, Illinois

CONTENTS

The primary biological effects of nonionizing electromagnetic radiation probably do not include direct genetic damage. More subtle and indirect modes of action, which include disruption of chronobiological relationships, however, are almost certainly involved. Disruption of temporal circadian relationships has been shown to have far-reaching biological consequences.

Such temporal disruptions can be caused by alterations in environmental cues (light, food, temperature), by social and behavioral cues (such as experienced during jet lag and shift work), and by numerous chemicals and drugs. Nonionizing electromagnetic radiation has been shown to influence entrained circadian rhythms by altering phase relationships of physiological parameters such as body temperature, oxygen consumption, and carbon dioxide production. A behavioral-phase measure, gross animal activity, is also changed by electromagnetic-radiation exposure. The phase shifts of these gross parameters caused by acute electric-field exposure are measures of the disruption occurring in fundamental biochemical and cellular processes.

Disruptions of established circadian rhythms are associated with alterations in the cell division cycle, which may lead to mutagenesis and oncogenesis. In entrained, synchronized metabolic and biochemical pathways, the disruptions of cyclic processes affect many biological functions, decreasing the effectiveness of therapeutic drug treatment, producing affective (mood) disorders, decreasing mental and physical well-being, and impairing immunocompetence. Electromagnetic exposure and the resultant disruption of chronobiological integrity are environmental factors that provoke growing concern as having potentially significant impacts on human health.

Until recently, the threshold limit values for safe exposure to previously unknown or unmeasured environmental factors have been determined by measuring damage to genetic materials (e.g., mutagenesis, carcinogenesis), or by determining the level at which a certain percentage of the test organisms die (LD_{50}). Seldom does the measurement parameter depend on so apparently innocent a property as cognitive function or overall physical health ("feeling good"). Newly developed measures of the damage to circadian regulatory systems (dyschronogenesis) and complex circadian behavioral responses are able to determine objectively if a small rodent or a man is "feeling good" or "feeling bad." These methods include the description of the 24-hr (circadian) waveforms of sensitive parameters (e.g., activity, respiration, hormone activity). Until recently, these methods have been used only in disciplines that were but distantly related to bioelectromagnetics. Because these chronobiological methods have now demonstrated highly distinctive and significant effects of the actions in living systems caused by electromagnetic fields, attention must be given to this discipline of circadian chronobiology and its relationship to the other integrating temporal disciplines.

A simple statement of a target object (typically used in classical toxicological studies, bioelectromagnetics, and radiation biology) at any one of the organizational levels leaves one with a static image of that object, frozen in space—until the relevant dynamic discipline is applied in the analysis of temporal interactions (Fig. 1). The disciplines that are concerned

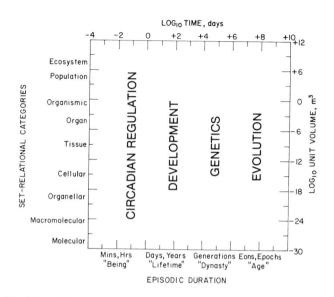

Fig. 1. The integrating disciplines of biological time structure. Evolution covers longest time spans (eons) and describes dynamics of population changes; genetics describes change and continuity on shorter time scale (generations); development describes changes within one lifetime; and chronobiology describes repetitive short-term physiological events within a lifetime (seconds, hours, days, weeks, years).

with domains of biological time are surprisingly few in number (four) and strike across the layers of levels of organizational complexity (Fig. 1). These four domains, in descending order of the biological time that they address (in providing rules, laws, and theorems of action) are evolution, genetics, development, and circadian regulatory biology. This last discipline studies the minutes, hours, days, weeks, and months within a lifetime, and is thus the one most closely allied with the maintenance of the health and function of the individual.

Mueller (1927) first showed that ionizing radiations most significantly damage living systems at the subcellular and macromolecular level where they cause gene mutations and chromosome damage. In sharp contrast, genetic damage does not appear to be involved as a primary biological effect of *nonionizing electromagnetic fields*. Rather, the primary effect appears to occur in chronobiological regulatory mechanisms that control temporal expressions in the ultradian (< 24 hr), circadian (~ 24 hr), and infradian (> 24 hr) domains (Fig. 2). An understanding of the importance of this newest of the integrating disciplines as it relates to the biological effects of electromagnetic field exposures requires a brief explanation of the discipline of circadian regulatory biology and its terms and rules.

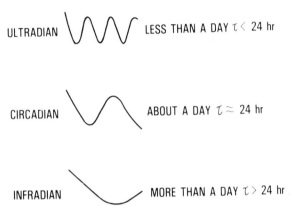

Fig. 2. Frequency domains. The frequency most thoroughly studied is about one period every 24 hr (circadian is defined from Latin *circa dies*, about 1 day). Ultradian rhythm periods are less than 1 day (high frequency), and infradian rhythm periods are more than 1 day (low frequency).

PROPERTIES OF CIRCADIAN RHYTHMS

Circadian oscillations are endogenous to all eukaryotic organisms and share common, definable characteristics (Edmunds and Laval-Martin, 1984). Their free-running period of about a day (circadian) is relatively temperature independent within normal physiological temperature ranges (temperature compensated). The phase of a circadian rhythm can be reset by the external environment (as seen in recovery from shift work and jet lag). The phases of these rhythms can be entrained to approximately 24-hr periods by a variety of environmental cues (Zeitgebers) such as 24-hr light:dark (LD) cycles, feed-starve (FS) cycles, and temperature (WC) cycles. In addition to these strong repetitive cycles, circadian rhythms can also be entrained by more subtle sound cues, social cues, and electromagnetic fields in mice (Dowse and Palmer, 1969; Duffy and Ehret, 1982; Ehret and Duffy, 1983) and in humans (Wever, 1979). When the entrainment cycle is removed (free-run), the entrained circadian rhythm continues with an underlying rhythm that had been hidden or "masked" during entrainment. The precise length of the free-running period (under constant conditions) is influenced by the quality of the synchronizing cycles responsible for entraining the organism before free-run. The qualities of the Zeitgebers (e.g., the intensity of a light cue) influence the length of the circadian period in a predictable manner in both nocturnal and diurnal animals (Aschoff, 1960). The descriptive parameters of the circadian waveform (period, amplitude, and mesor) are shown in Fig. 3.

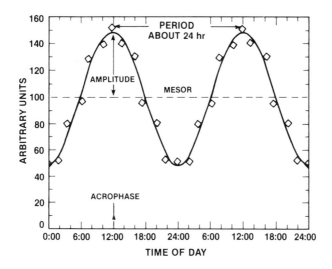

Fig. 3. Parameters of rhythms. The 24-hr cosine wave often used to model a circadian waveform has four parameters sufficient to describe cosine least-squares fit to a sinusoidal circadian rhythm: period, about 24 hr; amplitude, one-half distance between peak and trough within period; mesor, mean value of wave; acrophase, location of crest within 24-hr cycle.

The loss of circadian coordination within an organism (dyschronism) has been shown to have dramatic health effects (Halberg, 1976; Levine, 1976; Moore-Ede et al., 1983). Physiological and medical abnormalities that are clearly correlated with dyschronism include some immune dysfunctions (Smolensky et al., 1979), hypertension (Germano et al., 1984), affective disorders (Thompson, 1984; Welsh et al., 1986), shift-work fatigue (Ehret, 1981), and psychoses (Wehr et al., 1979).

Many experimental protocols in circadian biology involve, first, entrainment (giving a baseline for phase and amplitude), then free-run (removal of temporal cues), and finally an intervention during early free-run (exposure to the chemical or physical agent either at a specific phase of the physiological cycle or chronically). After a predetermined lag (frequently 7–10 days), the phases of the control animals (sham interventions) and of the exposed animals are compared to each other and to their relative phases during entrainment.

An example of a typical circadian study is seen in Figs. 4 and 5. The aim of this experiment was to determine if the methyl xanthine theoph-

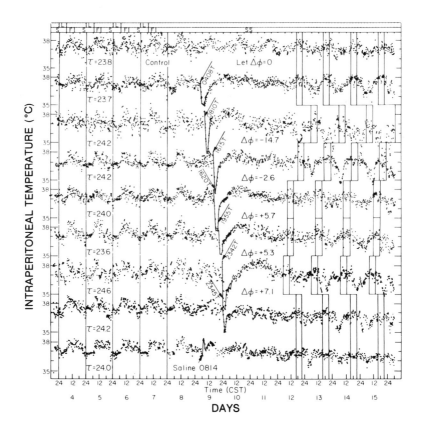

Fig. 4. Male rats were placed in light:dark (LD) cycles with 8 hr light and 16 hr darkness (LD 8:16), and feed-starve (FS) cycles of 8:16, during entrainment (through day 7; Fig. 4). Rats were then placed in free-run (days 8–16) with neither food (SS) nor light (DD). The circadian rhythm, which had a measured period length, tau (τ)=24.4 hr, continued unabated throughout free-run. Top animal is control. Other animals were given phase-specific injections of theophylline.

(Reprinted with permission from Ehret et al. 1975. Chronotypic action of theophylline and of phenobarbital as circadian Zeitgebers in the rat. *Science* 188:1212–1215.)

ylline is a Zeitgeber. The experimental design required punctate administration of the drug at predetermined intervals during early free-run (e.g., days 8–9 in Fig. 4). After this intervention, free-run continued. Some animals were indistinguishable from the controls, whereas others were either phase delayed or phase advanced. The circadian rhythm, which had a measured period length of 24.4 hr, continued unchanged throughout free-run. When results of this kind of experiment are plotted to show the magnitude of either delay or advance (phase shift) as a function of time of delivery (injection time), a *phase-response curve* (PRC) is developed. The

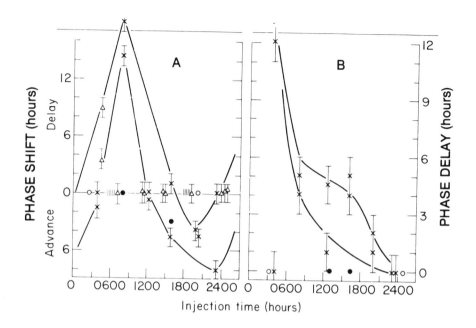

Fig. 5. Results of Fig. 4 are plotted in phase-response curve (PRC) to show magnitude of delay or advance (phase shift) as function of delivery time of theophylline (injection time) **(A)**. **(B)** Similar experiments with pentobarbital. PRCs show strong dependency of magnitude and direction of induced phase shift on injection time.
(Reprinted with permission from Ehret et al. 1975. Chronotypic action of theophylline and of phenobarbital as circadian Zeitgebers in the rat. *Science* 188:1212–1215.)

PRC for this theophylline experiment is given in Fig. 5A; a similar experiment with pentobarbital as the potential Zeitgeber gave the PRC in Fig. 5B. These PRCs show that phase shift depends strongly on the time of injection, both for the magnitude and the direction of the induced shift.

The circadian central pacemakers, which measure phase, can also be unaffected by a superficially positive circadian treatment (Fig. 6). The accuracy with which the proper (orthochronal) phase was recovered by the rats given a single exposure to lithium (Readey et al., in preparation) suggests that the master circadian pacemaker(s) controlling core body temperature was not disrupted by the punctate injection of lithium. That is to say, the circadian oscillators of the animals remained undisturbed and continued to keep track of the orthochronal phase relationships even during a strong transient disturbance. Treatments with this kind of chronobiotic agent [i.e., one that strikes at driven (or "slave") rather than at driving (or

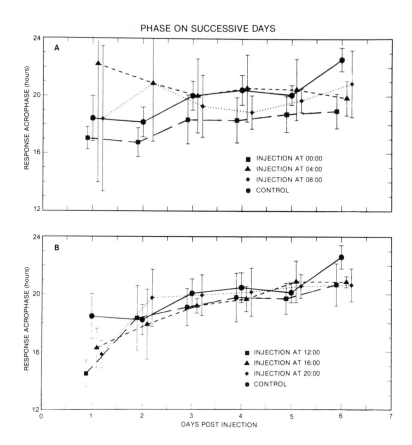

Fig. 6. Decay of treated animals to control phase. Rats treated with single exposure to LiCl change phase in chronotypic pattern; magnitude and direction of change (phase response) depend on time of administration (**A**). Phase change after single exposure to lithium (unlike that to theophylline and pentobarbital) is transient; periods immediately after exposure either shorten or lengthen phase change until altered phases are indistinguishable from phases of control animals. (**B**) Because rats were able to regain their preexposure phase, this result suggests that lithium does not target a master clock.

"master") oscillators] result in a transient resetting of the circadian clock with each dose.

PRCs demonstrate that a number of widely used pharmaceuticals are chronobiotics, for example, theophylline (Ehret et al., 1975), diazepam (Turek and Losee-Olson, 1986), barbiturates (Ehret et al., 1975; Peraino et al., 1980; Groh et al., 1988), and catecholamine-synthesis inhibitors (Cahill and Ehret, 1982). By definition, the response of an organism to a chronobiotic is highly dependent on the time of exposure. These temporal effects manifest themselves in health care, both as risks and as aids to good health.

CIRCADIAN DYSCHRONISM AND RELATED HEALTH RISKS

Indoleamine/Catecholamine Hypothesis and Circadian Addiction Physiology

The "biogenic amine" pathways commonly associated with the circadian clock are the indoleamine and catecholamine pathways (Figs. 7 and 8). These two pathways interact to form pacemakers for the millions of cellular clocks of the body. In a consistent dayworker, the catecholamine levels peak in the early morning and produce an "awakening" response. The indoleamines peak in the evening and trigger restfulness and sleep. Together they influence virtually every circadian property of the body (Fig. 9). These two pathways also are involved in the pharmacology of numerous addiction responses, including alcohol (Muller et al., 1980; Hunt and Dalton, 1981), opioids or narcotic analgesics (DiPalma and Sample, 1976), carbohydrates (Cahill and Ehret, 1982), barbiturates (Owasoyo et al., 1979; Owasoyo and Walker, 1980; Owasoyo, 1982), and amphetamines (DiPalma and Sample, 1976). Many of the addictive and dependency-forming drugs represented in these families are also chronobiotically active agents. The diazapam derivative triazolam (Turek and Losee-Olson, 1986), the methyl xanthines (Ehret et al., 1975; Ehret and Dobra, 1977), the amphetamines (Ruis et al., 1988), the opioids (Maestroni et al., 1987), and the barbiturates (Ehret et al., 1975; Peraino et al., 1980; Nishikawa et al., 1987a,b; Groh et al., 1988) have striking PRCs and other chronobiological properties.

Ethanol-addictive Wistar rats differ from nonethanol-addictive Sprague-Dawley rats in the effects of ethanol on their serotonin level; it is increased in the addictive animals and not in the nonaddictives (Muller et al., 1980; Hunt and Dalton, 1981). Norepinephrine levels appear to be lowered by chronic ethanol exposure in rat pups but not in adults (Muller et al., 1980). During intoxication, chronically treated mice show significantly depressed acetylcholine levels; withdrawal stages coincide with an increase and return to control levels of acetylcholine receptors (Tabakoff et al., 1980). Blood ethanol clearance rates are circadian in nature, with a fixed, endogenous relationship to core body temperature rhythms (Sturtevant and Garber, 1988), thus causing differential ethanol intoxication at different times of day. The risk of ethanol abuse may also be exacerbated in shift workers and other persons with poor chronohygiene because of the decreased mesor of the rate of ethanol clearance following a phase shift in exogenous Zeitgebers (Sturtevant and Garber, 1988).

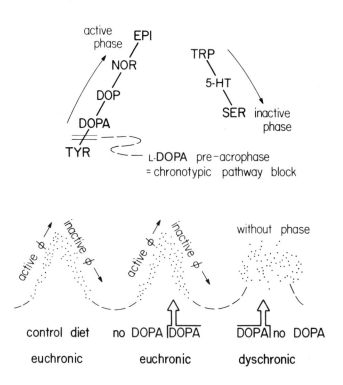

Fig. 7. Catecholamine-indoleamine model for circadian regulation shows negative feedback loop between stimulatory catecholamines and rest-inducing indoleamines. Circulating levels of these two families of hormones bring temporal information to the internal cellular clocks and thus coordinate them. In day-active animals such as humans, catecholamines are stimulated by light, increases in temperature, exercise, protein, social cues, and possibly electromagnetic fields; indoleamines are stimulated by darkness, carbohydrates, reduced temperature, and removal of social activity. If the endogenous clock sends these hormonal signals at appropriate times of day, eukaryote will be euchronic. If timing of signals is badly out of phase (eg., by addiction or induction of DOPA at wrong phase of circadian cycle), a eukaryote will become dyschronic from loss of temporal coordination among cells and organ systems. Abbreviations: TYR, tyrosine; DOPA, dihydroxyphenylalanine; DOP, dioctyl phthalate; NOR, norepinephrine; EPI, epinephrine; TRYP, tryptophan; 5-HT, 5-hydroxytryptophan; SER, serotinin.

Pentobarbital dependency alters circadian activity and feeding patterns in rats, shifting the bulk of their daily caloric intake to the early active phase. The late-active feeding peak is virtually eliminated by the drug dependency (Peraino et al., 1980). The turnover rate of serotonin also in-

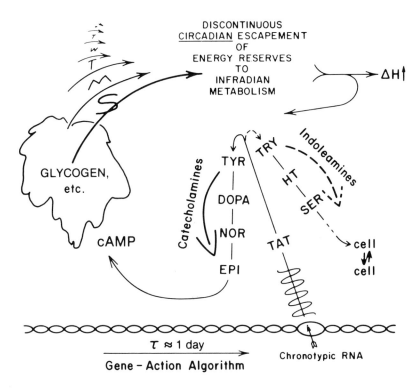

Fig. 8. Internal control of cell timing by catecholamine-indoleamine model. Biogenic amines interact with modulating parameters, such as liver glycogen reserves (which stimulate indoleamines and suppress catecholamines) to maintain consistent temporal environment. Any pharmaceutical that interferes with these control pathways would necessarily affect the clock. Abbreviations: cAMP, adenosine cyclic phosphate; EPI, epinephrine; NOR, norepinephrine; DOPA, dihydroxyphenylalanine; TYR, tyrosine; TRY, tryptophan; HT, hydroxytryptophan; SER, serotonin.
(Reprinted with permission from Ehret, C. F. 1974. The sense of time: evidence for its molecular basis in the eukaryotic-gene action scheme. *Adv. Biol. Med. Phys.* 15:44–77.)

fluences the severity of convulsions during barbital withdrawal (Tagashira et al., 1982). Figures 10 through 12 (Groh et al., 1988) demonstrate the chronobiological profile of phenobarbital addiction through habituation (Fig. 10), addiction (Fig. 11), withdrawal (Fig. 12), and recovery. During the full addiction phase, the circadian waveform of the core body temperature of the addicted animals is stronger than that of the control animals. Addicted animals have a clearer circadian pattern, less ultradian variation, and an increased amplitude. This pattern suggests that the

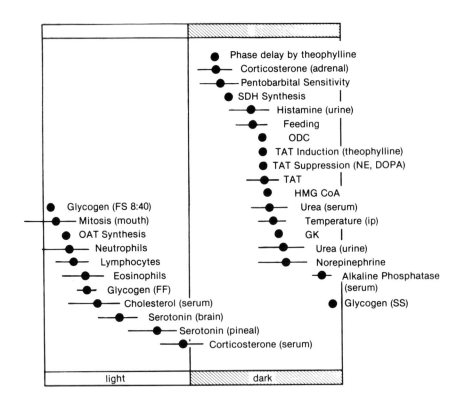

Fig. 9. Circadian chronotype of rat. These chronotypic traits depict acrophase or peak levels of physiological parameters against approximate time of relative light:dark cycle where that peak occurs. Lines represent 95% confidence interval for acrophase. Abbreviations: FS, feed-starve; OAT, ornithine amino transferase; FF, feed-feed; SDH, serine dehydratase; ODC, ornithine decarboxylase; TAT, tyrosine aminotransferase; HMG, β-hydroxy-β-methylglutaryl; ip, intraperitoneal; GK, glucokinase; SS, starve-starve.

dependent animals slept more deeply (lower body temperature) and had greater wakefulness (correlate of higher body temperature) during the active phase than the control animals.

Drugs that enhance the waveform in this manner would produce a circadian "high," or an artificially elevated "mood" (affect) in the animal. A person who has a reduced or disturbed circadian waveform, such as a shift worker (Cahill and Ehret, 1984; Reinberg et al., 1984), a worker under high job stress, a person with an affective disorder (Pflug et al., 1976), or one who is unemployed and thus without circadian social cues may therefore be at greater risk of dependency than a person with a similar genetic and social background but with a strong circadian waveform.

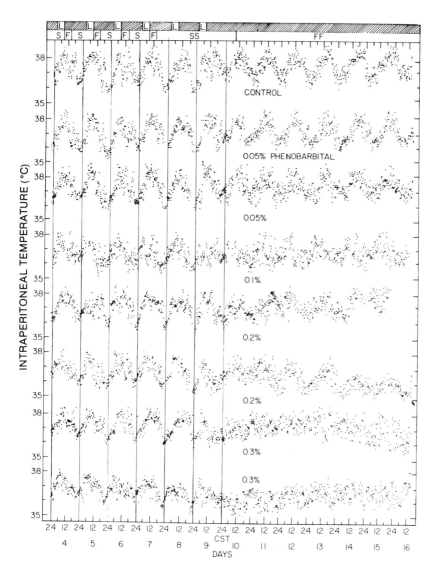

Fig. 10. Dyschronism as result of phenobarbital administration (habituation). Rats in free-run were given various percentages of phenobarbital in their food. At all levels tested, partial dyschronism resulted. Between 0.2% and 0.3% phenobarbital, waveform became visual blur. At 0.25%, phenobarbital acts as promoter of liver cancer (Peraino et al., 1980).

(Reprinted with permission from Ehret, C. F., Dobra, K. W. 1977. Oncogenic implications of chronobiotics in the synchronization of mammalian circadian rhythms: barbiturates and methylated xanthines. In: *Proceedings of the Third International Symposium on the Detection and Prevention of Cancer,* Nieburgs, H. E., ed., pp. 1101–1114. New York: Marcel Dekker.)

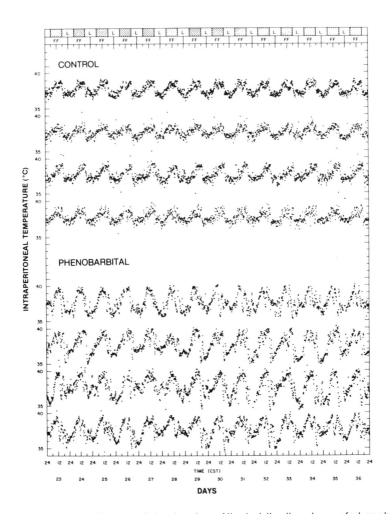

Fig. 11. Addiction. Once past dyschronism of the habituation stage of phenobarbital dependence, addicted rats develop stronger, more clearly defined circadian waveforms than nonaddicted controls. These plots of intraperitoneal temperature, a strong correlate of mood, indicate rats have physiologically adapted to drug and are functioning better than controls, as judged by this indicator.
(Reprinted with permission from Groh et al. 1988. Circadian manifestation of barbiturate habituation, addiction, and withdrawal in the rat. *Int. J. Chronobiol.* 5:153–166.)

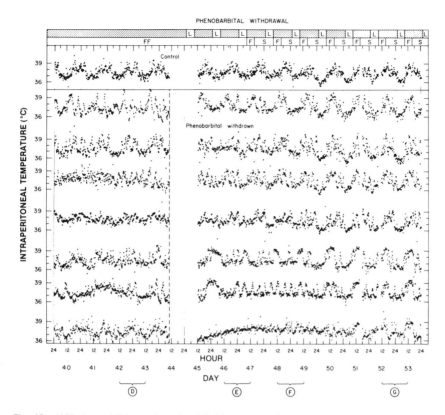

Fig. 12. Withdrawal. When phenobarbital is removed from addicted animals, resulting withdrawal syndrome can be characterized by partial or full loss of circadian frequency of biological waveform, even in presence of external LD cycle. Withdrawal lasts about 4–5 days. Not surprisingly, phenobarbital withdrawal and classical dyschronism share common characteristics, including disturbed sleep, acute eating disorders, acute affective disorders, and hallucinations.

(Reprinted with permission from Groh et al. 1988. Circadian manifestation of barbiturate habituation, addiction, and withdrawal in the rat. *Int. J. Chronobiol.* 5:153–166.)

Affective Disorders

Common clinical features correlating affective disorders with a disturbed circadian system (Pflug et al., 1976; Kripke et al., 1978; Giedke and Pflug, 1979; Jenner and Damas-Mora, 1979; Wehr and Goodwin, 1981) include premature (phase-advanced) awakening and increased diurnal variation in depressed patients (Weitzman, 1981) and remission of depressive symptoms following one night's sleep deprivation (Pflug and Tolle, 1971).

Depressed patients experience a premature acrophase of many circadian rhythms (Wehr et al., 1983) and a shortened free-run periodicity (Wehr et al., 1983); manic-depressives experience a preference for easterly (phase-advancing) air travel (Kripke et al., 1978). During the change from a depressive to a manic phase, a bipolar patient can experience from 1 to 10 sleep-wake cycles of 48 hr each (Wehr et al., 1982). Similar 2-day rhythms can occur in persons without an affective disorder if they are in temporal isolation (Chouvet et al., 1974; Wever, 1979), suggesting that the 2-day sleep-wake rhythm can be present but suppressed by temporal cues. In addition, dyschronogenic conditions can induce affective symptoms (Kripke et al., 1978; Graeber, 1982; Jauher and Weller, 1982; Reinberg et al., 1984).

Both synchronization (Richter, 1960) and desynchronization (Wever, 1979) of two or more endogenous oscillators (Aschoff, 1965; Pittendrigh, 1974; Wever, 1979; Moore-Ede et al., 1976; Monk et al., 1983; Moore-Ede, 1983) have been suggested as mechanisms underlying cyclic psychoses. One of these mammalian circadian pacemakers appears to be located in the suprachiasmatic nucleus (SCN) of the hypothalamus, and the second in an as-yet-unidentified anatomical area (Moore-Ede, 1983). In these anatomical models, dependent on the biochemical pathways discussed, the underlying strong and weak oscillators either couple or uncouple (Pflug et al., 1976; Kripke et al., 1978), changing their relative phase angle and thus causing the cyclic mood disorder. The strong oscillator is thought to phase advance more often and to a greater extent than the weak oscillator (Wehr et al., 1983). Highly variable, abnormal circadian-phase relationships are thus associated with mood disorders.

Intercellular communication of temporal information is essential to a strong circadian rhythm. Dyschronisms such as phase-angle disorders, loss of circadian waveform, and depressed amplitudes may result from either improper pacemaking or the breakdown of the communication of temporal information among the individual cells (clocks). The ability of dyschronism to induce acute affective symptoms in individuals with no prior history of such disorders suggests that either a transient, shortened period or a displacement of the phasing of environmental and endogenous cues may play a role in some affective disorders (Kripke et al., 1978; Graeber, 1982; Jauher and Weller, 1982; Reinberg et al., 1984). The symptoms of dyschronism such as those experienced during jet lag and shift work, including the transient affective symptoms, can be alleviated by the use of multiple Zeitgebers in the proper (orthochronal) phase relationships (Ehret et al., 1975, 1980; Ehret, 1983).

Chronohygiene and Affective Disorders

The practice of good circadian chronohygiene can control and indeed prevent some symptoms of dyschronism. Chronohygienic practices, including the control of dietary, social, and lighting cycles (Wever, 1979; Wehr et al., 1985; Lewy and Sack, 1986), may aid in the control of recurring dyschronism symptoms, including those seen in bipolar disorders, shift workers, and individuals suffering from jet lag (Ehret et al., 1978, 1980; Ehret, 1981). Multiple controlled orthochronal time cues can be given alone or in combination with an orthochronally administered program of phase-shifting or period-lengthening medications such as Li^{2+} (Jenner, 1970; Lingjaerde, 1983), with sleep-deprivation therapy (van Bemmel and van den Hoofdakker, 1981; Dessauer et al., 1985; Baxter et al., 1986; Borbely, 1987), or with light therapy (Czeisler et al., 1986; Lewy and Sack, 1986; Lewy et al., 1987).

Sleep deprivation and phototherapy, like lithium therapy, are punctate in their effects and need to be repeated episodically to achieve therapeutic effects. Sleep-deprivation therapy also resembles lithium therapy in that it appears to be effective in treating dyschronism-related bipolar disorders (Nosachev, 1985; Kuhs and Tolle, 1986; Levine, 1986). An orthochronal regime, given by itself or in combination with punctate chronobiotics, may serve to improve the long-term remission of endogenous and exogenous dyschronism disorders. The orthochronal approach should work directly on the pacemaker system if the problem is one of phasing. If the period is shortened in a given individual, then a period-lengthening drug such as lithium could be coupled to a 24-hr orthochronal regime to help the person to adapt to his newfound circadian environment.

Circadian Physiology and Carcinogenesis

Many chronobiologists now suspect a link between the loss of the timing of cellular events (dyschronism) and cancer promotion (Ehret and Dobra, 1977). Circadian dyschronism is the disruption of the circadian integrity of the temporal physiological organization that is present from the cellular to the organismic levels. By definition, dyschronism at the cellular level interferes with the timing and function of cell division and transcription/translation cycles. Biochemical and environmental agents that disrupt cell cycles are frequently oncogenic, teratogenic, or mutagenic.

Dyschronism at the organismic level is associated with dyschronism at the cellular level. Light, a strong circadian Zeitgeber, also can synchronize populations of slowly dividing (infradian mode) cells (Meinert et al., 1975) and of exponentially growing (ultradian mode) cells (Readey, 1986, 1987). In these populations of cells, the events leading to cell division are restricted to specific phases of the cycles of the synchronizing Zeitgeber cycle. If cell division is blocked during the phase of the cell cycle when division events are possible, the excess DNA that is synthesized is maintained and eliminated on release from the block by repeated, rapid divisions until the proper DNA complement is reestablished.

The important Zeitgebers for humans include visible light (Czeisler et al., 1986), social interactions (Wever, 1979), eating schedules (Ehret et al., 1980), and pharmacological agents (Ehret et al., 1975; Groh et al., 1988). Individuals at risk for chronic circadian dyschronism include those who have a poor or marginal circadian waveform and who are further perturbed by phase-specific exposures to chronobiotic agents, such as persons who work odd hours, do shift work, or have irregular exposure to or phase ratios of the major Zeitgebers.

The phase relationships in which the Zeitgebers and chronobiotics interact in moderately to severely dyschronic individuals also differ from normal diurnal patterns. As a consequence, the dyschronism experienced by these people is further enhanced at both organismic and cellular levels (Czeisler et al., 1982; Cahill and Ehret, 1984). Although most models of circadian rhythmicity involve the cell cycle, to date no studies have tested repeated resetting of the biological clock for the oncogenic effects that are consequences of these models. Rhythmicity in carcinogenesis (Moller, 1984) has not been adequately explored to determine the extent of circadian variations in cell proliferation in neoplastic cells. Although a disturbed temporal structure is characteristic of a variety of cancers, including ovarian and breast cancers and leiomyosarcoma (Bailleul et al., 1986), less is known about the contributions of dyschronism to oncogenesis. Several theories of the production of circadian rhythms (chronon: Ehret and Trucco, 1967; cytochron-cytogene: Edmunds and Adams, 1981; quantal: Klevecz, 1976) suggest that oncogenesis is a potential danger and result of dyschronism.

Dyschronism interferes with the timing and function of the cell division and cell transcription cycles (Readey, 1987) and thus could be oncogenic, teratogenic, and/or mutagenic. One chronobiological approach to cancer therapy relies on the fact that antineoplastics, and the cytotoxicity of antineoplastics, are both often cell cycle dependent in their mode of action (Hrushesky, 1983, 1985; Klevecz and Braly, 1986, 1987). Although removal of the complex of Zeitgebers has been implicated in both dyschronism and oncogenesis, the correlation between dyschronism and oncogenesis has not yet been experimentally tested.

The data on interaction between human neoplastic cells and circadian rhythms that do exist are fragmentary (Voutilainen, 1953; Scheving, 1959; Mauer, 1965; Garcia-Sainz and Halberg, 1966). Recent data (Klevecz and Braly, 1986, 1987) indicate fundamental differences in the temporal organization of both human and mouse ovarian cancers from normal ovarian tissues. Although normal ovarian tissue shows a strong 24-hr periodicity, ovarian tumors in both humans and mice show a severely reduced amplitude for their circadian rhythms (dyschronism) and strengthened 12-hr and higher frequency rhythms of proliferation. Similar results have been seen in mouse melanoma (Garcia-Sainz and Halberg, 1966).

Both the rate of cell proliferation in the tissues and the phase in the cell cycle influence the effectiveness of a carcinogen (Laerum, 1976). Proliferating tissues are generally considered at higher risk for carcinogenesis than nonproliferating tissues (Columbano et al., 1981; Rajewsky, 1986). Studies have linked specific cell cycles and circadian-phase tumor yield after phase-specific induction of carcinogenesis both in vivo and in vitro (Bertram et al., 1975; Iverson and Kauffman, 1980; Iverson, 1982a,b; Clausen et al., 1984).

Hepatic carcinogenesis occurs most frequently when the tissue is exposed during the early S phase of the cell cycle (Rabes et al., 1986). Nonproliferating cells maintain their circadian physiology and continue to cycle, passing through a modified S phase about every 24 hr (Klevecz et al., 1975). It is not clear at this time if the lower rate of carcinogenesis in nonproliferating cells truly results from gained physiological resistance, or if the targeted phases of the cell cycle for initiation or induction (such as S phase) simply occupy a relatively smaller portion of the cycle and are therefore harder to hit.

There is some evidence that suppression of the immune system leads to an increased suceptibility to cancer (Blome and Larko, 1984). Again, immune dysfunction (Smolensky et al., 1979) may be a consequence of dyschronism and thus another, possibly secondary, factor for the onset of carcinogenesis in dyschronic individuals. To be effective, a chronobiotically active carcinogen or cocarcinogen would need to be administered at the proper phase angle, even in an immunologically compromised individual. Note, however, that the link between immunosuppression and carcinogenesis is weak, and only clearly established for lymphatic cancers and Kaposi's sarcoma.

Humans become progressively more likely to develop cancer with age, and indeed the incidence of cancer rises sharply toward the end of the life span in numerous species (Anisimov, 1983; Holmes, 1983). This increase in cancer is correlated with the breakdown of circadian regulation in older animals (Sacher and Duffy, 1978; Duffy et al., 1987).

Circadian Regulation and Immune Responses

Anatomically, the immune system and the circadian regulatory system share many components. Lymphocyte rhythms of activity and proliferation are influenced by the hypothalamic–pituitary–adrenal–cortical axis (Riley, 1981). The suprachiasmatic nuclear region of the hypothalamus is strongly implicated in the regulation of the vertebrate clock (Kawamura et al., 1982; Rusak, 1982). In the adrenal cortex, catecholamine release, a central circadian regulatory component, is itself modulated by the glucocorticosteroids (Weiner and Taylor, 1985). In this context, the synthetic corticosteroid dexamethasone is another powerful Zeitgeber for which elegant circadian PRCs are known (Horseman and Ehret, 1982). Corticosteroids also modulate immune functions, possibly by controlling the expansion of lymphocyte populations by inhibiting T-cell growth factors (Hayashi and Kikuchi, 1982; Indiveri et al., 1985). Epinephrine may also play a role in immunoregulation (Jankovic and Isakovic, 1973).

Indoleamine (serotonin and melatonin) regulation is directed by the pineal gland. Melatonin levels help transmit information about the temporal environment to the various organs and their endogenous cellular clocks. This transmission includes information about light and temperature (Axelrod et al., 1982; Reiter, 1984) as well as the electromagnetic environment (Wilson et al., 1981; Olcese et al., 1988). At most times of day, inhibition of melatonin synthesis reduces the humoral and cellular immune responses. The only time when this result is not obtained is when synthesis is inhibited in the early light phase, when melatonin levels normally decrease. Similarly, treatment with exogenous melatonin in the early dark phase (during the phase when melatonin levels normally increase to the maximal active phase levels) increased the primary responses of mice to an antigen (Maestrioni et al., 1987). As predicted by Ehret et al. (1979), the circadian rhythms of both the indoleamines and catecholeamines are altered by exposure to 60-Hz electric fields (Vasquez et al., 1988). The corresponding 25% inhibition of allogenic cytotoxicity of target cells by cytotoxic T lymphocytes is also seen after a 4-hr exposure to a 60-Hz field (Lyle et al., 1988).

The strength of the immune response in mice and hamsters varies seasonally, with the highest resistance to antigens occurring during the long days of summer. This increase again correlates well with the decrease in melatonin during the long photophases of summer. The corresponding increase in melatonin occurs during the winter, when long dark scotophases are the norm. In small rodents, this seasonal difference is mainly expressed in decreased weight of the spleen in winter, although no difference in antibody production is observed (Petterborg et al., 1981; Brainard et al., 1987). These types of changes have also been correlated with the intensity of the light (number of delivered photons: Vriend and Lauber, 1973). This seasonality is supplemented by an endogenous annual rhythm that persists under constant light:dark (LD) ratios (Kim et al., 1980; Brock, 1983). In beagles, the number of cells undergoing lectin-induced lymphocyte transformation varies seasonally (Shifrine et al., 1980, 1982a). In the limited human studies to date, it is unclear if the peak in the number of circulating lymphocytes occurs in summer (Rocker et al., 1980) or in winter (Reinberg et al., 1977, 1978). The studies of Bratescu and Teodorescu (1981) indicate that human T lymphocytes peak in late fall and B cells peak during winter. Transformation of lymphocytes also indicate a circannual (endogenous yearly) cycle of cell-mediated immune responsiveness (Shifrine et al., 1982b).

In a favorable circadian environment, a healthy individual has circadian rhythms in lymphocyte subsets (Hahn et al., 1980; Haus et al., 1983; Ritchie et al., 1983; Miyawaki et al., 1984; Signore et al., 1985). Felder et al. (1985) failed to show such a rhythm in lymphocyte stimulation in patients with rheumatoid arthritis, an autoimmune dysfunction; it is possible that the periodicity was present, but at a much reduced amplitude with smaller changes in T:B ratios.

During dyschronism induced by light:dark inversions of an L:D 12:12 cycle, the circadian rhythms of corticosterone levels adjusted to the new phase relationships during a 5- to 6-day interval (Hayashi and Kikuchi, 1982). During that time, the mesor corticosterone level also increased above noninverted LD phase controls. This rise was correlated with a decrease in T-lymphocyte levels, which are normally suppressed by increases in the endogenous corticosterone levels (Shek and Sabiston, 1983). It is not known if the ratios of T-cell subpopulations are also altered. This change is, in turn, correlated with a decrease in the immune response in mice to both thymus-dependent and thymus-independent antigens (Hayashi and Kikuchi, 1982, 1985; Kort and Weijma, 1982). Nakaio et al. (1982) and Aizawa et al. (1982) showed that rotating shift work and permanent night work suppressed the immunofunction of T cells. This depression of the immune response was differentially altered by different rotation systems.

Although a separate phenomenon, sleep deprivation is often associated with dyschronism, and mild cases of these two conditions often occur in tandem under normal social conditions. Total sleep deprivation results in a decreased uptake of interperitoneally administered antigens by peritoneal cells and spleen, but with a corresponding increase in uptake by the liver (Casey et al., 1973).

Circadian Rhythms and Bioelectromagnetics

Because the central nervous system is intimately involved in the regulation of circadian rhythms, and because biogenic amines have a central role in neurological control, Zeitgebers and chronobiotics that are administered at an inappropriate time of the day result in chronotypic changes in levels of neurotransmitters, such as indoleamines and catecholamines, that coordinate the individual cellular clocks within the body (Cahill and Ehret, 1981). The loss of intracellular temporal coordination/communication induces dyschronism, which increases the risk of genetic damage, oncogenesis, and mutagenesis.

In this context, the relationship between electromagnetic-field exposure and carcinogenesis is not immediately obvious. It is difficult to reconcile the largely negative results from traditional laboratory approaches such as assays for mutagenicity and carcinogenicity with the claims of a few epidemiological investigations, many of which appear to be positive. The

epidemiological results, if accepted, appear to show a link between electromagnetic-field exposure and increased incidence of cancer, especially in
children living close to power lines. This discrepancy is not easily explained
in terms of classical mechanisms. This chapter explains how this gulf can
be bridged if certain chronobiological properties of living systems are taken
into account.

The electromagnetic spectrum, especially the visible light region, has
long been known to have a strong synchronizing influence on chronobiological rhythms (LD cycle) (Aschoff, 1965), although the receptors and
mechanism(s) have not been completely elucidated. At the cellular level,
the action spectra of the porphyrins and nucleic acids were shown to be
involved in circadian clock resetting (Ehret, 1959). A typical approach to
the identification of a photoreceptor pigment(s) is to determine an action
spectrum for a given chronobiological process and look for a corresponding absorption spectrum of some putative substance present in the cell
or tissue. However, different photoreceptor systems appear to underlie
different circadian biological responses to light, even in the same species.
For example, 10-fold less energy of blue light (442 nm) was required for
phase shifting in the early subjective-night phase of the free-running circadian rhythm of eclosion in *Drosophila pseudoobscura* than was necessary in the late subjective-night phase, which could suggest that two different pigments or primary processes are responsible for light absorption
at the two times (Chandrashekaran and Engelmann, 1973).

Different spectral sensitivities (presumably different photoreceptors)
exist for entrainment and photoperiodic induction in the house sparrow
Passer domesticus (Menaker and Eskin, 1967). Noninductive green light
can entrain circadian rhythms even in blinded birds (Menaker, 1968) without affecting testicular potency. This wavelength was used to phase the
rhythms of motor activity and of sensitivity to light so that perturbations
by potentially inductive white light could fall at different circadian times.
As anticipated, the white-light exposures were either inductive or noninductive depending on the circadian time (CT) at which the signals were imposed. Winfree (1970) also has reported that this same circadian system exhibits two qualitatively different types of response, that is, phase-resetting
behavior, in response to the length of the light pulse administered (type 0,
"fast"; type 1, "weak," gradual transition).

A study of the clock receptor in the hamster retina (Takahashi et al.,
1984) showed that the rhodopsin-containing photoreceptor in the cones
has a high threshold (about 1 million times higher than that for rod vision). The reciprocity between intensity and duration of light stimuli is maintained for pulses as long as 45 min. This system appeared to serve as a
photon integrator. Ninnemann (1979) compiled a comprehensive and critical review of photoreceptors for circadian rhythms in both plants and animals. Given the complexity of the problem as just discussed, it perhaps

is not surprising that Ninnemann concluded that there was no specific universal light-absorbing pigment in organisms that effectively initiates, phase shifts, or inhibits circadian rhythmicity across species. Action spectra for the phase-shifting of circadian rhythms in *Paramecium* (Ehret and Wille, 1970), *Neurospora, Gonyaulax, Drosophila*, the moth *Pectinophora*, and the hamster *Mesocricetus* were all different. Just as no specific universal structure has been developed as the anatomical locus for circadian oscillators, no universal photoreceptor or pigment has evolved.

CIRCADIAN EFFECTS OF ELECTRIC-FIELD EXPOSURE

Exposure to nonionizing electromagnetic fields has been known for some time to alter circadian rhythms. One of the most sensitive circadian responses to an electric field was seen by Wever (1970) in humans. A 2.5-V/m, 10-Hz square-wave field caused internal desynchronization of body temperature and activity and shortened the circadian period for activity.

Although most research into the effects of electromagnetic radiation has centered on 60 Hz, the absence of identification of a receptor or solid mechanism for electromagnetic radiation effects suggests either multiple receptors and/or mechanisms. Blackman (1988) found discrete "windows" of frequency sensitivity for the efflux of calcium ions from chicken brain tissue. Harmonic frequency stimulation and amplification of the weak, nonionizing nature of electromagnetic radiation have been considered necessary to explain the observed biological effects of electromagnetic radiation (Adey, 1988). Semm and Demaine (1986) found peaks in sensitivity to magnetic fields when pigeons were simultaneously exposed to 503- and 582-nm light. These few examples suggest that the receptor(s) and mechanism(s) for electromagnetic effects on biological organisms may be specific not only for frequency and electric or magnetic field but also for species, as well as specific for the biological effect under observation.

Many circadian studies require facilities in which many animals can be continuously monitored for long periods of time without disturbing ongoing entrained ultradian and circadian rhythms and that permit simultaneous exposure to multiple environmental and biochemical time cues (Zeitgebers), with the monitoring of subsequent response(s) in a noninvasive and disruptive manner (Runge et al., 1974; Ehret et al., 1979). Facilities that monitor and record circadian and ultradian metabolic and behavioral responses provide the most sensitive measures that can be obtained of environmental perturbations of the nervous system. In our laboratory, microcomputer-controlled data acquisition systems collect circadian data from organisms that range from populations of eukaryotic cells to individually housed small mammals (rats and mice) (Jaroslow and Eisler, 1968; Runge et al., 1974; Ehret et al., 1979; Cahill and Ehret, 1982; Horseman and Ehret, 1982; Rosenberg et al., 1983; Groh et al., 1987).

Because of the variation among individual animals, large sample sizes for each experimental condition are helpful, especially when weak Zeitgebers such as electric fields are being studied. The timing of the selected Zeitgeber and its mode of administration are also critical to the magnitude and direction of the response. Long-term, simultaneous monitoring of multiple metabolic parameters such as energy metabolism, animal activity, and body temperature have revealed underlying ultradian rhythms that are dependent on the inherent circadian phase and the circadian perturbations of environmental time cues (Sacher and Duffy, 1978). Simultaneous measurement of several chronobiological physiological and behavioral parameters in two of these microcomputer-controlled data acquisition systems has been used to investigate the effect(s) of horizontal, floating, and vertical, grounded 60-Hz electric fields.

In one facility (Groh et al., 1987), mice were tested in an eight-cage respiration system (CO_2 production and O_2 consumption) that additionally monitors animal activity and deep body temperature in the presence of a vertical, grounded 60-Hz electric field. The white-footed mouse *Peromyscus leucopus* is typically used as the test subject, but several other rodents have been examined. The effects of electric-field exposures (individual, multiple, or continuous) have been examined after light-cycle entrainment [light:dark (LD)]. When mice are exposed to a 60-Hz electric field, an immediate "startle" response in activity and increased respiration is observed (Fig. 13) (Sacher and Duffy, 1978; Rosenberg et al., 1983). The animals respond to the presence of the electric field at a threshold field strength of 35–50 kV/m. Below this range, observed responses are few and inconsistent. Above 50 kV/m, there is no increase in the percentage of the population of mice responding when they are exposed to the electric field.

A typical longitudinal plot of circadian LD entrainment followed by exposure to an electric field can be seen in Fig. 14. For as long as 14 days, mice were entrained to a light:dark (LD 12:12) cycle followed by 8 days of free-run, where all environmental cues were removed or held constant (DDFF). A 60-Hz electric field was then applied (and the absence of exogenous time cues was maintained). Protocols of constant or multiple, intermittent exposures to increase the circadian effect have been explored. As can be seen in Fig. 14, there was an immediate "startle" response, with an increase in activity and respiration when the electric field was first established, followed later (during the first exposure) by a generalized increase in total activity and respiration. Exposures at the same circadian time on subsequent days continued to stimulate a response to the electric field for several days; this was followed by a generalized decrease in the amplitude of response by the end of the experiment (day 10). After the third electric-field exposure, circadian torpor or "mini-hibernation" was

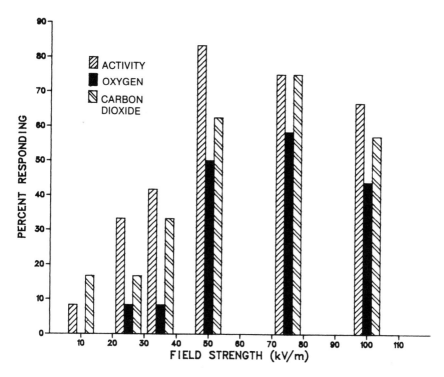

Fig. 13. Percentage of mice that exhibit arousal or "startle" response to 60-Hz electric field between 10 and 100 kV/m, first hour of exposure, grounded. Activity, oxygen consumption, and carbon dioxide production are indicated by histograms.
[Reprinted with permission from Groh et al. 1980. Biomedical effects associated with energy transmission systems: effects of 60-Hz electric fields on circadian and ultradian physiological and behavioral functions on small rodents. In: *Energy Conservation,* Tech. Rep. DOE/TIC/1027653 (DE81027653). Washington, DC: U. S. Department of Energy.]

observed in this animal, as evidenced by both oxygen consumption and carbon dioxide production. The acrophase of the three measured circadian parameters remained at midnight, at the midpoint of the active, dark phase, and its position was not altered by the multiple electric-field exposures.

Another circadian response to 60-Hz electric fields that we have observed is "splitting" in which the single, circadian peak separates or "splits" into two peaks of activity and respiration (Fig. 15). The control, sham-exposed animals did not show any splitting during free-run (Fig. 16). One exposure was sufficient to cause this response, and the magnitude of the splitting often approached the classical, 12-hr, 180° out-of-phase response also known to be caused by lesions of the suprachiasmatic nucleus of the

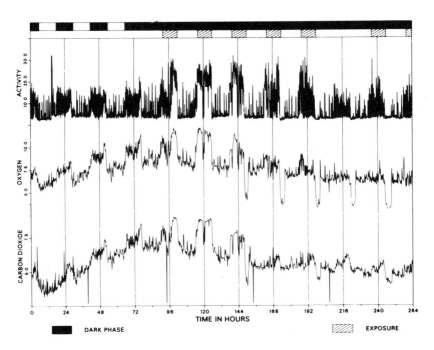

Fig. 14. Circadian effect of electric-field exposure on animal activity (top), oxygen consumption (middle), and carbon dioxide production (bottom). Representative mouse was entrained for 14 days on light:dark (LD) 12:12 (upper schedule, top), with food available ad lib (FF). On third day, LD entrainment ceased; rest of experiment was free-run (DD). At onset of entrained dark phase of LD cycle, mouse was exposed to 100-kVm electric field for 9 hr on 7 subsequent days (lower schedule, top). (See Fig. 15 for parameters.)

hypothalamus (Rusak, 1977). Pittendrigh and Daan (1976) identified two distinct peaks of activity in hamsters, the first at lights-on (morning) and the other at lights-off (night). In free-run (DD), when an electric-field exposure occurs at one of these boundaries, the electric field appears to mimic the Zeitgeber effect of light and cause a similar separation of the single activity peak into two distinct bursts of activity. Pittendrigh and Daan (1976) proposed a two-oscillator model for the circadian pacemaker that responds to different light intensities and appears to also respond to electric-field exposure.

One of the strongest circadian responses to 60-Hz electric-field exposure is a phase shift in the entrained circadian rhythm of activity and respiration (Fig. 17). Phase shifts to both an earlier (advance) and to a later time of day (delay) have since been observed in mice and rats. A single, 4-hr exposure above a threshold field strength of 35 kV/m is sufficient to

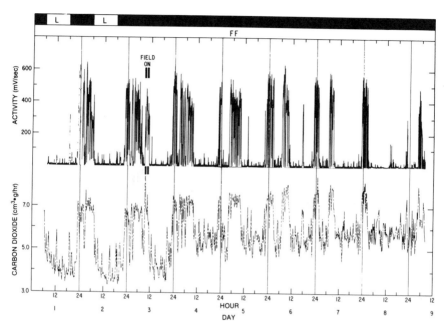

Fig. 15. Circadian "splitting" following short electric-field exposure. Experimental conditions for representative mouse as described in Fig. 14, except 2- to 30-min, 100-kV/m exposures, separated by 30 min, were given on third day during early inactive (light) phase of circadian cycle.

cause a phase shift in the acrophase of activity and respiration. As noted earlier, such phase shifts can be associated with changes in mood, immune function, or reliability of cell replication process. Whether these purely circadian effects are beneficial or harmful to an organism depends on their ability to strengthen or weaken the circadian structure of that organism.

A convenient procedure to represent these phase shifts is the cosinor method of Halberg et al. (1972). A linear least-squares sinusoidal fit to the data is typically graphed on a clock face, with error measures of the fit. In Fig. 18, a phase delay of 4.6 hr in the carbon dioxide circadian acrophase was observed after this entrained animal was exposed to a 60-Hz electric field for 9 hr. Phase shifts as great as 8 hr (advance and delay) have been induced by exposure to electric fields. The magnitude and direction of a circadian-phase shift depend on the phase relationship between the exposure to the electric field and the entraining light:dark cycle. The magnitude of the arousal or "startle" response and the magnitude of the subsequent phase shift are also positively correlated (Fig. 19). The phase delay seen after an electric-field exposure is greater when the animal also

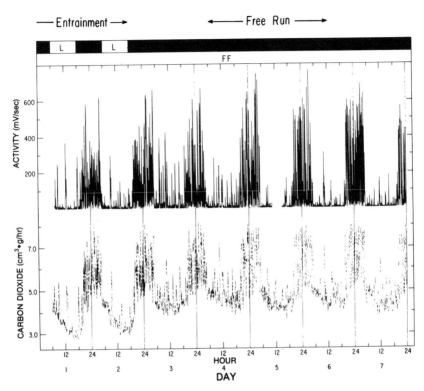

Fig. 16. Control, sham exposure. Mouse was entrained and entered free-run (DDFF) as did animals represented in Figs. 14 and 15, but no electric-field exposure was given on third represented day. Switch to dark:dark (DD) also demonstrated the "light masking effect," which was observed as baseline depression of carbon dioxide production during light (L) portion of LD cycle. No phase shift is seen in activity (top) or carbon dioxide (bottom).

shows a large startle response (Duffy et al., 1987; Groh et al., in manuscript).

In a number of animal species, and in man, 60-Hz electric fields are circadian synchronizers. The circadian effects seen in this laboratory have been primarily phase shifts (both advances and delays) in animal body temperature, activity, and respiration. These endpoints were chosen to permit noninvasive, sensitive monitoring of the circadian effects of electric-field exposure. As is shown in later chapters in this volume, however, these endpoints are not the only circadian effects that have been observed as a result of electromagnetic-field exposure. It is clear from these studies

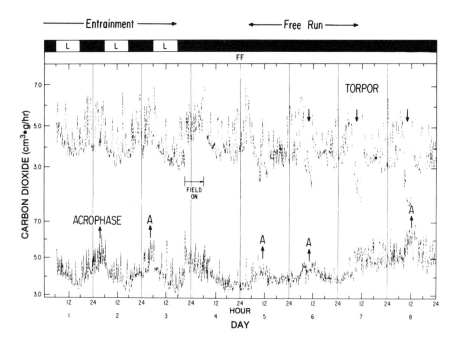

Fig. 17. Circadian phase delay after electric-field exposure. Entrainment and free-run as in Figs. 14, 15, and 16. Exposure to 60-Hz electric field at 100 kV/m for 9 hr on day 3 during active (dark) phase caused phase delay in carbon dioxide (bottom of figure) and stress response (torpor) (top of figure) in mouse.

that the circadian pacemaker is one of the main targets of electromagnetic fields, with wide-ranging potential impact on biological organisms.

ACKNOWLEDGMENTS

We thank Mr. J. Nicholas for statistical help, Ms. C. Fox for software assistance, Mr. W. Eisler, Jr., and Mr. D. LeBuis for design and maintenance of the experimental facilities, Ms. S. Barr for editorial assistance, and Dr. P. Klingensmith, Mr. G. Grose, Ms. S. Kramer, Ms. Y. Costellanos, Ms. K. McArdle, Ms. K. Supalo, Ms. J. Bannister, and Ms. M. Jirka for experimental assistance. This work was supported by the U.S. Department of Energy, Conservation and Renewable Energy, Office of Energy Storage and Distribution, under contract No. W–31–109–ENG–38.

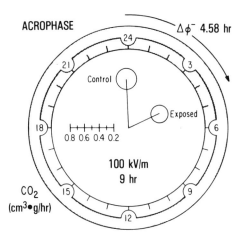

Peromyscus leucopus

Fig. 18. Cosine graph of phase delay of CO_2 in the mouse shown in Fig. 17. Linear least-squares fit to longitudinal circadian data is followed by error measurements of quality of fit. This information is then displayed on clock face; position of acrophase and amplitude of rhythm are indicated by position and length of "clock" hand, respectively. Error estimates of acrophase and amplitude are shown by circle surrounding tip of line ("hand"). By convention, any significant displacement from center indicates presence of circadian rhythm.

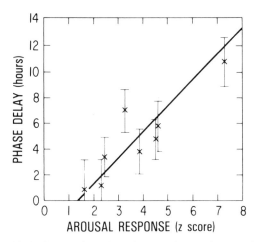

Fig. 19. Relationship between circadian phase delay and amount of activity arousal induced by 60-Hz electric-field exposure. A measure for arousal was calculated by comparing activity during exposure and activity at same circadian time on previous day and computing resultant z score. Mice were entrained to light:dark (LD) 12:12 as in Figs. 14 through 17, and exposed during inactive phase to 100-kV/m, 60-Hz electric field for 30 min. At end of free-run, cosine plots of phase shift (delay) compared amount of phase delay to z score for each individual mouse.

REFERENCES

Adey, W. R. 1988. Physiological signalling across cell membranes and cooperative influences of extremely low frequency electromagnetic fields. In: *Biological Coherence and Response to External Stimuli*, Frohlich, H., ed., pp. 148–170. Heidelberg: Springer.

Aizawa, Y., Hitosugi, M., Takata, T., Aono-Nakamura, K., Mori, K. 1982. Effects of night work on immunological circadian rhythm. *J. Hum. Ergol. (Tokyo)* 11:121–129 (Suppl.).

Anisimov, V. N. 1983. Carcinogenesis and aging. In: *Advances in Cancer Research*, Vol. 40, Klein, G., Weinhouse, S., eds., pp. 365–415. New York: Academic Press.

Aschoff, J. 1960. Exogenous and endogenous components in circadian rhythms. *Cold Springs Harbor Symp. Quant. Biol.* 25:11–28.

Aschoff, J. 1965. Circadian rhythms in man. *Science* 148:1427–1432.

Axelrod, J., Fraschini, F., Velo, G. P. 1982. The pineal gland and its neuroendocrine role. In: *Proceedings of the NATO Advanced Study*, Erice, Italy. New York: Plenum Press.

Bailleul, F., Levi, F., Reinberg, A., Mathe, G. 1986. Interindividual differences in the circadian hematologic time structure of cancer patients. *Chronobiol. Int.* 3:47–54.

Baxter, L. R. Jr., Liston, E. H., Schwartz, J. M., Altshuler, L. L., Wilkins, J. N., Richeimer, S., Guze, B. H. 1986. Prolongation of the antidepressant response to partial sleep deprivation by lithium. *Psychiatry Res.* 9:17–23.

Bertram, J. S., Person, A. R., Heidelberg, C. 1975. Chemical oncogenesis in cultured mouse embryo cells in relation to the cell cycle. *In Vitro* 11:97–106.

Blackman, C. F. 1988. Stimulation of brain tissue in vitro by extremely low frequency, low intensity sinusoidal electromagnetic fields. *Prog. Clin. Biol. Res.* 257:107–117.

Blome, I., Larko, O. 1984. Premalignant and malignant skin lesions in renal transplant patients. *Transplantation* 37:165–167.

Borbely, A. A. 1987. The S-deficiency hypothesis of depression and the two-process model of sleep regulation. *Pharmacopsychiatry* 20:23–29.

Brainard, G. C., Knobler, R. L., Podolin, P. L., Lavasa, M., Lublin, F. D. 1987. Neuroimmunology: modulation of the hamster immune system by photoperiod. *Life Sci.* 40:1319–1326.

Bratescu, A., Teodorescu, M. 1981. Circannual variations in the B cell/T cell ratio in normal human peripheral blood. *J. Allergy Clin. Immunol.* 68:273–280.

Brock, M. A. 1983. Seasonal rhythmicity in lymphocyte blastogenic responses of mice persists in a constant environment. *J. Immunol.* 130:2586–2587.

Cahill, A. L., Ehret, C. F. 1981. Circadian variations in the activity of tyrosine aminotransferase and tryptophan hydroxylase: relationship to catecholamine metabolism. *J. Neurochem.* 37:1109–1115.

Cahill, A. L., Ehret, C. F. 1982. Alpha-methyl-*p*-tyrosine shifts circadian temperature rhythms. *Am. J. Physiol.* 243:R218–222.

Cahill, A. L., Ehret, C. F. 1984. Chronobiological consequences of various shift work schedules. In: *Proceedings of the 15th International Conference of the International Society of Chronobiology*, pp. 200–206. New York: Karger.

Casey, F. B., Eisenberg, J., Peterson, D., Pieper, D. 1973. Altered antigen uptake and distribution due to exposure to extreme environmental temperatures or sleep deprivation. *J. Reticuloendothel. Soc.* 15:87–95.

Chandrashekaran, M. K., Engelmann, W. 1973. Early and late subjective night phases of the *Drosophila pseudoobscura* circadian rhythm require different energies of blue light for phase shifting. *Z. Naturforsch. Sect. B Biosci.* 28C:750-753.

Chouvet, G., Mouret, J., Coindet, J., Siffre, M., Jouvet, M. 1974. Periodicite bicircadiane du cycle veille-sommeil dans les condition hors du temps. *Electroencephalogr. Clin. Neurophysiol.* 3:367–380.

Clausen, O. P. F., Iversen, O. H., Thorad, E. 1984. Circadian variation in the susceptibility of mouse epidermis to tumor induction by methyl-nitrosourea. *Virchows Arch. B Cell Pathol.* 45:325–329.

Columbano, A., Rajalakshini, S., Sarma, O. S. R. 1981. Requirement of cell proliferation for the initiation of liver carcinogenesis as assayed by three different procedures. *Cancer Res.* 42:2079-2083.

Czeisler, C. A., Moore-Ede, M. C., Coleman, R. M. 1982. Rotating shift work schedules that disrupt sleep are improved by applying circadian principles. *Science* 217:460-462.

Czeisler, C. A., Allan, J. S., Strogatz, S. H., Ronda, J. M., Sanchez, R., Rios, C. D., Freitag, W. O., Richardson, G. S., Kronaur, R. E. 1986. Bright light resets the human circadian pacemaker independent of the timing of the sleep-wake cycle. *Science* 233:667-670.

Dessauer, M., Goetze, U., Tolle, R. 1985. Periodic sleep deprivation in drug-refractory depression. *Neuropsychobiology* 13:111-116.

DiPalma, J. R., Sample, R. G., eds. 1976. *Basic Pharmacology in Medicine.* New York: McGraw-Hill.

Dowse, H. B., Palmer, J. D. 1969. Entrainment of circadian activity rhythms in mice by electrostatic fields. *Nature (London)* 222:564-566.

Duffy, P. H., Ehret, C. F. 1982. Effects of intermittent 60-Hz electric field exposures: circadian phase shifts, splitting, torpor, and arousal response in mice. In: *Abstracts, Fourth Annual Meeting of the Bioelectromagnetics Society.* (Available from the Bioelectromagnetics Society, P. O. Box 3729, Frederick, Maryland 20878.)

Duffy, P. H., Feuers, R. J., Hart, R. W. 1987. Effect of age and torpor on the circadian rhythms of body temperature, activity, and body weight in the mouse (*Peromyscus leucopus*). In: *Progress in Clinical and Biological Research. Advances in Chronobiology B*, Pauly, J. E., Scheving, L. E., eds., pp. 111–120. New York: Alan R. Liss.

Edmunds, L. N. Jr., Adams, K. J. 1981. Clocked cell cycle clocks. *Science* 211:1002-1013.

Edmunds, L. N. Jr., Laval-Martin, D. L. 1984. Cell division cycles and circadian oscillators. In: *Cell Cycle Clocks*, Edmunds, L. N. Jr., ed., pp. 295–324. New York: Marcel Dekker.

Ehret, C. F. 1959. Induction of phase shift in cellular rhythmicity by far ultraviolet and its restoration by visible radiant energy. In: *Photoperiodism and Related Phenomena in Plants and Animals*, Withrow, R. B., ed., pp. 541–550. Washington, DC: American Association for the Advancement of Science.

Ehret, C. F. 1974. The sense of time: evidence for its molecular basis in the eukaryotic-gene action scheme. *Adv. Biol. Med. Phys.* 15:44–77.

Ehret, C. F. 1981. New approaches to chronohygiene for the shift worker in the nuclear power industry. In: *Night and Shift-Work: Biological and Social Aspects,* Reinberg, A., Vieux, N., Andlauer, P., eds. New York: Pergamon Press.

Ehret, C. F. 1983. Future perspectives for the application of chronobiological knowledge in occupational work scheduling. Biological clocks and shift work scheduling. In: *Hearings Before the Subcommittee on Investigations and Oversight Committee on Science and Technology,* House of Representatives, 98th Congress, First Session, March 23, 24 (No. 7), pp. 321–355.

Ehret, C. F., Dobra, K. W. 1977. Oncogenic implications of chronobiotics in the synchronization of mammalian circadian rhythms: barbiturates and methylated xanthines. In: *Proceedings of the Third International Symposium on the Detection and Prevention of Cancer,* Nieburgs, H. E., ed., pp. 1101–1114. New York: Marcel Dekker.

Ehret, C. F., Duffy, P. H. 1983. High strength electric fields are circadian zeitgebers in mice. *Chronobiologica* 1:24 (Abstr.).

Ehret, C. F., Trucco, E. 1967. Molecular models for the circadian clock: I. The chronon concept. *J. Theor. Biol.* 15:240-262.

Ehret, C. F., Wille, J. J., Jr. 1970. The photobiology of circadian rhythms in protozoa and other eukaryotic microorganisms. In: *Photobiology of Microorganisms,* Halldal, P., ed., pp. 369–416. New York: Wiley.

Ehret, C. F., Groh, K. R., Meinert, J. C. 1978. Circadian dyschronism and chronotypic ecophilia as factors in aging and longevity. *Adv. Exp. Med. Biol.* 108:185–213.

Ehret, C. F., Groh, K. R., Meinert, J. C. 1980. Considerations of diet in alleviating jet lag. In: *Chronobiology: Principles and Applications to Shifts in Schedules,* Scheving, L. E., Halberg, F., eds., pp. 109–125. Netherlands: Sijthoff and Noordhoff.

Ehret, C. F., Potter, V. R., Dobra, K. W. 1975. Chronotypic action of theophylline and of phenobarbitol as circadian zeitgebers in the rat. *Science* 188:1212–1215.

Ehret, C. F., Sacher, G. A., Langsdorf, A., Lewis, R. N. 1979. Exposure and data collection facilities for circadian studies of electric field effects upon behavior, thermoregulation, and metabolism in small rodents. In: *Biological Effects of Extremely Low Frequency Electromagnetic Fields,* Proceedings of the 18th Annual Hanford Biology Symposium, Phillips, R. D., Gillis, M. F., Kaune, W. T., Mahlum, D. D., eds., pp. 198–224. CONF-781016. Springfield, Virginia: National Technical Information Service.

Felder, M., Dore, C. J., Knight, S. C., Ansell, B. M. 1985. In vitro stimulation of lymphocytes from patients with rheumatoid arthritis. *Clin. Immunol. Immunopathol.* 37:253–261.

Garcia-Sainz, M., Halberg, F. 1966. Mitotic rhythms in human cancer, reevaluated by electronic computer programs – evidence for chronopathology. *J. Natl. Cancer Inst.* 37:279–292.

Germano, G., Damiani, S., Ciavarella, M., Appolloni, A., Ferrucci, A., Corsi, V. 1984. Detection of a diurnal rhythm in arterial blood pressure in the evaluation of 24-hour antihypertensive therapy. *Clin. Cardiol.* 7:525–535.

Giedke, H., Pflug, B. 1979. Diurnal rhythms in manic-depressive disorders. In:

Neuro-Psychopharmacology, Saletu, B., Berner, P. L., Hollister, L., eds., pp. 221–231. Oxford: Pergamon Press.

Graeber, R. C. 1982. Alterations in performance following rapid transmeridian flight. In: Rhythmic Aspects of Behavior, Brown, F. M., Graeber, R. C., eds. Hillsdale, New Jersey: Erlbaum.

Groh, K. R., Ehret, C. F., Eisler, W. J., Jr., LeBuis, D. A. 1987. Multiparameter data acquisition systems for studies of circadian rhythms. In: Chronobiotechnology and Chronobiological Engineering, Scheving, L. E., Halberg, F., Ehret, C. F., eds., pp. 397–405. NATO ASI Series E, Applied Sci. No. 120. Boston: Martinus-Nijhoff.

Groh, K. R., Ehret, C. F., Peraino, C., Meinert, J. C., Readey, M. A. 1988. Circadian manifestations of barbiturate habituation, addiction and withdrawal in the rat. Chronobiol. Int. 5:153–166.

Hahn, B. H., McDermott, R. P., Jacobs, S. B., Pletscher, L. S., Beale, M. C. 1980. Immunosuppressive effects of low doses of glucocorticoids: effects on autologous and mixed leukocyte reactions. J. Immunol. 124:2812–2817.

Halberg, F. 1969. Chronobiology. Annu. Rev. Physiol. 31:675–725.

Halberg, F. 1976. Some aspects of chronobiology relating to the optimization of shift work. In: Shift Work and Health: A Symposium, HEW Publ. (NIOSH) 76-203, Rentos, P. G., Shepard, R. D., eds., pp. 13–47. Washington, DC: Department of Health, Education and Welfare.

Halberg, F., Johnson, E. A., Nelson, W., Runge, W., Sothern, R. 1972. Autorhythmometry–procedures for physiologic self-measurements and their analysis. Physiol. Teach. 1:1–11.

Haus, E., Lakatua, D. J., Swoyer, J., Sackett-Lundeen, L. 1983. Chronobiology in hematology and immunology. Am. J. Anat. 168:467–517.

Hayashi, O., Kikuchi, M. 1982. The effects of the light-dark cycle on humoral and cell-mediated immune responses of mice. Chronobiologia 9:291–300.

Hayashi, O., Kikuchi, M. 1985. The influence of phase shift in the light-dark cycle on humoral immune responses of mice to sheep red blood cells and polyvinylpyrrolidone. J. Immunol. 134:1455–1461.

Holmes, F. F. 1983. Aging and Cancer. Recent Results in Cancer Research, Vol. 87. New York: Springer-Verlag.

Horseman, N. D., Ehret, C. F. 1982. Glucocorticosteroid injection is a circadian zeitgeber in the laboratory rat. Am. J. Physiol. 243:R373-R378.

Hrushesky, W. J. M. 1983. The clinical aplication of chronobiology to oncology. Am. J. Anat. 168:519–542.

Hrushesky, W. J. M. 1985. Circadian timing of cancer chemotherapy. Science 228:73–75.

Hunt, W. A., Dalton, T. K. 1981. Neurotransmitter-receptor binding in various brain regions in ethanol dependent rats. Pharmacol. Biochem. Behav. 14:733–739.

Indiveri, F., Pierri, I., Rogna, S., Poggi, A., Montaldo, P., Romano, R., Pende, A., Morgano, A., Barabino, A., Ferrone, S. 1985. Circadian variations of autologous mixed lymphocyte reactions and endogenous cortisol. J. Immunol. Methods 82:17–24.

Iverson, O. H. 1982a. Enhancement of methylnitrosourea skin carcinogenesis by inhibiting cell proliferation with hydroxyurea or skin extracts. Carcinogenesis 3:881–889.

Iverson, O. H. 1982b. Hydroxyurea enhances methylnitrosourea skin tumorigenesis when given shortly before, but not after, the carcinogen. *Carcinogenesis* 3:891–894.

Iverson, O. H., Kauffman, S. L. 1980. Circadian variation in the susceptibility of mouse epidermis to chemical carcinogens. *Int. J. Chronobiol.* 8:49–62.

Jankovic, B. D., Isakovic, K. 1973. Neuroendocrine correlates of immune response. I. Effects of brain lesions on antibody production. Arthus reactivity and delayed hypersensitivity in the rat. *Int. Arch. Allergy Appl. Immunol.* 45:360-372.

Jaroslow, B., Eisler, W. Jr. 1968. *Telemetry of Hibernation,* Argonne National Laboratory Report (Publ. 7525), pp. 103–105. Argonne, Illinois: Argonne National Laboratory.

Jauher, P., Weller, M. P. I. 1982. Psychiatric morbidity and time zone changes: a study of patients from Heathrow Airport. *Br. J. Psychiatry* 140:231–235.

Jenner, F. A. 1970. The physiology and biochemistry of periodic psychoses including periodic catatonia. In: *Biochemistry, Schizophrenias, and Affective Illnesses,* Himwich, H. E., ed., pp. 29–42. Baltimore: Williams & Wilkins.

Jenner, F. A., Damas-Mora, J. 1979. Cyclic process and abnormal behavior. In: *Handbook of Biological Psychiatry: Part I – Disciplines Relevant to Biological Psychiatry,* van Praag, H. M., Lader, M. H., Rafaelsen, O. J., Sachar, E. J., eds., pp. 229–272. New York: Marcel Dekker.

Kawamura, H., Inouye, S., Ebihara, S., Noguchi, S. 1982. Neurophysiological studies of the SCN in the rat and in the Java Sparrow. In: *Vertebrate Circadian Systems,* Aschoff, J., Daan, S., Groos, G. A., pp. 106–111. New York: Springer-Verlag.

Kim, Y., Pallansch, M., Carandente, F., Reissmann, G., Halberg, E., Halberg, F., Halberg, F. 1980. Circadian and circannual aspects of the complement cascade – new and old results, differing in specificity. *Chronobiologia* 7:189–204.

Klevecz, R. R. 1976. Quantized generation time in mammalian cells as an expression of the cellular clock. *Proc. Natl. Acad. Sci. USA* 73:4012–4016.

Klevecz, R. R., Braly, P. S. 1986. Synchronous waves of proliferation in human ovarian cancers. *Annu. Rev. Chronopharmacol.* 3:176–178.

Klevecz, R. R., Braly, P. S. 1987. Circadian and ultradian rhythms of proliferation in human ovarian cancer. *Chronobiol. Int.* 5:513–524.

Klevecz, R. R., Keniston, B. A., Deaven, L. L. 1975. The temporal structure of S phase. *Cell* 5:195–203.

Kort, W. J., Weijma, J. M. 1982. Effect of chronic light-dark shift stress on the immune response of the rat. *Physiol. Behav.* 29:1083–1087.

Kripke, D. F., Mullaney, D. J., Atkinson, M., Wolf, S. 1978. Circadian rhythms disorders in manic depressives. *Biol. Psychiatry* 13:335–351.

Kuhs, H., Tolle, R. 1986. Schlafentzug (Wachtherapie) als Antidepressivum. *Fortsch. Neurol. Psychiatr.* 54:341–355.

Laerum, O. D. 1976. Possible influences of circadian rhythms in experimental carcinogenesis. *Arch. Toxicol.* 36:247–259.

Levine, H. 1976. Health and work shifts. In: *Shift Work and Health: A Symposium,* Rentos, P. G., Shepard, R. D., eds., pp. 57–69. HEW Pub. (NIOSH) 76–203. Washington, DC: National Institute of Occupational Safety and Health.

Levine, S. 1986. The management of resistant depression. *Acta Psychiatr. Belg.* 86:141–151.

Lewy, A. J., Sack, R. L. 1986. Light therapy and psychiatry. *Proc. Soc. Exp. Biol. Med.* 183:11–18.

Lewy, A. J., Sack, R. L., Miller, S., Hoban, T. M. 1987. Antidepressant and circadian phase-shifting effects of light. *Science* 235:352–354.

Lingjaerde, O. 1983. The biochemistry of depression: a survey of monoaminergic, neuroendocrinological, and biorhythmic disturbances in endogenous depression. *Acta Psychiatr. Scand. (Suppl.)* 302:36–51.

Lyle, D. B., Ayotte, R. D., Sheppard, A. R., Adey, W. R. 1988. Suppression of T-lymphocyte cytotoxicity following exposure to 60-Hz sinusoidal electric fields. *Bioelectromagnetics* 9:303–313.

Maestroni, G. J. M., Conti, A., Pierpaoli, W. 1987. The pineal gland and the circadian opiatergic, immunoregulatory role of melatonin. In: *Neuroimmune Interactions: Proceedings of the Second International Workshop on Neuroimmunomodulation,* Jankovic, B. D., Markovic, B. M., Spector, N. H., eds. *Ann. N.Y. Acad. Sci.* 496:67–77.

Mauer, A. M. 1965. Diurnal variation of proliferative activity in the human bone marrow. *Blood* 26:1–7.

Meinert, J. C., Ehret, C. F., Antipa, G. A. 1975. Circadian chronotypic death in heat-synchronized infradian mode cultures of *Tetrahymena pyriformis* W. *Microb. Ecol.* 2:201–214.

Menaker, M. 1968. Extraretinal light perception in the sparrow, I. Entrainment of the biological clock. *Proc. Natl. Acad. Sci. USA* 80:6119–6121.

Menaker, M., Eskin, A. 1967. Circadian clock in photoperiodic time measurement: a test of the Bünning hypothesis. *Science* 157:1182–1185.

Miyawaki, T., Taga, K., Nagaoki, T., Seki, H., Suzuki, Y., Taniguchi, N. 1984. Circadian changes of T-lymphocyte subsets in human peripheral blood. *Clin. Exp. Immunol.* 55:618–622.

Moller, U. 1984. Diurnal rhythmicity used in experimental cancer research. In: *Cell Cycle Clocks,* Edmunds, L. N., Jr., ed., pp. 501–254. New York: Marcel Dekker.

Monk, T. H., Weitzman, E. D., Fookson, J. E., Moline, M. L., Kronauer, R. E., Gander, P. H. 1983. Task variables determine which biological clock controls circadian rhythms in human performance. *Nature (London)* 304:543–545.

Moore-Ede, M. C. 1983. The circadian timing system in mammals: two pacemakers preside over many secondary oscillators. *Fed. Proc.* 42:2802–2806.

Moore-Ede, M. C., Czeisler, C. A., Richardson, G. S. 1983. Circadian timekeeping in health and disease. Part 1. Basic properties of circadian pacemakers. *N. Engl. J. Med.* 309:530-536.

Moore-Ede, M. C., Kass, D. A., Herd, J. A. 1976. Internal organization of the circadian timing system in multicellular animals. *Fed. Proc.* 35:2333–2338.

Mueller, H. J. 1927. Artificial transmutation of the gene. *Science* 66:84–87.

Muller, P., Britton, R. S., Seem, P. 1980. The effects of long-term ethanol on brain receptors for dopamine acetylcholine, serotonin and noradrenaline. *Eur. J. Pharmacol.* 65:31–37.

Nakaio, Y., Miura, T., Hara, I., Aono, H., Miyano, N., Miyajima, K., Tabuchi, T., Kosaka, H. 1982. Effect of shift work on cellular immune function. *J. Hum. Ergol. (Suppl.)* 11:131–137.

Ninnemann, H. 1979. Photoreceptors for circadian rhythms. *Photochem. Photobiol. Rev.* 4:207–265.

Nishikawa, J., Kast, A., Albert, H. 1987a. Circadian rhythms of the liver of male rats dosed with phenobarbital. I. Organ weight, cellular structures, glycogen contents and mitotic activity. *Chronobiol. Int.* 4:161–173.

Nishikawa, J., Yabe, T., Kast, A., Albert, H. 1987b. Circadian rhythm of the liver of male rats pretreated with phenobarbital. II. Hexobarbital sleeping time and lipids content in liver and serum. *Chronobiol. Int.* 4:175–182.

Nosachev, G. N. 1985. [Treatment of endogenous depressions by sleep deprivation.] Terapiia endogennykh depressii deprivatsei sna (English summary). *Zh. Nevropatol. Psikhiatr. IM. S. S. Korsakova* 85:565–570 .

Olcese, J., Reuss, S., Semm, P. 1988. Geomagnetic field detection in rodents. *Life Sci.* 42:605–613.

Owasoyo, J. O. 1982. Circadian rhythm in brain histamine alterations by phenobarbital pentylene tetrazole and picrotoxin. *J. Interdiscip. Cycle Res.* 13:29–36.

Owasoyo, J. O., Walker, C. A. 1980. The effects of sodium phenobarbital on the circadian levels of norepinephrine and serotonin in rat brains. *J. Interdiscip. Cycle Res.* 11:95–101.

Owasoyo, J. O., Walker, C. A., Whitworth, U. G. 1979. Diurnal variation in the dopamine level of rat brain areas: effect of sodium phenobarbital. *Life Sci.* 25:119–122.

Peraino, C., Ehret, C. F., Groh, K. R., Meinert, J. C., D'Arcy-Gomez, G. 1980. Phenobarbital effects on weight gain and circadian cycling of food intake and body temperature. *Proc. Soc. Exp. Biol. Med.* 165:473–479.

Petterborg, L. J., Richardson, B. A., Reiter, R. J. 1981. Effect of long or short photoperiod on pineal melatonin content in the white-footed mouse, *Peromyscus leucopus. Life Sci.* 29:1623–1627.

Pflug, B., Tolle, R. 1971. Disturbance of the 24-hr rhythm in endogenous depression and treatment of endogenous depression by sleep deprivation. *Int. Pharmacopsychiatry* 6:187–196.

Pflug, B., Erikson, R., Johnsson, A. 1976. Depression and daily temperature, a long-term study. *Acta Psychiatr. Scand.* 54:254–266.

Pittendrigh, C. S. 1974. Circadian oscillations in cells and the circadian organization of multicellular systems. In: *The Neuroscience Third Study Program,* Schmitt, F. O., Worden, F. G., eds., pp. 437–458. Cambridge: MIT Press.

Pittendrigh, C. S., Daan, S. 1976. A functional analysis of circadian pacemakers in nocturnal rodents. V. Pacemaker structure: a clock for all seasons. *J. Comp. Physiol.* 106:333–355.

Rabes, H. M., Mueller, L., Hartmann, A., Kerler, R., Schuster, C. 1986. Cell cycle-dependent initiation of adenosine triphosphate-deficient populations in adult rat liver by a single dose of *N*-methyl-*N*-nitrosourea. *Cancer Res.* 46:645–650.

Rajewsky, M. F. 1986. Tumorigenesis by exogenous carcinogens: role of target-cell proliferation and state differentiation (development). In: *Age-Related Factors in Carcinogenesis,* Likachev, A., Anisimov, V., Montesano, R., eds., pp. 215–224. IARC Scientific Publication 58. Lyon, France: International Agency for Research on Cancer.

Readey, M. A. 1986. *A Comparison of the Ultradian and Infradian Modes of Growth in the Ciliate Protozoan Tetrahymena.* Doctoral dissertation, University of Toronto, Toronto, Ontario, Canada.

Readey, M. A. 1987. Ultradian photosynchronization in *Tetrahymena pyriformis*

GLC is related to modal cell generation time: further evidence for a common timer model. *Chronobiol. Int.* 4:195–208.

Reinberg, A., Andlaner, P., De Prins, J., Malbecq, W., Vieux, N., Bourdeleau, P. 1984. Desynchronization of the oral temperature circadian rhythm and intolerance to shift work. *Nature (London)* 308:272–274.

Reinberg, A., Schuller, E., Delasnerie, N., Clench, J., Helary, M. 1977. Rhythmes circadiens et circannuels des leucocytes, proteines totales, immunoglobulines A, G, et M. Etude chez 9 adultes jeunes et sains. *Nouv. Presse Med.* 6:3819–3823.

Reinberg, A., Lagoguey, M., Cesselin, F., Touitou, Y., Legrand, J. C., Delassalle, A., Antreassian, J., Lagoguey, A. 1978. Circadian and circannual rhythms and hormones and other variables of five healthy young human males. *Acta Endocrinol.* 88:417–427.

Reiter, R. J., ed. 1984. *The Pineal Gland.* New York: Raven Press.

Richter, C. P. 1960. Biological clocks in medicine and psychiatry: shock-phase hypothesis. *Proc. Natl. Acad. Sci. USA* 46:1506–1530.

Riley, V. 1981. The psychoneuroendocrine influences on immunocompetence and neoplasia. *Science* 212:1100–1109.

Ritchie, A. W. S., Oswald, I., Micklem, H. S., Boyd, J. E., Elton, R. A., Jazwinskee, E., James, K. 1983. Circadian variation of lymphocyte subpopulations: a study with monoclonal antibodies. *Br. J. Virol.* 286:1773–1775.

Rocker, L., Feddersen, H. M., Hoffmeister, H., Junge, B. 1980. Jahreszeitliche Veranderungen diagnostisch wichterger Blutbestandteile. *Klin. Wochenschr.* 58:769–778.

Rosenberg, R. S., Duffy, P. H., Sacher, G. A., Ehret, C. F. 1983. Relationship between field strength and arousal response in mice exposed to 60-Hz electric fields. *Bioelectromagnetics* 4:181–191.

Ruis, J. F., Cambras, T., Buys, J. P., Rietveld, W. J. 1988. Methamphetamine-induced internal desynchronization in rats. In: *Program and Abstracts for the First Meeting of the Society for Research on Biological Rhythms,* Abstr. 46.

Runge, W., Lange, K., Halberg, F. 1974. Some instruments for chronobiologists developed or used in systems at the University of Minnesota. *Int. J. Chronobiol.* 2:327–341.

Rusak, B. 1977. The role of the suprachiasmatic nuclei in the generation of circadian rhythms in the golden hamster, *Mesocricetus auratus. J. Comp. Physiol.* 118:145–164.

Rusak, B. 1982. Physiological models of the rodent circadian system. In: *Vertebrate Circadian Systems,* Aschoff, J., Daan, S., Groos, G. A., eds. New York: Springer-Verlag.

Sacher, G. A., Duffy, P. H. 1978. Age changes in the rhythms of energy metabolism, activity, and body temperature in *Mus* and *Peromyscus. Adv. Exp. Med. Biol.* 108:105–124.

Scheving, L. E. 1959. Mitotic activity in the human epidermis. *Anat. Rec.* 135:7–20.

Semm, P., Demaine. C. 1986. Neurophysiological properties of magnetic cells in the pigeon's visual system. *J. Comp. Physiol.* 159:619–625.

Shek, P. N., Sabiston, B. H. 1983. Neuroendocrine regulation of immune process: changes in circulating corticosterone levels induced by the primary antibody response in mice. *Int. J. Immunopharmacol.* 5:23–33.

Shifrine, M., Taylor, N., Rosenblatt, L. S., Wilson, F. 1980. Seasonal variation in cell-mediated immunity of clinically normal dogs. *Exp. Haematol.* 8:318–326.

Shifrine, M., Rosenblatt, L. S., Taylor, N., Hetherington, N. W., Mathews, V. J., Wilson, F. D. 1982a. Seasonal variations in lectin-induced lymphocyte transformation in beagle dogs. *J. Interdiscip. Cycle Res.* 13:151–165.

Shifrine, M., Garsd, A., Rosenblatt, L. S. 1982b. Seasonal variation in immunity in humans. *J. Interdiscip. Cycle Res.* 13:157–165.

Signore, A., Cugini, P., Letizia, C., Lucia, P., Murano, G., Pozzilli, P. 1985. Study of the diurnal variation of human lymphocyte subsets. *J. Clin. Lab. Immunol.* 17:25–28.

Smolensky, M. H., Reinberg, A., McGovern, J. P., eds. 1979. *Recent Advances in the Chronobiology of Allergy and Immunology. Vol. 28. Advances in the Biosciences.* New York: Pergamon Press.

Sturtevant, R. P., Garber, S. L. 1988. Circadian rhythm of blood ethanol clearance rates in rats: response to reversal of the L/D regimen and to continuous darkness and continuous illumination. *Chronobiol. Int.* 5:137–148.

Tabakoff, B., Melchior, C., Urwyler, S., Hoffman, P. L. 1980. Alterations in neurotransmitter function during the development of ethanol tolerance and dependence. In: *Alcohol and Brain Research*, Idestrom, C., ed. *Acta Psychiatr. Scand.* 62:153–160.

Tagashira, E., Hiramori, T., Urano, T., Nakao, K., Yanaura, S. 1982. Participant of serotonin turnover rate in the brain on barbital withdrawal convulsions (sic). *Jpn. J. Pharmacol.* 32:159–167.

Takahashi, J. S., DeCouresy, P. J., Bauman, L., Menaker, M. 1984. Spectral sensitivity of a novel photoreceptive system mediating entrainment of mammalian circadian rhythms. *Nature (London)* 308:186–188.

Thompson, C. 1984. Circadian rhythms and psychiatry. *Br. J. Psychiatry* 145:204–206.

Turek, F. W., Losee-Oleson, S. 1986. A benzodiazepine used in the treatment of insomnia phase-shifts the mammalian circadian clock. *Nature (London)* 321:167–168.

van Bemmel, A. L., van den Hoofdakker, R. H. 1981. Maintenance of therapeutic effects of total sleep deprivation by limitation of subsequent sleep. A pilot study. *Acta Psychiatr. Scand.* 63:453–462.

Vasquez, B. J., Anderson, L. E., Lowery, C. I., Adey, W. R. 1988. Diurnal patterns in brain biogenic amines of rats exposed to 60-Hz electric fields. *Bioelectromagnetics* 9:229–236.

Voutilainen, A. 1953. Uber die 24-Stunden-rhythmic der mitosenfrequencz in malignen Tumoren. *Acta Pathol. Microbiol. Scand. (Suppl.)* 99:1–104.

Vriend, J., Lauber, J. K. 1973. Effects of light intensity, wavelength and quanta on gonads and spleen of the deer mouse. *Nature (London)* 244:37–38.

Wehr, T. A., Goodwin, F. K. 1981. Biological rhythms and psychiatry. In: *American Handbook of Psychiatry, 2nd Ed., Vol. VII: Advances and New Directions*, Arieti, S., Brodie, H. K. H., eds., pp. 46–74. New York: Basic.

Wehr, T. A., Sack, D. A., Rosenthal, N. E. 1985. Antidepressant effects of sleep deprivation and phototherapy. *Acta Psychiatr. Belg.* 85:593–602.

Wehr, T. A., Wirz-Justice, A., Goodwin, F. K. 1979. Phase advance of the circadian sleep-wake cycle as an antidepressant. *Science* 206:710-713.

Wehr, T. A., Goodwin, F. K., Wirz-Justice, A., Craig, C., Breitmeier, J. 1982. 48-hour sleep-wake cycles in manic-depressive illness: naturalistic observations and sleep-deprivation experiments. *Arch. Gen. Psychiatry* 39:559–565.

Wehr, T. A., Sack, D., Rosenthal, N., Duncan, W., Gillin, J. C. 1983. Circadian rhythm disturbances in manic-depressive illness. *Fed. Proc.* 42:2809–2814.

Weiner, N., Taylor, P. 1985. Neurohumoral transmission: the autonomic and somatic motor nervous systems. In: *The Pharmacological Basis of Therapeutics*, Gilman, A. G., Goodman, L. S., Rall, T. W., Murad, F., eds., pp. 66–99. New York: Macmillan.

Weitzman, E. D. 1981. Sleep and its disorders. *Annu. Rev. Neurosci.* 4:381–417.

Welsh, D. K., Nino-Murcia, G., Gander, P. H., Keenan, S., Dement, W. C. 1986. Regular 48-hour cycling of sleep duration and mood in a 35-year-old woman: use of lithium in time isolation. *Biol. Psychiatry* 21:527–537.

Wever, R. A. 1970. Zur Zeitgeber-Starke eines Licht-Dunkel-Wechsels für die circadiane Periodic des Menschen. *Pfluegers Arch. Gesamte Physiol. Menschen Tiere* 321:133–142.

Wever, R. A. 1979. *The Circadian System of Man: Results of Experiments Under Temporal Isolation.* New York: Springer-Verlag.

Wilson, B. W., Anderson, L. E., Hilton, D. I., Phillips, R. D. 1981. Chronic exposure to 60-Hz field effects on pineal function in the rat. *Bioelectromagnetics* 2:371–380.

Winfree, A. T. 1970. Integrated view of resetting a circadian clock. *J. Theor. Biol.* 28:327–374.

5 Effects of Light and Stress on Pineal Function

RUSSEL J. REITER
Department of Cellular and Structural Biology
University of Texas Health Science Center
San Antonio, Texas

CONTENTS

This chapter serves as an introduction to pineal gland physiology. The influence of light and stress on the pineal gland is reviewed, and some endocrine implications of this organ are briefly discussed. The pineal gland exerts a wide-ranging influence on the endocrine system. In chronobiological terms, light acts as a Zeitgeber to the pineal. Stress, inappropriate light exposure, and exposure to electric or magnetic fields can alter normal pineal rhythms. In this chapter, reproduction is the principal paradigm illustrating the effects on pineal physiology from these external stimuli.

The pineal gland is a neuroendocrine transducer: it converts a neuronal signal into an endocrine output. The neural signal in question originates

in the suprachiasmatic nuclei of the hypothalamus. Because the suprachiasmatic nuclei generate a number of rhythms within the organism, the pineal
rhythms depend on intact suprachiasmatic nuclei. Activities of the suprachiasmatic nuclei, at least those relative to the pineal gland, are synchronized by light and darkness as perceived by the lateral eyes in mammals.
In nonmammalian vertebrates, the pineal gland is directly photosensitive,
responding directly to light and darkness. But in mammals, including man,
it is the perception of light by the eyes that is of import. Neural connections between the suprachiasmatic nuclei and the pineal gland are relatively
well defined, at least peripherally (Ariens Kappers, 1960). Preganglionic
sympathetic neurons from the thoracic cord go to the superior cervical
ganglia, and postganglionic fibers then carry the information to the pineal
(Fig. 1). Neural connections between the suprachiasmatic nuclei and the
intermediolateral cell column are somewhat ambiguous. We know that they
exist: when transections of the cervical cord or lesions in the hypothalamus

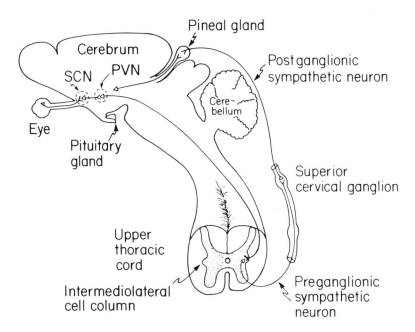

Fig. 1. Neural connections between mammalian eyes and the mammalian pineal
gland.
(Reprinted with permission from Reiter, R.J. 1981. The mammalian pineal gland: structure and function. *Am. J. Anat.* 162:287–313.)

such as those caused by tumors interrupt these pathways, they also interfere with all known pineal rhythms. The same thing happens on removal of the superior cervical ganglia or sympathetic innervation. Certain fibers in the central nervous system have axons that project up the stalk into the pineal gland (Korf and Møller, 1984). The functional relationship of these neurons to the pineal remains to be determined. However, the anatomy of the pathway itself is quite well described.

Within the pineal gland, the postganglionic sympathetic fibers terminate near the pinealocytes, the endocrine elements or the functional elements of the gland. The relationships of these two cells is reasonably well known. In all mammals, norepinephrine is released from the postganglionic sympathetic neurons during darkness. It acts on β-adrenergic receptors on the pinealocyte membrane to stimulate a series of events that culminate in the production of several hormones or putative hormones.

CIRCADIAN RHYTHMS IN THE PINEAL GLAND

Melatonin, the most-often studied of the pineal hormones, is synthesized from serotonin and is among the primary tryptophan metabolites. Tryptophan is taken up from the blood and converted to serotonin in a twostep process (Fig. 2). Serotonin is acted upon by *N*-acetyltransferase, with the resultant formation of *N*-acetyl serotonin. This compound is then O methylated by the enzyme hydroxyindole-*O*-methyltransferase to form melatonin. This pathway is extremely active at night, and there is an obvious 24-hr cycle in melatonin production.

Within the nerve endings, there are also several metabolic cycles that can be measured. For example, tyrosine hydroxylase, the enzyme that determines norepinephrine production, also shows a distinct rhythm in the pineal. The pineal gland in general has a highly rhythmic output (Fig. 3).

There are other putative hormones of pineal gland origin. For example, Wilson et al. (1978) showed that 5-methoxytryptophol is also produced in a circadian manner in the pineal. More recently, 5-methoxytryptamine has been espoused as a pineal secretory product. Interestingly, melatonin, 5-hydroxytryptophol, and 5-methoxytryptophol require the same enzyme for synthesis. Hydroxyindole-*O*-methyltransferase converts *N*-acetyl serotonin to melatonin. The same enzyme converts 5-hydroxytryptophol to 5-methoxytryptophol, and the O-methylation of serotonin directly, by the same enzyme, produces 5-methoxytryptamine. This enzyme within the pineal gland shows either a very weak or no rhythm in activity, whereas the acetylating enzyme that results in synthesis of melatonin shows a very dramatic 24-hr cycle (Fig. 3).

Figure 3 compares pineal rhythms in three species: the common laboratory rat, the Syrian hamster, and the cotton rat (Reiter, 1981). Each of these species is nocturnal, but whether a species is nocturnal or diurnal,

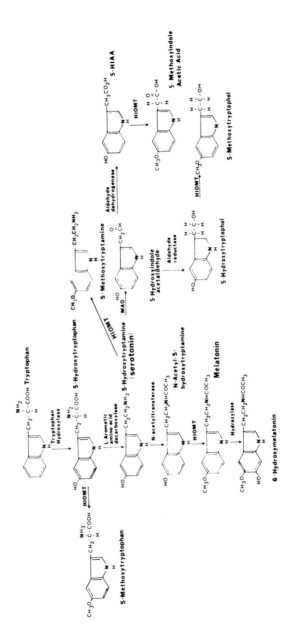

Fig. 2. Tryptophan metabolism in the pineal gland.

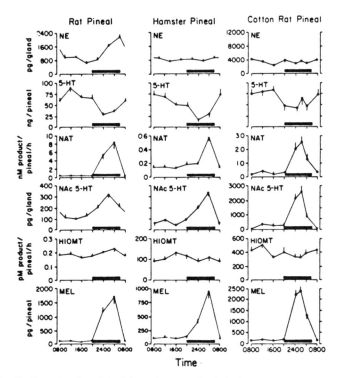

Fig. 3. Rhythms in pineal indoleamines and related enzymes.
(Reprinted with permission from Reiter, R.J. 1986a. Pineal melatonin production: photoperiodic and hormonal influences. In: *Advances in Pineal Research*, Vol. I, Reiter, R.J., Karasek, M., eds., pp. 77–87. London: John Libbey.)

the patterns of melatonin production are very similar. Figure 3 shows a 24-hr period with a black bar representing the daily period of darkness. In all species, the serotonin concentration within the pineal gland is highest during the light phase of the light:dark cycle. With the onset of darkness, a decrease in the amount of serotonin within the pineal gland can be measured in all species in which it has been studied. This drop in serotonin is presumably partially related to the large increase in the acetylating enzyme, which increases at night in every species investigated. The magnitude of change varies according to the species (Rudeen et al., 1975); for example, in the gerbil this increase is only about twofold, but in the albino rat it may be as much as a hundredfold.

The rise is believed to be a consequence of the release of norepinephrine and its action on β-, and possibly α-, adrenergic receptors to stimulate *N*-acetyl transferase activity. This causes the nightly accumulation of *N*-acetyl serotonin in the pineal gland. Although the methylating enzyme shows either a very weak rhythm or none throughout the 24-hr period, by mass action N-acetyl serotonin is converted to melatonin and, almost without exception, melatonin increases within the pineal gland in a very obvious manner during every dark period (Reiter, 1986a).

Tyrosine hydroxylase controls norepinephrine synthesis within the pineal gland and also shows a circadian rhythm. Tyrosine hydroxylase activity, as measured by the accumulation of dihydroxyphenylalanine (DOPA) after inhibition of DOPA decarboxylase, shows a large increase at night (Craft et al., 1984), a uniform finding among mammalian species.

An apparent additional rhythm is in the actual number of β receptors on the pinealocyte membrane. Using iodocyanopindolol (ICYP), which specifically binds β receptors, during the dark phase of the light:dark cycle a marked nocturnal increase in IYCP binding suggests larger numbers of β receptors on rat pinealocytes (Reiter et al., 1985). Thus, the increase in melatonin production at night is a consequence of the synthesis or release of norepinephrine, but complementing this is a rise in the number of β receptors available to the norepinephrine as well.

PATTERNS OF PINEAL GLAND MELATONIN SECRETION

Patterns of melatonin production vary among animals (Reiter, 1987). In all species the light phase is associated with low melatonin; during the first several hours of darkness there may not be an immediate rise in melatonin production. In the hamster a very discrete short-term melatonin peak occurs under this specific light:dark cycle (14:10 light:dark). This is typical of the Syrian hamster, with which we have worked extensively, and also common in the Mongolian gerbil. In many species, the onset of darkness is associated with a gradual melatonin rise, with a peak somewhere near the middle of the dark period and then an anticipatory (of light onset) drop in melatonin production. This is typical of the albino rat, Richardson ground squirrel, thirteen-lined ground squirrel, and human.

In some species the pattern of melatonin production appears to resemble a square wave. Very soon after the onset of darkness (within 30 min), melatonin production reaches a peak and remains elevated during the entire dark period. This is typical of the white-footed mouse, the Djungarian hamster, and also several breeds of sheep. Whether these patterns have any physiological significance remains unknown, but the patterns seem to differ among the species.

Humans generally fall into the second category, although there are many individual variations in humans that are not as apparent in experimental animals. In a group of individuals, the peak is near the middle of the dark period although there are variations as to the time when peak occurs. In animals, the standard errors are very small in melatonin measurements; animals are much more homogeneous genetically. Humans are much more heterogeneous and hence one would anticipate more variability.

Another common characteristic of the melatonin rhythm is that as the duration of the dark period to which animals are exposed increases, the duration of the melatonin peak increases (Reiter, 1987). In other words, if animals are exposed to 8 hr of darkness, the pattern in elevated melatonin is short. If darkness is increased to 10 or 12 hr, the melatonin peak is prolonged accordingly. It is as if the pineal in this case is measuring the duration of the dark period. This may be the signal to the organism that it must adjust physiologically to a new photoperiodic environment. The duration hypothesis has become very popular as an explanation of how the pineal gland signals the organism in reference to the light:dark cycle. It is not the only hypothesis, however.

Melatonin produced in the pineal gland seems to be very rapidly released, and as a consequence blood levels or cerebrospinal fluid levels of melatonin follow very closely a pattern similar to that within the pineal (Wilkinson et al., 1977). Figure 4, for example, shows the nighttime rise of pineal and plasma melatonin in the hamster. The relatively rapid release of melatonin produced in the pineal gland causes a similar rise in melatonin within the plasma. Thus, levels of melatonin usually are taken

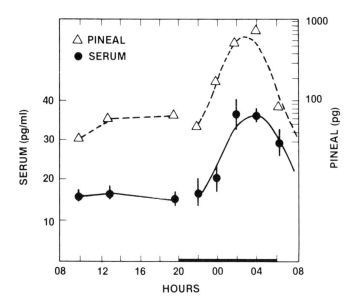

Fig. 4. Close parallelism of pineal and serum melatonin in the hamster.
(Reprinted with permission from Vaughan, G.M. et al. 1986. Serum melatonin after a single aqueous subcutaneous injection in Syrian hamsters. *Neuroendocrinology* 42:124–127.)

as an index of the amount being produced by the pineal gland within that same approximate time frame. This is important because other organs also produce melatonin. The Harderian gland, a large compound tubular alveolar gland around the eye, may contain massive amounts of melatonin, especially in females of certain strains of hamsters (Menendez-Pelaez et al., 1987). The retina itself produces melatonin. It is also possible that red blood cells, and perhaps the gastrointestinal tract, produce small amounts of melatonin. The 24-hr pattern of melatonin in blood, however, derives primarily, if not exclusively, from the pineal gland.

Melatonin does not appear to be taken up in large quantities by the Harderian gland. Small amounts of melatonin may be taken up by the retina, and perhaps by some other organs. No reduction in melatonin occurs within the other organs as the result of pinealectomy; synthesis at such sites seems to be normal. In the female hamster Harderian gland, for example, there is a large amount of melatonin because the gland is so large, thousands of times larger than the pineal gland, and thus all the melatonin could not be of pineal origin. In a study done by Panke and colleagues, removal of the Harderian gland slightly depressed the amount of melatonin within the pineal gland at one time point during darkness (Panke et al., 1979a). Injected, radiolabeled melatonin concentrates within the pineal, suggesting that the pineal gland takes up melatonin.

After removal of the pineal gland itself, or removal of the superior cervical ganglia, or destruction of the sympathetic innervation, there is no nocturnal rise in plasma melatonin (Reiter, 1986b). There may be, however, quantities of melatonin within the blood that are recognizable using a melatonin-specific antibody. Others have argued that evidence from a GC/MS assay indicates melatonin essentially disappears from the blood after pinealectomy. In all mammalian species in which it has been examined, the nocturnal melatonin rise is clearly a consequence of pineal secretion.

EFFECT OF LIGHT ON PINEAL MELATONIN SYNTHESIS

Light exposure is an important Zeitgeber for the circadian rhythm in melatonin production. If animals are in constant light, providing the light is bright enough, the circadian rhythm of melatonin production is eliminated (Reiter, 1986a). If the animals are held in constant darkness, the melatonin rhythm free-runs with a period slightly greater than 24 hr. This free-run has been measured in one or two species at about 24.7 hr (Lewy and Newsome, 1983). In summary, the melatonin rhythm free-runs under constant darkness, and is eliminated by constant light.

If one merely leaves the lights on in the animal room, the melatonin rhythm is indeed totally suppressed (Panke et al., 1979b). In this case, the light had an irradiance, or brightness, of about 40 μW/cm^2. Artificial room lighting is usually of the order of 20 μW/cm^2, and is sufficient to inhibit melatonin production in rats. If during the night the animals are exposed to a sufficiently bright light, melatonin levels will drop precipitously (Brainard et al., 1982).

In species in which the pineal melatonin levels have been measured after light onset in the middle of the night, the time required to reduce blood concentrations in half has been of the order of 8–10 min. The exception was the cotton rat, in which the half-time was in fact only 2 min (Thiele et al., 1983). It would appear that a half-time after acute light exposure at night is ≤10 min. Melatonin concentrations return to daytime levels within 15–20 min after nighttime light exposure.

Using this experimental paradigm, we defined the minimal intensity of light that was required to inhibit pineal gland function in various species. Syrian hamsters (Fig. 5) were exposed to various intensities of light during the middle of the dark phase. Many intensities were tested, and only some data are shown in the figure. The dimmest light normally encountered indoors under artificial lighting, for example, is considerably more than

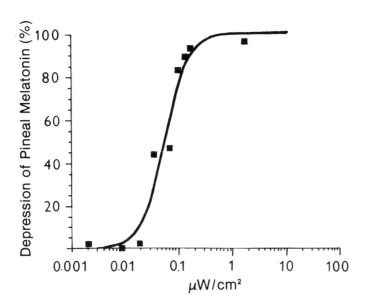

Fig. 5. Dose–response curves of sensitivity of hamster pineal melatonin to light exposure at $p < .001$.

(Reprinted with permission from Brainard, G.C., Richardson, B.A., King, T.A., Matthews, S.A., Reiter, R.J. 1983. The suppression of pineal melatonin content and N-acetyltransferase activity by different light irradiances in the Syrian hamster: a dose-response relationship. *Endocrinology* 113:293–296.)

3.5 µW/cm². This dim light, however, leads to precipitous decline in melatonin (Brainard et al., 1983). When light levels were decreased to approximately 0.2 µW, there was a similar decline in hamster pineal melatonin. Inhibition of melatonin production in these species seems to be an all-or-nothing phenomenon. If melatonin is inhibited by acute light exposure, it is totally inhibited. If it is not inhibited, that is, if the light is not bright enough, there is no reduction.

In these experiments, the light level was further reduced to approximately 0.02 µW, which was not effective in reducing melatonin levels. Any electric light provides more than this. In a dose–response study using a large number of animals, we found that, for the Syrian hamster, about 0.1 µW/cm² will inhibit the pineal gland (Fig. 5). Thus, this light level appears to be on the threshold for this species. Some animals responded with 100% inhibition and some showed no response. Any light brighter than 0.1 µW totally inhibited melatonin production, and any light of lesser intensity was ineffective.

Every species seems to have a different threshold for light inhibition of the pineal gland. Six species have been casually examined. As noted, about 0.1 µW of light will inhibit the pineal gland of the Syrian hamster. The albino rat pineal gland is exquisitely sensitive, responding to an irradiance of 5×10^{-2} µW of light (Webb et al., 1985). The retina responds to lower irradiances of light than this. Although the retina senses lights of this intensity, however, the pineal gland does not necessarily respond.

The cotton rat has an irradiance threshold of about 0.01 µW/cm². The thirteen-lined ground squirrel requires at least 925 µW, and the Richardson ground squirrel as much as 1,850 µW/cm² (Reiter et al., 1984). Although this seems like very bright light, it is not very bright relative to sunlight. On a clear day at high noon, sunlight can have an irradiance of 50,000 µW/cm², which is to say none of the artificial lights tested are very bright. Note here primarily that there is much species variation in terms of sensitivity to light.

We have calculated from studies by Lewy et al. (1980) that the 2500 lux used in their light experiments with humans is roughly 150 µW/cm². It appears that the previous lighting history may be critical in determining the sensitivity of the pineal gland to light perceived by the retina and the pineal gland may act as a comparator. Also, the retinas of diurnal and nocturnal animals are very different; nocturnal animals have primarily rod-dominated retinas with relatively few cones. This factor may also be critical in determining the sensitivity of the pineal gland to light. Further, if an animal is acutely exposed to light at night, the duration of exposure needed to elicit a response to the light can be very short. For the hamster, 1 sec of light, provided it exceeds 0.1 µW/cm², totally inhibits melatonin production until 30 min later (Fig. 6) (Reiter et al., 1986). This occurs regardless of when light is administered during the dark period. With short acute

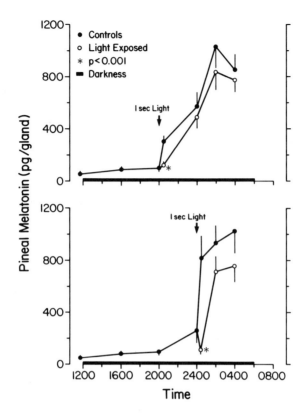

Fig. 6. Pineal melatonin inhibition in hamsters after a 1-sec light exposure at night. (Reprinted with permission from Reiter, R.J., Joshi, B.N., Heinzeller, Th., Nurnberger, F. 1986. A single 1 or 5 second light pulse at night inhibits hamster pineal melatonin. *Endocrinology* 118:1906–1909.)

exposures, melatonin depression may be transient, and melatonin levels may again increase later in the dark phase.

Light wavelength or color also appears to have an effect on melatonin inhibition. Ultraviolet light generally has no inhibitory effect, but some evidence suggests that ultraviolet light of approximtely 260 nm may inhibit melatonin. Red light and yellow light are ineffective in suppressing melatonin; green light suppresses melatonin production to some extent. The most effective light clearly is blue (Brainard et al., 1984). This same light, blue light, also inhibits melatonin production in the human (Brainard et al., 1985), which may imply that rhodopsin is the retinal mediator of the effects of light on the pineal gland. Rhodopsin is activated by light at about 502 nm. Blue light, or light in this range, seems to be the most effective in inhibiting melatonin in both the pineal gland and the blood. The studies in the human were much more refined in terms of the spectral band width of the light that was administered.

EFFECTS OF STRESS ON PINEAL MELATONIN PRODUCTION

Effects of stress on the pineal gland are extremely complex (Reiter, 1988). Norepinephrine, released from sympathetic neural endings within the pineal gland, normally stimulates melatonin production. Very high circulating levels of norepinephrine can occur, such as in stress responses in which blood levels of norepinephrine may increase severalfold. If norepinephrine circulates to the pineal gland, why does it not act on these receptors and stimulate melatonin production even during the day? The theory has been that circulating norepinephrine arriving at the pineal is actively taken up, that is, nerve endings sequester all excess norepinephrine and as a consequence daytime stress does not increase melatonin production (Kvetnansky et al., 1979). This idea was espoused because if the superior cervical ganglia are removed, these nerve endings degenerate and stress then causes a rise in melatonin production (Parfitt and Klein, 1976). If norepinephrine uptake is blocked with tricyclic antidepressants, or reuptake inhibitors, then stress also has a marked stimulatory effect on melatonin production.

There are, however, certain stressful situations during the day that clearly stimulate melatonin production within the pineal gland. The best example is insulin-induced hypoglycemia (Lynch et al., 1973), which causes a very marked rise in melatonin production. The general consensus has been that if stress has any effect on the pineal gland, it in fact elevates melatonin production.

About a year ago we did an experiment at night wherein we selected two stresses, either hind-leg injection of saline or swimming for 10 min. We thought one would be relatively mild, and one would be more severe; swimming seemed to be much more severe. After swimming for 10 min, the rats were very exhausted. In the case of hind-leg injection, they seemed to settle down relatively rapidly after the treatment. These experiments were conducted during the day and again at night, and the nighttime effects were unexpected. We anticipated a subsequent rise in melatonin production. Upon injection of saline at night, melatonin levels and N-acetyltransferase activity dropped precipitously (Troiani et al., 1988). This response was prevented if we removed the adrenal gland. It appeared as though some substance of adrenal origin in the circulation shuts off the pineal gland at night. We think this factor derives from the adrenal cortex rather than from the medulla.

When the experiment was repeated using the 10-min swim, a more prolonged and sustained stress, melatonin levels also dropped precipitously, even more dramatically, but *N*-acetyltransferase was totally unaffected (Troiani et al., 1987). In fact, *N*-acetyltransferase activity continued to rise. We next measured blood levels of melatonin, which increased during swimming. It appears that, in this case, melatonin production was not inhibited. Instead, there appeared to be a massive release of all residual quantities of melatonin from the pineal. The release was dramatic, causing blood levels to rise and pineal levels to drop. This effect was not prevented by adrenalectomy or by hypophysectomy. If we removed the superior cervical ganglia to prevent the nighttime rise in melatonin, then gave isoproterenol, a β agonist, at the time of darkness to stimulate melatonin production artificially, and then caused the animals to swim, melatonin levels still dropped (Wu et al., 1987). This response to swimming is a totally different response from the hind-leg saline injection. It is not related to the adrenal, because adrenal gland removal does not affect the massive release of melatonin.

In summary, hind-leg saline injection is a mild stress, and swimming is a severe stress. Hind-leg injection causes a drop in *N*-acetyltransferase and melatonin that is prevented by adrenalectomy. Swimming causes a drop in melatonin apparently because of a massive release that does not influence synthesis of melatonin. *N*-acetyltransferase levels continue to increase, and that response is not influenced by the removal of the adrenal gland. Apparently swimming speeds up the secretory mechanism, whatever that may be. We do not know how melatonin leaves the pinealocyte, but presume that it simply diffuses out of the cell.

PINEAL GLAND EFFECTS ON REPRODUCTIVE PHYSIOLOGY

The pineal gland in some photoperiodic species has a very profound influence on reproductive physiology, and reproduction is only one of many neuroendocrine systems influenced by the pineal gland (Reiter, 1980). Syrian hamsters, a highly photosensitive species, when placed in an environment where days are short (less than 12.5 hr of light), will cease to be reproductively competent. Under long days, Syrian hamsters have large functional testes, high levels of luteinizing hormone (LH) in the blood, high levels of follicle-stimulating hormone (FSH), large amounts of prolactin, and large amounts of testosterone. Clearly, under long-day conditions (more than 12.5 hr of light), these animals are reproductively functional. In short days, however, the testes show dramatic atrophy, the accessory sex organs regress, LH levels fall, FSH levels are diminished, prolactin levels plummet, and testosterone levels likewise drop (Fig. 7). Changing the animals from long days to short days causes this total collapse of the reproductive system. Removal of the pineal gland or the superior cervical ganglia blocks these changes. Thus, the testes remain large and functional, the accessory organs do not atrophy, and LH, FSH, prolactin, and testosterone all remain normal. In other words, the regressive response of the reproductive

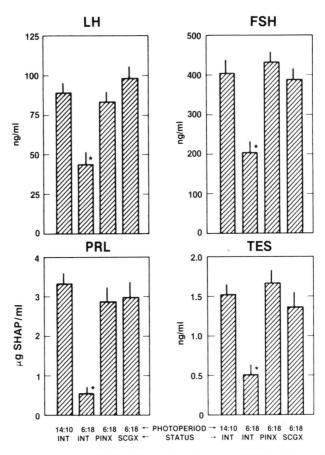

Fig. 7. Changes in plasma LH, FSH, PRL, and testosterone (TES) levels in male hamsters exposed to short days and the effects of pinealectomy; SHAP, standard hamster anterior pituitary.
(Reprinted with permission from Reiter, R.J. 1980. The pineal and its hormones in the control of reproduction in mammals. *Endocrinol. Rev.* 1:109–131.)

organs of this species to short days, which requires about 6 to 10 weeks, is clearly a pineal-mediated phenomenon.

This response can be modified by other parameters, particularly environmental temperature. At cooler temperatures, these responses are much faster. We have just completed a study wherein we kept animals at 33° versus 23°C. This warmer temperature markedly delays the atrophic response of the reproductive organs (Li et al., 1987; Reiter et al., 1988). It appears that temperature changes do not affect the melatonin signal. We believe there is a change in the sensitivity of the animal to the melatonin, depending on change in the ambient temperature. The interaction of environmental variables is complex in this regard. The pineal gland is sending out a signal, and the sensitivity of the animal is adjusted by other environmental factors.

Under natural daily photoperiods, reproductive physiology changes dramatically in the hamster. These effects are not necessarily unique to hamsters; many species show these responses. Reproductive responses are very complex and variable on a species-to-species basis. All that is necessary to show the effects of the daylength on the pineal gland and in turn on reproductive competence under these conditions is to move the animal to short days. For the hamster, anything less than 12.5 hr of light per day is perceived as a short day. It is an all-or-nothing phenomenon; whether the animal is exposed to 10, 8, 6, or 2 hr of light makes no difference. If exposure is less than 12.5 hr of light per day, gonads atrophy. Pinealectomy totally reverses this effect.

What importance does all this information have? Daylength changes as a function of season. Exclusive of man, all animals in their natural habitats are exposed to seasonal variations in daylength. We as humans control our photoperiodic environment, and yet we may be seasonal in many respects. The longest days occur at the time of summer solstice, and the shortest days at winter solstice. These variations are inverted in the northern and southern hemisphere. Near the equator these seasonal variations are slighter. At the extremes of latitude they become very great, such that within the Arctic or Antarctic circles there are actually periods of constant light and periods of constant dark. Animals must adjust their physiology accordingly. They rely on this photoperiodic information; the pineal transduces this photoperiodic information into usable hormones that signal the animal to make the appropriate physiological adjustments.

Thus, we know the pineal gland in some of these species regulates seasonal reproduction (Fig. 8). In hamsters kept in San Antonio, Texas (29°33″N), where the longest days are 14 hr of light in summer and shortest days are 10 hr of light in winter, these extremes overlap the 12.5-hr critical photoperiod. During the summertime these animals are reproductively competent. As fall approaches the gonads atrophy; they remain atrophic through the winter, eventually regenerate, and are totally competent again during the summer months (Reiter, 1973). This constitutes the annual cycle of reproduction. Removal of the pineal gland in these animals renders them sexually competent at all times, that is, they show no seasonality with regard to reproductive physiology whatsoever. The annual cycle of reproduction, in this and other species, is mediated by the pineal gland. Winters in San Antonio are not particularly harsh, but in Edmonton, Alberta, Canada, or even in Seattle, Washington, the conditions are harsher in the winter. Seasonal breeding animals cannot afford to have young during the middle of the winter, so they have developed mechanisms to avoid producing

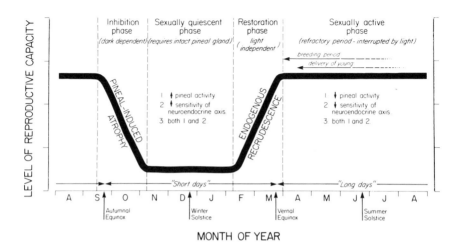

MONTH OF YEAR

Fig. 8. Seasonal changes in reproduction mediated by the pineal gland.
(Reprinted with permission from Reiter, R.J. 1978. Interaction of photoperiod, pineal and seasonal reproduction as exemplified by findings in the hamster. In: *The Pineal and Reproduction*, Reiter, R.J., ed., pp. 169–190. Basel: Karger.)

young at this time of year. These animals rely on photoperiod, signaled to them by means of the pineal gland, to seasonally adjust their reproductive competence (Fig. 8) (Reiter, 1974).

INTERNAL COINCIDENCE AND DURATION MODELS OF PINEAL SIGNAL PERCEPTION

How is the photoperiod information signal transferred to the organism? The two theories that I present are not mutually exclusive, and there may be many other explanations. The common theory is that under long days and short nights, a melatonin peak occurs with a finite duration of high melatonin production. Increasing the duration of darkness, in fact, prolongs the melatonin peak. This may be the means by which the animal is apprised of the photoperiodic environment by the pineal gland. This I refer to as the duration hypothesis (Fig. 9).

At least one other theory, termed an internal coincidence model, has some support. This hypothesis states that under short-day photoperiods, melatonin concentrations peak at night but the animal's sensitivity to melatonin is temporally displaced (Reiter, 1987) and occurs at a different time. The melatonin peak and melatonin receptor populations are not coincident under long days. When the animals are on short days, these rhythms

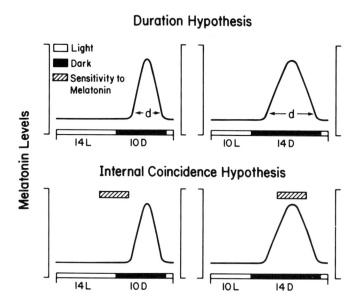

Fig. 9. Duration and internal coincidence hypotheses explaining effect of melatonin on the endocrine systems.
(Reprinted with permission from Reiter, R.J. 1986b. The pineal gland: an important link to the environment. *News Physiol. Sci.* 1:202–205.)

become coincident. Thus, when the melatonin peak occurs simultaneously with the sensitivity of the animal to the melatonin peak there is ample response (Fig. 9). This theory is attractive because every organ could have its own sensitivity phase in a different temporal place. For example, during the time that melatonin affects reproduction it may not have an effect on the pancreas. At another time, by contrast, melatonin may be released and influence the pancreas but not the reproductive system.

The fact that pineal melatonin sometimes influences reproduction, and not another system, and then sometimes influences another system and not reproduction, is explicable in terms of this internal coincidence model (Reiter, 1987). It is very important that these rhythms occur simultaneously for the animal to be properly synchronized. Each organ may have a different rhythm of sensitivity to melatonin, in that one melatonin rhythm sends out a basic signal and the organ determines how and when it uses this signal. The pineal gland in itself is relatively passive.

In mammals, the eye provides light perception for the pineal gland (Fig. 10). In nonmammalian species the pineal gland is directly photosensitive.

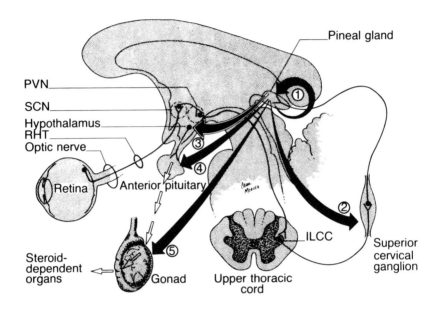

Fig. 10. Summary of interrelationships of visual system, pineal, and reproductive organs.

Information about the light:dark environment is transferred initially to the suprachiasmatic nuclei, which are very important to the pineal gland. These nuclei generate the melatonin rhythm and many other rhythms within the gland. Pathways between the hypothalamus and the upper thoracic cord remain ambiguous. Clearly, however, the peripheral sympathetic nervous system carries information from the thoracic cord outside the skull back into the brain, and here this information, in the form of a neural signal, is transduced, changed, altered, and comes out as a chemical: melatonin, 5-hydroxytryptophol, 5-hydroxytryptamine, miscellaneous peptides, and maybe many other hormone secretions. Certainly we know a great deal about melatonin; it is released and enters the blood in very copious quantities. It is also found in the cerebrospinal fluid with a rhythm similar to that in the blood. Where it acts, however, we do not really know, but we think it has strong neural effects on the hypothalamus.

Melatonin regulates the release of hypothalamus-releasing hormones and, in the case of prolactin, an inhibitory hormone. Melatonin also regulates the anterior pituitary and gonads. It is important to keep in mind that it is not the function of the pineal gland to regulate reproduction. The bulk of my work during the past 20 years has been relative to the reproductive system, and I have come to the conclusion that the pineal gland does not regulate reproduction per se. Its scope is much broader than this. The function of the pineal gland is to keep the animal in appropriate synchrony with its external environment, which may only coincidentally mean determining seasonal reproduction. But it may also involve migration, brown fat physiology, adrenal metabolism, insulin regulation, and so forth; any number of different systems are influenced by melatonin and the pineal gland.

Thus, the pineal gland must synchronize the organism. To do so, it must have predictive value to the organism. It is not sufficient for the animal to recognize days are short on December 25th; the animal must know in September that days will be short on December 25th so it can make the physiological adjustments in advance of that time. This is what the pineal gland does: It continually apprises the animal of the environmental state and adjusts its physiology accordingly. Further, I am convinced it has these same functions in man. The signal that emanates from the human pineal gland has many effects that we simply do not understand.

CONCLUSION

The pineal gland is a highly active organ of internal secretion and secretes a number of potential hormones. The best known is melatonin, which has a gamut of physiological effects. Some are relegated to the neuroendocrine system, but there are also effects on brain neurotransmitter metabolism that may be completely unrelated to reproductive or thyroid consequences of this gland. The pineal is an ubiquitously acting and active organ of internal secretion, and the function of the gland is to synchronize the physiology of the organism with the prevailing environmental conditions.

REFERENCES

Ariens Kappers, A. 1960. Development, topographical relations and innervation of epiphysis cerebri in albino rat. *Z. Zellforsch. Mikrosk. Anat.* 52:163–215.

Brainard, G. C., Richardson, B. A., Petterborg, L. J., Reiter, R. J. 1982. The effect of different light intensities on pineal melatonin content. *Brain Res.* 233:75–81.

Brainard, G. C., Richardson, B. A., King, T. S., Reiter, R. J. 1984. The influence of different light spectra on the suppression of pineal melatonin content in the Syrian hamster. *Brain Res.* 294:333–339.

Brainard, G. C., Richardson, B. A., King, T. A., Matthews, S. A., Reiter, R. J. 1983. The suppression of pineal melatonin content and N-acetyltransferase activity by different light irradiances in the Syrian hamster: a dose-response relationship. *Endocrinology* 113:293–296.

Brainard, G. C., Lewy, A. J., Menaker, M., Frederickson, R. H., Miller, L. S., Weleber, R. G., Cassone, V., Hudson, D. 1985. Effect of light wavelength on the suppression of nocturnal plasma melatonin in normal volunteers. *Ann. N.Y. Acad. Sci.* 453:376–378.

Craft, C. M., Morgan, W. W., Reiter, R. J. 1984. 24-hour changes in catecholamine synthesis in rat and hamster pineal glands. *Neuroendocrinology* 38:193–198.

Korf, H., Møller, M. 1984. The innervation of the mammalian pineal gland with special reference to central pinealopetal projections. *Pineal Res. Rev.* 2:41–86.

Kvetnansky, R., Kopin, I. J. Klein, D. C. 1979. Stress increases pineal epinephrine. *Commun. Psychopharmacol.* 3:69–74.

Lewy, A. J., Newsome, D. A. 1983. Different types of melatonin secretory rhythms in some blind subjects. *J. Clin. Endocrinol. Metab.* 56:1103–1107.

Lewy, A. J., Wehr, T. A., Goodwin, F. K., Newsome, D. A., Markey, S. P. 1980. Light suppresses melatonin secretion in humans. *Science* 210:1267–1269.

Li, K., Reiter, R. J., Vaughan, M. K., Oaknin, S., Troiani, M. E., Esquifino, A. I. 1987. Elevated ambient temperature retards the atrophic response of the neuroendocrine-reproductive axis of male Syrian hamsters to either daily afternoon melatonin injections or to short photoperiod exposure. *Neuroendocrinology* 45:356–362.

Lynch, H. J., Eng, J. P., Wurtman, R. J. 1973. Control of pineal indole biosynthesis by changes in sympathetic tone caused by factors other than environmental lighting. *Proc. Natl. Acad. Sci. USA* 70:1704–1707.

Menendez-Palaez, A., Howes, K. A., Gonzalez-Brito, A., Reiter, R. J. 1987. *N*-acetyltransferase activity, hydroxyindole-*O*-methyltransferase activity and melatonin levels in the Harderian glands of female Syrian hamsters: changes during the light:dark cycle and the effect of 6-parachlorophenylalanine administration. *Biochem. Biophys. Res. Commun.* 145:1231–1238.

Panke, E. S., Reiter, R. J., Rollag, M. D. 1979a. Effect of removal of the Harderian glands on pineal melatonin concentrations in the Syrian hamster. *Experientia* 35:1405–1406.

Panke, E. S., Rollag, M. D., Reiter, R. J. 1979b. Pineal melatonin concentrations in the Syrian hamster. *Endocrinology* 104:194–197.

Parfitt, A. G., Klein, D. C. 1976. Sympathetic nerve endings in the pineal gland protect against acute stress-induced increase in *N*-acetyltransferase activity. *Endocrinology* 99:840–844.

Reiter, R. J. 1973. Pineal control of a seasonal reproductive rhythm in male golden hamsters exposed to natural daylight and temperature. *Endocrinology* 92:423–430.

Reiter, R. J. 1974. Circannual reproductive rhythms in mammals related to photoperiod and pineal function: a review. *Chronobiologia* 1:365–367.

Reiter, R. J. 1978. Interaction of photoperiod, pineal and seasonal reproduction as exemplified by findings in the hamster. In: *The Pineal and Reproduction*, Reiter, R. J., ed., pp. 169–190. Basel: Karger.

Reiter, R. J. 1980. The pineal and its hormones in the control of reproduction in mammals. *Endocrinol. Rev.* 1:109–131.

Reiter, R. J. 1981. The mammalian pineal gland: structure and function. *Am. J. Anat.* 162:287–313.

Reiter, R. J. 1986a. Pineal melatonin production: photoperiodic and hormonal influences. In: *Advances in Pineal Research*, Vol. 1, Reiter, R. J., Karasek, M., eds., pp. 77–87. London: John Libbey.

Reiter, R. J. 1986b. The pineal gland: an important link to the environment. *News Physiol. Sci.* 1:202–205.

Reiter, R. J. 1987. The melatonin message: duration versus coincidence hypotheses. *Life Sci.* 46:2119–2131.

Reiter, R. J. 1988. Pineal responses to stress: implications for reproductive physiology. In: *Biorhythms and Stress in Physiopathology of Reproduction,* Pancheri, P., Zichella, L., eds., pp. 215–226. New York: Hemisphere.

Reiter, R. J., Joshi, B. N., Heinzeller, Th., Nurnberger, F. 1986. A single 1 or 5 second light pulse at night inhibits hamster pineal melatonin. *Endocrinology* 118:1906–1909.

Reiter, R. J., Esquifino, A. I., Champney, T. H., Craft, C. M., Vaughan, M. K. 1985. Pineal melatonin production in relation to sexual development in the male rat. In: *Paediatric Neuroendocrinology,* Gupta, D., Borrelli, P., Attanasio, A., eds., pp. 190-202. London: Croam Helm.

Reiter, R. J., Hurlbut, E. C., Brainard, G. C., Steinlechner, S., Richardson, B. A. 1984. Influence of light irradiance on hydroxyindole-*O*-methyltransferase activity, serotonin-*N*-acetyltransferase activity, and radioimmunoassayable melatonin levels in the pineal gland of the diurnally active Richardson's ground squirrel. *Brain Res.* 288:151–157.

Reiter, R. J., Li, K., Gonzalez-Brito, A., Tannenbaum, M. G., Vaughan, M. K., Vaughan, G. M., Villanua, M. 1988. Elevated environmental temperature alters the responses of the reproductive and thyroid axes of female Syrian hamsters to afternoon melatonin injections. *J. Pineal Res.* 5:301–315.

Rudeen, P. K., Reiter, R. J., Vaughan, M. K. 1975. Pineal serotonin *N*-acetyltransferase in four mammalian species. *Neurosci. Lett.* 1:225–229.

Thiele, G., Holtorf, A., Steinlechner, S., Reiter, R. J. 1983. The influence of different light irradiances on pineal *N*-acetyltransferase activity and melatonin levels in the cotton rat, *Sigmodon hispidus. Life Sci.* 33:1543–1547.

Troiani, M. E., Reiter, R. J., Vaughan, M. K., Oaknin, S., Vaughan, G. M. 1987. Swimming depresses nighttime melatonin content without changing *N*-acetyltransferase activity in the rat pineal gland. *Neuroendocrinology* 47:55–60.

Troiani, M. E., Reiter, R. J., Vaughan, M. K., Gonzalez-Brito, A., Herbert, D. C. 1988. The depression in rat pineal melatonin production after saline injection at night may be elicited by corticosterone. *Brain Res.* 450:18–24.

Vaughan, G. M., Mason, A. D., Jr., Reiter, R. J. 1986. Serum melatonin after a single aqueous subcutaneous injection in Syrian hamsters. *Neuroendocrinology* 42:124–127.

Webb, S. M., Champney, T. H., Lewinsi, A. K., Reiter, R. J. 1985. Photoreceptor damage and eye pigmentation: influence on the sensitivity of rat pineal *N*-acetyltransferase activity and melatonin levels to light at night. *Neuroendocrinology* 40:205–209.

Wilkinson, M., Arendt, J., Bradtke, J., de Ziegler, D. 1977. Determination of a dark-induced increase in pineal *N*-acetyltransferase activity and simultaneous radioimmunoassay of melatonin in pineal, serum and pituitary tissue of the male rat. *J. Endocrinol.* 72:243–249.

Wilson, B. W., Lynch, H. J., Ozaki, Y. 1978. 5-Methoxytryptophol in rat serum and pineal: detection, quantitation and evidence for daily rhythmicity. *Life Sci.* 23:1019–1024.

Wu, W., Reiter, R. J., Troiani, M. E., Vaughan, G. M. 1987. Elevated daytime rat pineal and serum melatonin levels induced by isoproterenol are depressed by swimming. *Life Sci.* 41:1473–1479.

6 Oncogenes and Leukemia

WILLIAM P. HAMMOND, IV
Division of Hematology, Department of Medicine
University of Washington School of Medicine
Seattle, Washington

CONTENTS

Leukemia is a family of neoplastic diseases of the hematopoietic system in which a clone of cells proliferates abnormally, replacing normal marrow, infiltrating other hematopoietic tissues, and leading to death. First recognized independently by Bennett in England and Virchow in Germany in 1845, it was quickly recognized by Virchow not to represent "suppuration of blood" but rather a primary abnormality in what he called "cellular pathology" (Wintrobe et al., 1981; Virchow, 1860). Virchow subsequently distinguished two forms of "leukemie" ("white blood"); in one, lymphadenopathy was marked (chronic lymphocytic leukemia), and in the second, splenic enlargement dominated (chronic myelogenous leukemia). In 1857, Friedreich first reported leukemia occurring in an acute form; after the development of Ehrlich's methods for differential staining of blood leukocytes, the various cytologic subtypes of leukemia were described. Today, the name "leukemia" is given to any of a broad group of malignant disorders in which a clonal population of cells derived from hematopoietic tissue infiltrates the blood and bone marrow, resulting in symptoms from either excessive numbers of abnormal cells in circulation or the failure of bone marrow production of normal cells (Wintrobe et al., 1981).

In normal hematopoiesis, a hierarchically organized developmental pathway proceeds from a pluripotent stem cell through a series of steps of commitment and differentiation to fully matured end cells (Fig. 1). Stem cells are characterized by extensive capacity for self-renewal and differentiation capacity for multiple cellular subtypes. At the other extreme, the far more numerous differentiated progeny have lost their self-renewal capacity and are committed to function in a particular cellular lineage. In normal states, the cells in the pluripotent and committed stem cell compartments comprise less than 0.01% of bone marrow cells, and the bone marrow appears to regulate its release of cells so that only fully differentiated and functionally mature cells are usually seen in blood. In normal

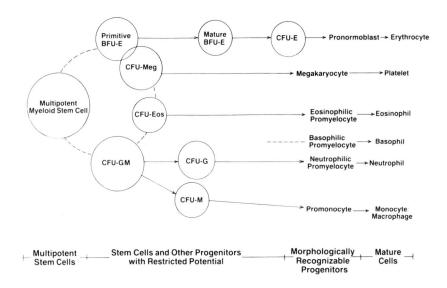

Fig. 1. A hierarchical model of progenitor cells in bone marrow. Stem cells possess both self-renewal capacity and capacity for differentiation into various cellular lineages; depending on which cell line they become "committed" to, different end cells (*far right*) are produced. Abbreviations: BFU-E, burst-forming unit erythroid; CFU-E, colony-forming unit erythroid; CFU-Meg, colony-forming unit megakaryocyte; CFU-Eos, colony-forming unit eosinophil; CFU-GM, colony-forming unit granulocyte macrophage; CFU-G, colony-forming unit granulocyte (neutrophil); CFU-M, colony-forming unit monocyte/macrophage.

individuals a very small proportion of circulating immature cells may be demonstrable, and thus it is either the appearance of increased total numbers or increased proportions of less mature cells that constitute the cardinal findings in leukemias.

In respect to regulation of these cellular compartments and movement from one to the next, a series of nonspecific as well as lineage-specific growth factors have been defined that are required for the expression of these cellular phenotypes (Adamson, 1984; Metcalf, 1984). In vitro culture in semisolid media (agar or methylcellulose) has been used extensively to demonstrate the existence of individual cells in marrow and blood that are capable of proliferation and differentiation to form a colony of one or another of the various differentiated blood cells. Recent developments in serum-free culture have defined requirements for the proliferation and differentiation of these cells, which include, as nonspecific growth factors, iron-saturated transferrin, insulin, and serum albumin (possibly containing other trace factors as yet unidentified) (Iscove et al., 1980; Migliacco and Migliacco, 1987). The presence of specific growth factors such as erythropoietin or granulocyte-macrophage colony-stimulating factor (GM-CSF) (Table I) are then required for the expression of the fully developed phenotype of erythroid or granulocyte- and macrophage-containing colonies, respectively. These hematopoietic growth factors appear to act at various levels in the system outlined (see Fig. 1); some factors, such as erythropoietin in the erythroid system, granulocyte colony-stimulating factor (G-CSF) in the granulocytic system, and interleukin-5 (IL-5) in the eosinophilic system, act relatively late ("mature cell factors"), and others act at earlier stages ("primitive cell factors") on cells of multiple lineages or cells in the stem cell compartment. Between 1983 and 1987, several of these growth factors were cloned using recombinant DNA technology, and their structural and functional relationships are just beginning to emerge (Jacobs et al., 1985; Kawasaki et al., 1985; Lee et al., 1985; Wong et al., 1985; Kaushansky et al., 1986; Metcalf, 1986; Nagata et al., 1986; Yang et al., 1986; Broudy et al., 1987). Certain of the factors are also capable of stimulating the release of other factors within this group [e.g., interleukin-1 (IL-1) stimulates increased production of GM-CSF and G-CSF by endothelial cells (Broudy et al., 1987)]. In addition, there are several reports of synergistic activity between various factors in the promotion of colony formation of any particular cell type. The physiology of this regulatory system continues to excite considerable research interest (Metcalf, 1986).

Current hypotheses for the mechanism of leukemogenesis suggest that the clonal proliferation of leukemic cells derives from a single transformation event in this hierarchy that confers growth advantage in association with a variable degree of abnormal differentiation on the affected cell. For example, a transformation event occurring at one of the stages

TABLE I.
Hematopoietic Growth Factors

Factor[a]	Target cells[b]
Mature Cell Factors	
Erythropoietin	BFU-E, CFU-E through retics
Granulocyte colony-stimulating factor (G-CSF)	CFU-G through neutrophils
Macrophage colony-stimulating factor (M-CSF)	CFU-M, monos, macrophages
Interleukin 5 (IL-5, eosinophil differentiating factor)	CFU-Eos, through eosinophils
Interleukin 2 (IL-2, T-cell growth factor)	T lymphocytes
B-cell growth factors (BCGF)	PreB lymphocytes, B lymphocytes
Primitive Cell Factors	
Granulocyte macrophage colony-stimulating factor (GM-CSF)	CFU-GM, BFU-E, CFU-G, CFU-M, neutrophils, monocytes, eosinophils
Interleukin 3 (IL-3)	CFU-GEMM, CFU-GM, B
Interleukin 1 (IL-1, "hemopoietin 1")	CFU-S, CFU-GEMM, ?CFU-GM
Interleukin 6 (IL-6, hybridoma growth factor)	CFU-S, CFU-GEMM

[a] Factors: usually named originally by function assays; these have all been cloned by recombinant DNA technology (Jacobs et al., 1985; Lee et al., 1985; Wong et al., 1985; Kawasaki et al., 1985; Kaushansky et al., 1986; Metcalf, 1986; Nagata et al., 1986; Yang et al., 1986; Broudy et al., 1987).
[b] Target cells: represent cells in the hierarchy of hematopoiesis (Fig. 1). CFU-GEMM represents the multipotent myeloid stem cell; "?" refers to uncertainty in data as to sites of action.

represented in the left-hand portion of Fig. 1 could then result in growth of a neoplastic clone. Depending on which particular cell was affected, one would expect a different phenotype such as erythroid leukemia, or monocytic leukemia, or a very primitive "undifferentiated" leukemia. The details of the mechanism underlying such transformation events constitute the topic of this review.

ONCOGENES

Cancer generally, and leukemia specifically, have been demonstrated to result from an alteration in the DNA of the neoplastic cell population. For example, in methylcholanthrene-induced fibrosarcomas of mice, the transforming principle was shown unequivocally to be DNA by serial transfer experiments in which tumorigenicity of in vitro transformed cells was demonstrated (Fig. 2) (Shih et al., 1979; Weinberg, 1983). From experiments such as these it was concluded that cancer does, indeed, have a "genetic" basis, and following the earlier proposal by Huebner and Todaro (1969), the altered genes were called oncogenes.

Oncogenes such as those reported with chemically induced cancers had also been previously demonstrated with polyoma virus-induced transformation of hamster fibroblasts capable of growing as tumors in mice (Tooze, 1973). This suggested that a stable, genetic transformation event can also be conferred by viral infection. Both DNA and RNA tumor viruses have been described, and the independent discoveries by Temin and Baltimore in 1970 of RNA-dependent DNA polymerase (reverse transcriptase) (Temin and Mizutani, 1970; Baltimore, 1970) provided a mechanism for the action of RNA tumor viruses via a stably altered DNA "provirus." Because the oncogenes of DNA tumor viruses appear to be intimately involved in viral replication as well as transformation, they have been difficult to study. The oncogenes of RNA tumor viruses appear not to be primarily involved in viral replication, and thus their functions and identity have been more readily approached experimentally (Varmus, 1987). An additional unusual biological property of the RNA tumor viruses is that their "transforming" DNA sequences are structurally similar to those of normal cellular genes. Hence, many of these oncogenes of viral origin have been defined and are termed "viral oncogenes" or simply *v-onc* genes, and have been named with three-letter abbreviations reflecting their viral origins (Tables II and III). The homologous "cellular" oncogenes, also called proto-oncogenes or just *c-onc* genes, have been extensively studied initially in avian and murine systems and more recently in man. Interestingly, there is extraordinary phylogenetic conservation of these proto-oncogenes, as demonstrated by highly homologous sequences in the fruit fly *Drosophila melanogaster* and in the yeast *Saccharomyces cerevisiae* (Varmus, 1987). From such data as these it has been concluded that understanding the relationship between proto-oncogenes and the transforming viral oncogenes is likely to reveal answers to fundamental questions about neoplasia.

CELL GROWTH AND SIGNALING

To begin to understand its regulation, we must recognize that cellular growth normally proceeds through an orderly process that has been referred to as the "cell cycle." Cell growth both in vitro and in vivo is characterized by a series of events beginning with a signal to initiate cell growth (G_1 phase); this is followed by DNA synthesis (S phase), a period of further growth (G_2 phase), and finally mitosis (M) (Fig. 3) (Varmus, 1987;

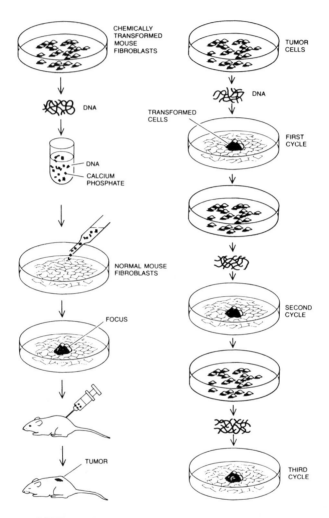

Fig. 2. Tumor cell DNA was shown to encode cancer traits by a gene-transfer experiment (*left*). DNA was extracted from mouse cells that had been transformed by a chemical carcinogen. The DNA was coprecipitated with calcium phosphate, which facilitated its entry into normal mouse cells. Transfected normal cells gave rise to foci of transformed cells. Injection of transformed cells into mice gave rise to a tumor. In a serial-transfer experiment (*right*), DNA extracted from human tumor cells transformed normal mouse cells. The transformed cells were grown into a large culture whose DNA was extracted in turn and again served to transform mouse cells. The process was repeated in a third cycle and even a fourth. Only short segments of DNA could survive the process of repeated extraction and coprecipitation, and so the successive transformations showed that the transforming agent must reside in a single segment.
(Reprinted with permission from Weinberg, R. A. 1983. A molecular basis of cancer. *Sci. Am.* 249:126–142.)

TABLE II.
Selected Viral and Cellular Oncogenes in Leukemia

Virus name	V-Onc	Origin	Animal disease
Rous sarcoma	src	Chicken	Sarcoma
Fujinami sarcoma	fes	Chicken	Sarcoma
Myelocytomatosis-29	myc	Chicken	Carcinoma, myeloid leukemia
Avian myeloblastosis	myb	Chicken	Myeloid leukemia
Avian E-26 myeloblastosis	ets	Chicken	Myeloid leukemia
Avian erythroblastosis-H	erb B	Chicken	Erythroleukemia
Avian erythroblastosis-ES4	erb A	Chicken	Erythroleukemia
Avian reticuloendotheliosis	rel	Turkey	Immature B-cell lymphoma
Abelson murine leukemia	abl	Mouse	PreB-cell lymphoma
Kirsten murine sarcoma	Ki-ras	Rat	Sarcoma, erythroleukemia
Harvey murine sarcoma	Ha-ras	Rat	Sarcoma, erythroleukemia
Moloney murine sarcoma	mos	Mouse	Sarcoma, leukemia
McDonough feline sarcoma	fms	Cat	Sarcoma
Simian sarcoma	sis	Monkey	Sarcoma
H-Z feline leukemia	kit	Cat	Sarcoma, leukemia
(?)	B-lym	Chicken	Lymphoma
Mouse mammary tumor	int-2	Mouse	Mammary carcinoma
(None)	N-ras	Human	—

TABLE III.

Activities and Associations of Viral Oncogene Products

Protein product/activity	V-Onc	Disease association in man
Tyr kinase (pp60$^{v\text{-}src}$)	src	—
Tyr kinase	fes	Acute leukemia
Nuclear protein	myc	Burkitt's lymphoma
Nuclear binding	myb	Acute leukemia
	ets	—
Truncated EGF receptor	erbB	—
Triiodothyronine receptor	erbA	—
Serine/threonine kinase	rel	—
Tyr kinase	abl	CML
GTP-binding protein	Ki-ras	AML
GTP-binding protein	Ha-ras	Non-Hodgkins lymphoma
Serine/threonine kinase	mos	—
M-CSF receptor homologue	fms	—
PDGF-B chain	sis	—
Related to EGF receptor	kit	—
Nuclear protein	B-lym	Burkitt's lymphoma
Related to FGF	int-2	—
GTP-binding protein	N-ras	Neuroblastoma, AML

Baserga, 1981). Cells may withdraw from the cell cycle during G_1 into a resting state (called G_0), and then be activated back into the cell cycle by various signals. Alternatively, after mitosis cells may undergo differentiation and depart from G_0 or G_1, ultimately losing their self-replicative capacity as they fulfill their function as mature end cells. Understanding the processes regulating transition between G_0 and G_1 and between G_0 or G_1 and the differentiated state are the central questions for our understanding of cellular growth control. The capacity to enter G_0 appears to be decreased or lost in certain malignant cell populations; alternatively, the transition from G_0 back to G_1 may be obligatory in certain abnormal settings. A third possible mechanism underlying neoplastic transformation could be a failure to depart G_0 or G_1 (and the replicative cell cycle) into a differentiated state. The phenotypic expressions of such a lack of growth control in leukemia generally include such abnormal in vitro parameters as unrestrained cell proliferation ("immortalization"), failure of differentiated phenotypes to develop in the normal sequence ("maturation arrest"), loss of contact inhibition ("focus formation"), and anchorage-independent growth.

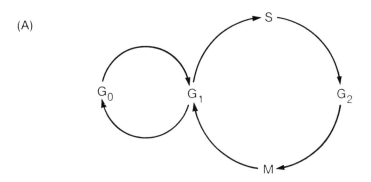

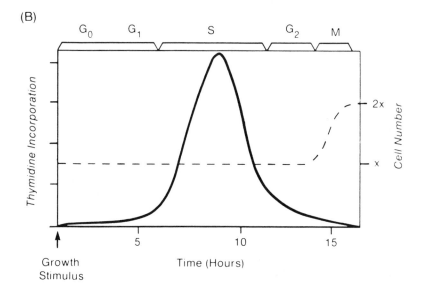

Fig. 3. **(A)** The cell growth cycle. The two major functions, replication of chromosomal DNA (S=DNA synthesis) and cell division (M=mitosis), are separated by two intervals (G₁ and G₂=gaps). Cells can escape from the cycle via early G₁ to enter a resting phase (G₀); resting cells can reenter the cycle via G₁. **(B)** Idealized representation of the response of resting (G₀) mammalian cell culture to a growth stimulus (eg., addition of growth factors). Entry of cells into S phase is detected by incorporation of radioactive thymidine into DNA (*continuous line*), and passage through M phase is measured by counting cells (*dashed line*). If cells are sufficiently synchronized to detect a second peak of DNA synthesis, the length of the cell cycle can be determined from the interval between S phases.

(Reprinted with permission from Varmus, H. 1987. Cellular and viral oncogenes. In: *The Molecular Basis of Blood Diseases*, Stamatoyannopoulos, G., Nienhuis, A. W., Leder, R., Majerus, P. W., eds., pp. 347–346. Philadelphia: Saunders.)

A second principle underlying our understanding of the regulation of cell growth is that extracellular factors interacting with specific membrane receptors prominently influence cell growth. As outlined briefly for the hematopoietic system, multiple growth factors for specific lineages have been identified and cloned, and rapid progress is being made in the identification and cloning of their respective receptors. These receptors are usually transmembrane glycoproteins expressed in varying numbers on the cell surface (from hundreds to millions of copies per cell) that transduce information into the cell either directly via such activities as the protein kinase of their intracytoplasmic domains or indirectly via endocytosis in clathrin-coated invaginations of the cell membrane (Berridge, 1985; Ellis et al., 1986; Varmus, 1987).

The mechanisms whereby extracellular factors stimulate cell growth are incompletely understood, although two separate classes of actions have been defined. Certain factors are poorly mitogenic in themselves but markedly augment the mitogenic activity of other factors. Recently, a new theoretical model for cell proliferation, called the competence progression model, has been proposed to explain the findings (Pledger et al., 1977; Stiles et al., 1979; Kelvin et al., 1986). Thus, certain factors are referred to as "competence factors" because they are required to activate a response; those factors that augment responses to stimulation by the competence factors are called "progression factors." In general, progression factors are weakly active (or inactive) on their own, whereas competence factors are capable of producing distinct cellular responses; when combined in sequence, a competence factor and a progression factor produce the maximal biological response.

Third, to understand regulation of cell growth we need some concept of these intracellular signal transduction pathways. These pathways are relatively few in number, generally consist of a series of events beginning at the cell membrane surface receptor, and involve a transducing/amplifying element, production of second messengers, internal effectors, and finally a cellular response (Fig. 4) (Rodbell, 1980; Gomperts, 1983; Nishizuka, 1984; Berridge, 1984, 1985; Berridge et al., 1984; Ellis et al., 1986; Ullrich et al., 1986). As shown, there may be both stimulatory and inhibitory modulators of the response, and an extended series of intermediary reactions may occur before the final cellular response. Of particular note, protein kinase C, designated here as C kinase, is the site of action of the tumor promoters known as phorbol esters, and with calcium-dependent calmodulin activation of another protein kinase appear to participate in the principal signal transduction pathway for multiple different mitogenic signals. The calcium ionophores capable of directly altering intracellular calcium concentrations and the phorbol esters and inhibitors of protein kinase C have become important reagents in the dissection of these signal transduction pathways.

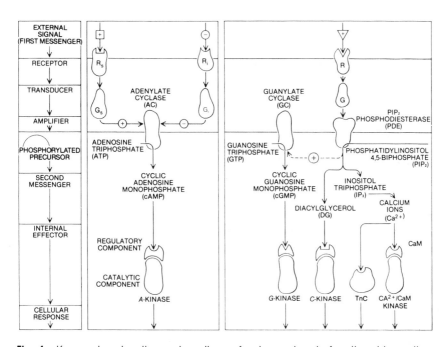

Fig. 4. Known signal pathways in cells are few in number. In functional terms they share a sequence of events (*left*). External messengers arriving at receptor molecules in the plasma membrane activate a closely related family of transducer molecules, which carry signals through the membrane, and amplifier enzymes, which activate internal signals carried by "second messengers." The pathway employing the second messenger cAMP (*middle*) has stimulatory receptors (R_s) and inhibitory ones (R_i), which both communicate with the amplifier adenylate cyclase (AC) by way of stimulatory and inhibitory transducers called G proteins because they require guanosine triphosphate (GTP) to function. Adenylate cyclase converts ATP into cAMP. The other major pathway (*right*) is not known to recognize inhibitory external signals. It employs a stimulatory G protein to activate its amplifier, a phosphodiesterase (PDE) enzyme. The enzyme makes phosphatidylinositol 4,5-biphosphate (PIP₂) into a pair of second messengers, diacylglycerol (DG) and inositol triphosphate (IP₃). In turn IP₃ induces the cell to mobilize still another messenger: calcium ions (Ca^{2+}). Moreover, the path somehow induces the amplifier guanylate cyclase (GC) to convert GTP into the second messenger cyclic guanosine monophosphate (cGMP). In general the second messengers bind to the regulatory component of a protein kinase, an enzyme that activates a cellular response such as contraction or secretion by adding phosphate (PO₄) groups to particular proteins. Calcium binds to a family of proteins including calmodulin (CaM) and troponin C(TnC). In turn CaM activates a protein kinase; TnC stimulates muscle contraction directly.

(Reprinted with permission from Berridge, M. J. 1985. The molecular basis of communication within the cell. *Sci. Am.* 253:142–152.)

HISTORICAL STUDIES OF ANIMAL MODELS OF NEOPLASIA

Unequivocal demonstration that viruses can transmit neoplastic diseases was reported in 1908 by Ellermann and Bang (Tooze, 1973), who demonstrated the transmission of leukemia in chickens by inoculating an extract from their leukemic cells that had been filtered to remove bacteria. Because leukemia was not then recognized as a cancer, this report from the Danish group did not produce the stir elicited by Peyton Rous at the Rockefeller Institute, who showed in 1911 that a spontaneous chicken sarcoma could likewise be transmitted via similar cell-free filtrates (Tooze, 1973). Subsequently, Rous isolated additional viruses from spontaneously occurring chicken tumors and suggested that viral agents might cause a substantial fraction of chicken tumors. The name Rous sarcoma virus (RSV) has been given to the original isolate as well as to additional independently isolated viruses that induce similar sarcomas; these are also given the generic name avian sarcoma virus (ASV). The chicken leukemia viruses originally described by the Danish group cause different subtypes of leukemia and have been collectively referred to as the avian leukemia viruses (ALV). The clinical presentation of these viruses may be (1) myeloblastosis, a disease characterized by large numbers of myeloblasts circulating in the blood; (2) erythroblastosis, a similar condition in which the cells are clearly of erythroid lineage; or (3) lymphoblastosis, a disorder in which lymphocytic infiltration of organs such as spleen, liver, and lung is most prominent. Certain strains of virus that characteristically produce one type of disease have been named accordingly, such as avian myeloblastosis virus (AMV) or avian erythroblastosis virus (AEV) (Tooze, 1973).

Knowing that avian leukemias were virally transmitted, at least two groups in the 1920s and 1930s began studies of inbred mouse strains characterized by very high frequencies of spontaneous leukemia. For example, in the AK strains, leukemia developed between 6 and 18 months of age, with as many as 85% of the mice developing disease before the time of expected death. Simultaneously, inbred strains with a very high incidence (90%) of mammary adenocarcinomas were noted to have extremely low leukemia frequencies (1%–2%). Consequently, a series of experiments were undertaken that attempted to pass leukemia between strains using cell-free extracts as was done by the Danish workers studying ALV (Tooze, 1973). Because mouse mammary adenocarcinomas had been shown to be transmissible in young (7- to 21-day-old) mice, Ludwig Gross in New York used 1-day-old suckling C3H mice (a strain with low leukemia incidence) to demonstrate transfer of leukemia using cell-free material from the AK leukemia cells (Gross, 1951). To demonstrate that the extracts were cell free, he showed that the leukemia cells had the determinants of recipient C3H mice rather than the specificities of donor AK cells, implying de novo transformation of C3H cells. Subsequent passages of this virus preparation have demonstrated more virulent growth patterns, and attempts at isolating viral agents capable of causing other murine cancers have resulted in identification of additional viruses causing phenotypically different leukemias. Among these murine leukemia viruses are those named the Moloney virus, the Friend virus, and the Rauscher virus, isolates that have stimulated numerous studies and many new insights into viral leukemogenesis (Friend, 1957; Moloney, 1960; Rauscher, 1962). The murine and avian leukemia viruses have provided sources for identification of both viral and cellular oncogenes, and studies comparing the structures of cloned sequences of both *v-onc* and *c-onc* genes have been crucial to this advancing knowledge (Bos et al., 1985; Neel et al., 1981). Although there is debate as to the explanation(s) for homologies between the *v-onc* and *c-onc* genes (Duesberg, 1987; Varmus, 1987; Weinberg, 1987), it is clear that the differences between these highly homologous genes are functionally significant (see subsequent section, Oncogene Products as Growth Modulators).

GENETIC STUDIES IN HUMAN LEUKEMIA

The initial demonstration of a consistently demonstrable genetic abnormality in human leukemias occurred in 1960 when Nowell and colleagues at the University of Pennsylvania identified a characteristic, very small chromosome 22 in a large proportion of their patients with chronic myelogenous leukemia (CML) (Nowell, 1960; Holt et al., 1987). Given the stability of the clinical picture in such patients, the appearance of this abnormality added evidence for the theory that stable genetic alterations underlie at least some forms of neoplasia. It was later shown that patients without the classical "Philadelphia" chromosome had a distinctly worse prognosis, and that patients in whom a second Philadelphia chromosome or additional chromosomal abnormalities developed had a high probability of transformation to acute leukemia and death (Champlin and Golde, 1985; Holt et al., 1987).

The introduction, in 1970, of reliable methods for identifying specific bands within the chromosomes using quinacrine or Giemsa stain techniques resulted in an explosion of studies demonstrating specific nonrandom chromosomal abnormalities in the majority of patients with hematological malignancies (Caspersson et al., 1970; Patil et al., 1971; Rowley, 1982). Although elimination or reduplication of chromosomal material may occur, many defects involve reciprocal translocations of genetic material from one chromosome to another. For example, the Philadelphia chromosome represents a truncated form of chromosome 22 because of a translocation between chromosomes 9 and 22. Similarly, in Burkitt's lymphoma there is a characteristic translocation between chromosomes 8 and 14, and in acute promyelocytic leukemia there is a translocation between chromosomes 15 and 17. In the Burkitt's translocation, we see the breakpoints identified on the long arms of chromosome 8 (8q 24.130) and chromosome 14 (14q 32.33) and the classical translocation appearance (Fig. 5). As increasingly sophisticated banding techniques have been applied to the various leukemias, it has been possible to map with remarkable consistency the sites at which translocations and/or deletions occur in various disease states (Fig. 6). This type of data further solidifies the conclusion that the transformation event(s) leading to leukemia consist of a stable genetic

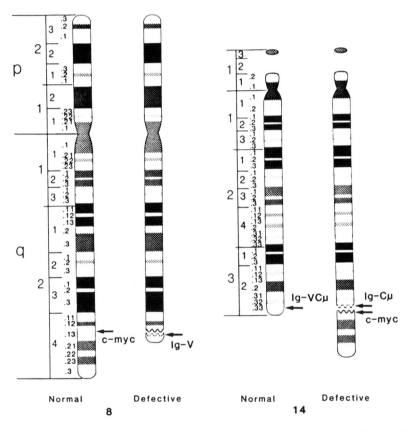

Fig. 5. Location of *c-myc* oncogene and heavy chain immunoglobulin variable (*V*) and constant μ (Cμ) genes on normal and defective chromosomes 8 and 14 in Burkitt's lymphoma, represented at the 1200 Giemsa band stage. The defective chromosome 8 loses the *c-myc* and gains *V* genes. The defective chromosome 14 gains *c-myc* from chromosome 8, becoming contiguous or near to Cμ. *Arrows* point to the normal and rearranged location of these genes. Broken ends of defective chromosomes indicate breakpoint sites.

(Reprinted with permission from Yunis, J. J. 1983. The chromosomal basis of human neoplasia. *Science* 221:227–236.)

alteration whose phenotypic expression is a function of both the precise location and the nature of the genetic change.

While these chromosomal studies were underway, the study of cellular proto-oncogenes had correspondingly expanded greatly and their precise chromosomal locations were identified. For instance, in the Burkitt's translocation, the *c-myc* proto-oncogene is moved from its normal location on chromosome 8 to a position immediately adjacent to the immunoglobulin heavy-chain constant region on chromosome 14 (see Fig. 5) (Dalla-Favera et al., 1982; Taub et al., 1982). It is hypothesized that this alters the normal regulation of *c-myc* expression, although the precise role of the immunoglobulin gene and its promoter is not as yet clear.

It is well accepted now that the same chromosomal abnormality may appear in various different clinical disorders, and that the same proto-oncogene moved to different locations will produce different diseases (Holt et al., 1987; Weinberg, 1987; Raskind et al., 1988). The latter finding suggests that the site "receiving" the displaced *c-onc* gene serves an important regulatory role, whereas the former finding suggests that alterations in clinical phenotype may reflect the presence of additional steps in the pathogenesis of the malignancy (Fialkow et al., 1981; Land et al., 1983; Slamon et al., 1984; Weinberg, 1987). The presence of a factor or factors other than the identifiable chromosomal abnormality thus may be required to explain the phenotypic difference in disease. The demonstration, using glucose-6-phosphate dehydrogenase (G-6-PD) isoenzymes, that clonal disease antedates the appearance of the chromosomal abnormality in CML (Fialkow et al., 1981) suggests that the chromosomal defect is not required for the initiation of the malignant phenotype. Interestingly, however, it may be a "necessary" step in the development of the abnormality of proliferation and hence a form of "progression" factor (see Cell Growth and Signaling).

Since the initial description of the Philadelphia chromosome in 1960, it has become possible to further characterize the genetic basis for chronic myelogenous leukemia. The proto-oncogenes *c-sis* and *c-abl* have been mapped to chromosome 22 and chromosome 9, respectively, and the gene located at the breakpoint of chromosome 22 has been identified and is now referred to as the breakpoint cluster region (*bcr*) gene (Fig. 7). The *c-abl* mRNA encodes a phosphoprotein of 145–150 kdaltons, which has homology to the tyrosine kinase family (see Table II; however, it lacks tyrosine kinase activity in vitro). The *bcr* gene encodes mRNAs of 4.5 and 6.7 kilobases and produces a peptide of approximately 190 kdaltons. Recently, CML cells and cell lines have been shown to contain a 210-kdalton phosphoprotein, which is a fusion product of the *bcr* and *abl* genes (Stam et al., 1985). Interestingly, this fusion protein has tyrosine kinase activity. It appears to be activated in a manner analogous to that occurring in *v-abl* of the Abelson murine leukemia virus, which is a fusion of viral *gag* and *c-abl* sequences.

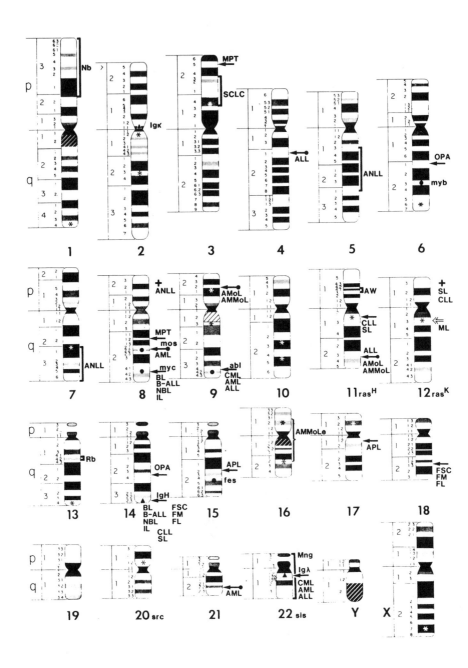

Fig. 6. See legend on facing page.

The significance of these genetic alterations is not yet totally clear. Because CML has multiple steps in its pathogenesis, it may well be that this translocation is a marker for the myeloid hyperplasia typical of CML, rather than of the initiating event. In this sense, it may be an expression of a critical "progression" factor rather than a "competence" factor. Comparisons of various alternative fusion products in CML (including both Philadelphia-positive and Philadelphia-negative variants) seems likely to help elucidate the pathophysiological role of tyrosine kinases in this form of leukemia.

The search for human leukemia viruses began in earnest in the early 1970s. Based on the earlier studies in avian, murine, and feline leukemias, it was hoped that direct demonstration of viral involvement would be possible. Gallo and co-workers demonstrated activities of a viral reverse transcriptase in cells from human leukemias and lymphomas, but extracellular virus was not demonstrable in these early studies (Reitz and Gallo, 1987). Demonstration that T lymphocytes require the presence of T-cell

Fig. 6. Human chromosome map of oncogenes (*dots*), fragile sites (*asterisks*), immunoglobulin genes (*triangles*), and consistent chromosome defects in human neoplasia. Deletions are represented with a bracket, inversion with a brace, trisomy with a plus sign, and reciprocal translocations with *solid arrows*. The karyotype represents Giemsa bands at the 400-band stage, according to the international nomenclature. Beginning with chromosome 1, abbreviations denote the following: Nb, neuroblastoma; Igκ, kappa light-chain immunoglobulin genes; MPT, mixed parotid gland tumor with t(3;8); SCLC, small cell lung cancer; ALL, acute "lymphocytic" leukemia with t(4;11); ANLL, acute nonlymphocytic leukemia; OPA, ovarian papillary adenocarcinoma with t(6;14); *mos*, Moloney sarcoma oncogene; AML, acute myelogenous leukemia with t(8;21); *myc*, myelocytoma oncogene; *BL, B-ALL, NBL*, and *IL*, Burkitt's lymphoma, B-cell type ALL, small noncleaved non-Burkitt's lymphoma, and immunoblastic lymphoma, respectively, with t(8;14); AMoL and AMMoL, acute monocytic and acute myelomonocytic leukemia with t(9;11); *abl*, Abelson oncogene; CML, chronic myelogenous leukemia with t(9;22); *ML* and *broken arrows*, not well defined malignant lymphoma associated with a t(12;14); AW, aniridia "Wilms" tumor syndrome; *CLL, SL*, chronic lymphocytic leukemia and small lymphocytic lymphoma, respectively; *ras*[H], *ras* Harvey oncogene identified at 11p; *ras*[K], Kirsten sarcoma oncogene identified on chromosome 12; *Rb*, retinoblastoma; *IgH*, heavy-chain immunoglobulin genes; *fes*, Snyder-Theilin feline sarcoma oncogene; *AMMoLe*, acute myelomonocytic leukemia with increased eosinophils and inversion 16; *FSC, FM, FL*, follicular small cleaved cell, follicular mixed, and follicular large cell lymphomas, respectively, with t(14;18); *src*, Rous sarcoma virus oncogene; *Mng*, meningioma; *sis*, Simian sarcoma oncogene; Igλ, immunoglobin light lambda chain genes. Igκ and Igλ are involved in the Burkitt's lymphoma variant with t(2;8) or t(8;22), respectively. Heritable fragile sites (*asterisks*) are found in Xq27, 2q11, 9p21, 10q23, 10q25, 11q13, 12q13, 16p124, 16q22, 17p12, and 20p11. Constitutional fragile sites occur in 1q44, 2q23, 3p14, 6q26, 7q31, 9q13, and 13q34.

(Reprinted with permission from Yunis, J. J. 1983. The chromosomal basis of human neoplasia. *Science* 221:227–236.)

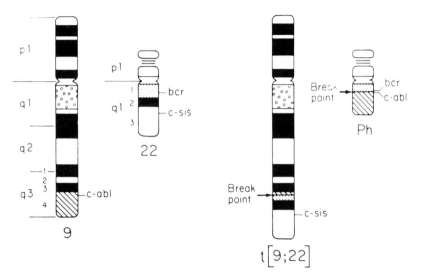

Fig. 7. Chromosomal rearrangements resulting in formation of the Ph chromosome. The location of the breakpoint cluster region (bcr) on chromosome 22, which is the site of the molecular breakpoints, is shown, as are the locations of the *c-sis* and *c-abl* proto-oncogenes. (Adapted from Champlin, R. E., Golde, D. W. 1985. Chronic myelogenous leukemia: recent advances. *Blood* 65:1039.)

(Reprinted with permission from Holt, J. T., Morton, C. C., Nienhuis, A. W., Leder, P. 1987. Molecular mechanisms of hematological neoplasms. In: *The Molecular Basis of Blood Diseases*, Stamatoyannopoulos, G., Nienhuis, A. W., Leder, P., Majerus, P. W., eds., pp. 341–376. Philadelphia: Saunders.)

growth factor (interleukin-2, or IL-2) for in vitro growth allowed the establishment of neoplastic T-cell lines and the subsequent discovery of C-type RNA viruses in these cultures. Because of its origin in cells derived from a patient with an aggressive cutaneous T-cell lymphoma, this first human leukemia virus was called human T-cell leukemia virus 1 (HTLV-1). HTLV-1 has a structure homologous to animal retroviruses, but differs in its core proteins and reverse transcriptase.

A search for known viral oncogenes has been made in HTLV-1-infected cells, but no consistent abnormality has been demonstrated. One provocative finding, however, is that all HTLV-1-infected cells express high levels of IL-2 receptor, and this receptor may differ qualitatively in its glycosylation sites. Although suggesting a role for an altered IL-2 receptor, these findings do not yet prove this is the mechanism for HTLV-1 transformation of T lymphocytes into malignant cells. These findings, however, do suggest that understanding the molecular relationships between various growth factors and oncogene expression will be required to elucidate the mechanism of certain leukemias.

ONCOGENE PRODUCTS AS GROWTH MODULATORS

The simultaneous reports in 1983 by Doolittle and colleagues from San Diego and by Waterfield and collaborators from London that the *v-sis* protein of simian sarcoma virus (SSV) was almost identical to the B chain of platelet-derived growth factor (PDGF) (Doolittle et al., 1983; Waterfield et al., 1983) was electrifying. By explicitly connecting the study of oncogenes and the study of growth factors, this discovery has precipitated a paradigm shift in cancer research (see Table III). Detailed structural and functional studies of these two genes reveal some differences, however. Nucleotide sequencing of the human gene most closely related to *v-sis* established that human *c-sis* does indeed encode the PDGF B chain (Josephs et al., 1984). The v-sis protein appears to represent a fusion protein combining part of the *c-sis* protein that is lacking its normal amino terminus and fused to an *env* protein derived from the SSV genome (Fig. 8) (Bishop and Varmus, 1985). Genetic studies suggest that some portion of the *env* domain (or a related sequence) is required to supply the activating membrane insertion signal, although the precise mechanism(s) whereby *v-sis* produces transformation remains to be proven.

Soon after this discovery, the epidermal growth factor (EGF) receptor and the protein product of the *v-erb B* gene were shown to contain related tryptic peptides and to share tyrosine kinase activity. This was substantiated by sequencing their cDNAs and demonstrating concordant assignment of both genes to chromosome 7 (Varmus, 1987). Detailed structural studies comparing chicken *v-erb B* and *c-erb B* genes suggested a new and potentially important mechanism for oncogenic activation, namely removal of the ligand-binding domain of a cell-surface receptor (see Fig. 8). The truncation of *c-erb B* removes the epidermal growth factor-binding domain, and is believed to result in both overproduction of the protein product and constitutive activation of the tyrosine kinase activity.

Still more recently, Sherr and colleagues provided evidence that the *c-fms* gene is the gene encoding the receptor for the macrophage colony-stimulating factor M-CSF (or CSF-1) (Scherr et al., 1985). This receptor is a surface glycoprotein of approximately 165 kdaltons with tryosine kinase activity. The *v-fms* of feline sarcoma virus (FeSV) encodes a protein product of about 120 kdaltons, and does not involve truncation of the amino terminus of *c-fms*. The *v-fms* protein appears to be a *gag-fms* fusion protein that retains its ability to bind M-CSF but has lost a tyrosine residue near the carboxyl terminus (Varmus, 1987). Interestingly, several variants of *v-fms* retain the capacity to transform NIH 3T3 cells (Roussel et al., 1987), but introduction of an *M-CSF* gene into an M-CSF-dependent cell line eliminates factor dependence yet fails to make the cells tumorigenic (Roussel et al., 1988). These data suggest that continued stimulation of the normal receptor is not in itself a sufficient cause for neoplastic transformation.

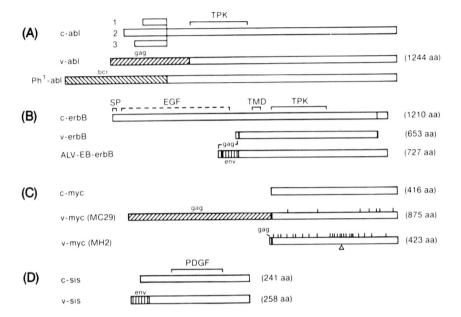

Fig. 8. Structural comparisons of the protein products of proto-oncogenes and mutant derivatives activated by retroviral transduction and other mutations. In each panel, the proto-oncogenic protein is shown above, with the activated proteins below; *shaded regions* indicate portions shared between proto- and viral oncogene proteins, and the sizes of the proteins are indicated, if known, in amino acids (aa). Each panel is drawn to a slightly different scale. (**A**) The products of mouse *c-abl*, the *v-abl* gene of Abelson murine leukemia virus, and the rearranged human *c-abl* gene in chronic myeloid leukemia cells with the Ph1 chromosome (t[9;22]). Alternative splicing can produce *c-abl* proteins with different aminotermini. The aminoterminus of the Ph₁ allele is encoded by the 5' end of a gene (*bcr*) from chromosome 22. TPK tyrosine protein kinase domain. (**B**) The products of human *c-erbB* (the EGF receptor gene), the *v-erbB* gene of avian erythroblastosis virus, strain ES4, and a chicken *c-erbB* gene activated by an avian leukosis virus insertion mutation in erythroblastosis. Designated regions of the proto-oncogene product include the signal peptide (SP), the extracellular ligand-binding domain (EGF), the hydrophobic transmembrane domain (TMD), and the tyrosine protein kinase domain (TPK). Both oncogenic proteins begin with six amino acids from the 5' end of *gag*; the viral protein lacks the normal carboxy-terminus, and the protein encoded by the insertionally mutated allele includes amino acids from the signal peptide of env. (**C**) The products of chicken *c-myc* and the *v-myc* genes of two avian leukemia viruses, myelocytomatosis-29 virus and Mill Hill-2 virus. *Short vertical lines* indicate positions of amino acid differences from *c-myc* protein; *asterisk* indicates a change common to the two viral proteins that is conjectured to be functionally important. The *triangle* indicates a four-codon deletion in MH-2 *myc*. (**D**) The products of human *c-sis* (believed to encode one subunit of PDGF) and the *v-sis* gene of simian sarcoma virus. The region of near identity with the sequenced amino-terminal portion of one of the chains of the mature PDGF heterodimer is indicated.

(Reprinted with permission from Varmus, H. 1987. Cellular and viral oncogenes. In: *The Molecular Basis of Blood Diseases*, Stamatoyannopoulos, G., Nienhuis, A. W., Majerus, P. W., eds., pp. 271–346. Philadelphia: Saunders.)

Extensive studies of the proto-oncogene family of nuclear DNA-binding proteins (of which *c-myc* and *c-fos* are prototypes) suggest their involvement in normal cellular responses to mitogenic stimulation. Nuclear factors capable of binding to enhancer and promoter regions of various genes are well described, and Rauscher et al. (1988) suggest that the protein product of the oncogene *c-fos* may bind to DNA at the site where transcription factor AP-1 binds. If confirmed, such interaction between defined nuclear binding proteins and these oncogene products might explain the complex responses of this family of genes to mitogenic stimuli. The exploration of these relationships between proto-oncogene products and intranuclear factors regulating gene expression has just begun.

Information available to date on regulation of *c-fos* suggests that it has a transcriptional regulatory control system. In this model system, type 1 genes are proposed to require both a protein kinase C signal and a calcium/calmodulin signal, whereas type 2 genes only require protein kinase C activation. In such a system, the cellular response of proliferation and/or differentiation may thus reflect responses to either a single signal or a precisely defined set of signals (as shown in Fig. 4). Because phorbol esters directly activate protein kinase C, stimulate increased expression of *c-fos* and *c-myc* mRNA, and are well characterized as tumor promoters in models of chemically induced neoplasms (Varmus, 1987), it seems likely that a genetic transformation event is possible at any point along their signal transduction pathways. To the degree that the early signals in the normal mitogenic response involve alterations in ionic fluxes (either across the cell membrane or intracellularly) (June et al., 1986; Rozengurt, 1986), these components of cellular signal transduction (Fig. 9) must also be considered important sites for regulatory control.

SUMMARY AND CONCLUSIONS

On the basis of the known characteristics of proto-oncogenes and their viral homologues, the transforming viral oncogenes, we may hypothesize at least three mechanisms whereby disturbed growth regulation leading to leukemia could occur (Fig. 10). In the first, an autocrine mechanism would be invoked wherein an excessive secretion of growth factor(s) produced by a cell promotes its own proliferation. A second possible mechanism would be a receptor alteration in which a faulty (e.g., truncated) receptor transmits stimulatory signals in the absence of ligand binding. A third possible mechanism would be an alteration in a "transducer element" in which interaction between a surface receptor and its signal transduction pathway is altered, again resulting in a permanently activated stimulatory signal and continued proliferation.

Many recent studies have suggested potential candidates for each of these mechanisms, some of which have already been discussed. For example,

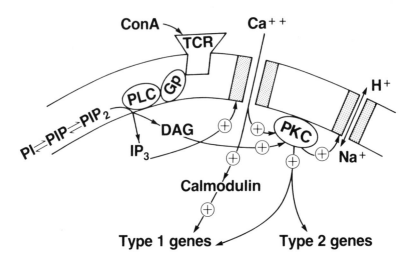

Fig. 9. A model for intracellular signal transduction that leads to turning on different classes of genes. Abbreviations: IP, inositol monophosphate; PIP, inositol diphosphate; PIP_2 (IP_3), inositol triphosphate; PLC, phospholipase C; G_p, guanine nucleotide-binding protein; ConA, concanavalin A; TCR, T-cell receptor; DAG, diacyl glycerol; PKC, protein kinase C.

Fig. 10. Oncogenes may provide independence from growth factors in at least three separate ways. The autocrine mechanism (*top*) is characterized by excessive secretion of growth factors that act on the tumor cell itself; an oncogene may indirectly stimulate expression of a growth factor gene (as shown), or may encode the growth factor and directly cause excessive secretion of the factor. A second mechanism involves production of altered growth factor receptors (*middle*); the oncogene may encode faulty receptors that transmit stimulatory signals to the cell even in the absence of extracellular growth factors. A third mechanism is characterized by interference with the signaling pathway that normally carries the stimulatory message from growth factor receptors to other sites in the cell (*bottom*). A normal gene may encode a protein that transduces signals from the receptor; the oncogenic version of this gene may produce a protein that sends out stimulatory signals even in the absence of signaling by the growth factor receptor.

(Reprinted with permission from Weinberg, R. A. 1960. Oncogenes and the mechanisms of carcinogenesis. In: *Scientific American Medicine*, Vol. 2, pp. 1–11. Washington, DC: Scientific American.)

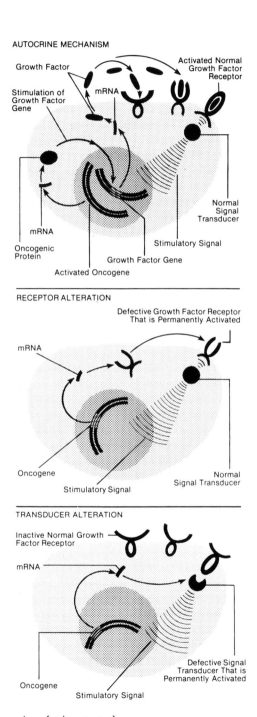

Fig. 10. (See legend on facing page.)

a receptor alteration is clearly the type of mechanism seen in the *v-erb B-* and *v-fms*-mediated transformation events described. Metcalf et al. have suggested an autocrine mechanism for one of the hematopoietic growth factors. They expressed the cDNA for GM-CSF in a factor-dependent cell line and demonstrated both autonomous growth in vitro and tumorigenicity in vivo (Lang et al., 1985). In contrast, a simple feedback loop of increased M-CSF failed to induce tumorigenicity, suggesting either additional altered regulatory domains were transferred with the GM-CSF cDNA or some specificity is present for types of growth factor-inducible transformation. The transducer alteration mechanism has been suggested as the mechanism whereby the *ras* proteins produce their effects. Viral-encoded *ras* genes produce altered guanine nucleotide-binding protein and thereby alter the intermediate components of the signal pathway (Varmus, 1987).

From these initial examples, one might predict certain candidates for "oncogene" status. The family of hematopoietic growth factors (see Table I) are obvious candidates for proto-oncogenes, as suggested by Metcalf and co-workers (Lang et al., 1985). Another candidate group for proto-oncogenes are those involved in regulating (intracellular) calcium fluxes and calmodulin-binding to its kinase, because these appear to be intimately related to the regulation of known oncogenes such as *c-myc*. A third class of factors that may be candidates for transformation events are the nuclear DNA binding proteins that serve to regulate both transcription and DNA replication.

The explosive growth of research in the areas of oncogenes, growth factors, and leukemogenesis is testimony to the interest in these candidate mechanisms for oncogenesis. Characterization of additional oncogenes, and elucidation of the pathways whereby growth factors regulate cell proliferation and differentiation, are likely to produce important insights into the genetic basis for leukemia. The next decade can be anticipated to contain significant new developments in this exciting area of research.

REFERENCES

Adamson, J. W. 1984. Analysis of hemopoiesis; the use of cell markers and in vitro culture techniques in studies of clonal hemopathies in man. *Clin. Haematol.* 13:489–502.

Baltimore, D. 1970. Viral RNA-dependent DNA polymerase. *Nature* 226:1209–1211.

Baserga, R. 1981. The cell cycle. *N. Engl. J. Med.* 304:453–459.

Berridge, M. J. 1984. Inositol triphosphate and diacylglycerol as second messengers. *Biochem. J.* 220:345–360.

Berridge, M. J. 1985. The molecular basis of communication within the cell. *Sci. Am.* 253(4):142–152.

Berridge, M. J., Heslop, J. P., Irvine, R. F., Brown, K. D. 1984. Inositol trisphosphate formation and calcium mobilization in Swiss 3T3 cells in response to platelet-derived growth factor. *Biochem. J.* 222:195–201.

Bishop, J. M., Varmus, H. 1985. Functions and origins of retroviral transforming genes. In: *RNA Tumor Viruses, 2nd Ed., Vols. 1 and 2*, Weiss, R., Teich, N., Varmus, H., Coffin, J., eds., Chapters 9 and 95. Cold Spring Harbor, New York: Cold Spring Harbor Laboratory.

Bos, J. L., Toksoz, D., Marshall, C. J., et al. 1985. Amino acid substitutions at codon 13 of the *N-ras* oncogene in human acute myeloid leukemia. *Nature (London)* 315:726.

Broudy, V. C., Kaushansky, K., Harlan, J. M., Adamson, J. W. 1987. Interleukin-1 stimulates human and endothelial cells to produce GM-CSF and G-CSF. *J. Immunol.* 139:464–468.

Caspersson, T., Zech, L., Johansson, C., Modest, E. J. 1970. Quinacrine mustard fluorescent banding. *Chromosoma* 30:215.

Champlin, R. E., Golde, D. W. 1985. Chronic myelogenous leukemia: recent advances. *Blood* 65:1039–1037.

Dalla-Favera, R., Bregni, M., Erickson, J., et al. 1982. Human *c-myc* oncogene is located on the region of chromosome 8 that is translocated in Burkitt lymphoma cells. *Proc. Natl. Acad. Sci. USA* 79:7824–7827.

Doolittle, R. F., Hunkapiller, M. W., Hood, L. E., DeVare, S. G., Robbins, K. C., Aaronson, S. A., Antoniades, H. N. 1983. Simian sarcoma virus *onc* gene, *v-sis*, is derived from the gene (or genes) encoding a platelet-derived growth factor. *Science* 221:275–277.

Duesberg, P. 1987. Cancer genes: rare recombinants instead of activated oncogenes (a review). *Proc. Natl. Acad. Sci. USA* 84:2117–2124.

Ellis, L., Morgan, D. O., Clauser, E., Edery, M., Jong, S.-M., Wang, L.-H., Roth, R. A., Rutter, W. J. 1986. Mechanisms of receptor-mediated transmembrane communication. *Cold Spring Harbor Symp. Quant. Biol.* 51:773–784.

Fialkow, P. J., Martin, P. J., Najfeld, V., Penfold, G. K., Jacobson, R. J., Hansen, J. A. 1981. Evidence for a multistep pathogenesis of chronic myelogenous leukemia. *Blood* 58:158–163.

Friend, C. 1957. Cell-free transmission in adult Swiss mice of a disease having the character of a leukemia. *J. Exp. Med.* 105:307.

Gomperts, B. D. 1983. Involvement of guanine nucleotide-binding protein in the gating of Ca^{2+} by receptors. *Nature (London)* 306:64–66.

Gross, L. 1951. "Spontaneous" leukemia developing in C3H mice following inoculation, in infancy, with Ak-leukemic extracts, or Ak-embryos. *Proc. Soc. Exp. Biol. Med.* 76:27.

Holt, J. T., Morton, C. C., Nienhuis, A. W., Leder, P. 1987. Molecular mechanisms of hematological neoplasms. In: *The Molecular Basis of Blood Disease*, Stamatoyannopoulos, G., Nienhuis, A. W., Leder, P., Majerus, P. W., eds., pp. 347–376. Philadelphia: Saunders.

Huebner, R. J., Todaro, G. 1969. Oncogenes of RNA tumor viruses as determinants of cancer. *Proc. Natl. Acad. Sci. USA* 64:1087–1094.

Iscove, N. N., Guilbert, L. J., Weyman, C. 1980. Complete replacement of serum in primary cultures of erythropoietin-dependent red cell precursors (CFU-E) by albumin, transferrin, unsaturated fatty acid, lecithin, and cholesterol. *Exp. Cell Res.* 6:121.

Jacobs, K., Shoemaker, C., Rudersdorf, R., Neill, S. D., Kaufman, R. J., Mufson, A., Seehra, J., Jones, S. S., Hewick, R., Fritsch, E. F., Kawakita, M., Shimizu,

T., Miyake, T. 1985. Isolation and characterization of genomic and cDNA clones of human erythropoietin. *Nature (London)* 313:806–810.

Josephs, S. F., Guo, C., Ratner, L., Wong-Staal, F. 1984. Human proto-oncogene nucleotide sequences corresponding to the transforming region of simian sarcoma virus. *Science* 223:487–491.

June, C. H., Ledbetter, J. A., Rabinovitch, P. S., Martin, P. J., Beatty, P. G., Hansen, J. A. 1986. Distinct patterns of transmembrane calcium flux and intracellular calcium mobilization after differentiation antigen cluster 2 (E rosette receptor) or 3 (T3) stimulation of human lymphocytes. *J. Clin. Invest.* 77:1224–1232.

Kaushansky, K., O'Hara, P. J., Berkner, K., Segal, G. M., Hagen, F. S., Adamson, J. W. 1986. Genomic cloning, characterization, and multilineage growth-promoting activity of human granulocyte-macrophage colony-stimulating factor. *Proc. Natl. Acad. Sci. USA* 83:3101–3105.

Kawasaki, E. S., Ladner, M. B., Wang, A. M., Arsdell, J. V., Warren, M. K., Coyne, M. Y., Schweikart, V. L., Lee, M. T., Wilsan, K. H., Boosman, A., Stanley, E. R., Ralph, P., Mark, D. F. 1985. Molecular cloning of a complementary DNA encoding human macrophage-specific colony-stimulating factor (CSF-1). *Science* 230:219.

Kelvin, D. J., Chance, S., Shreeve, M., Axelrad, A. A., Connolly, J. A., McLeod, D. 1986. Interleukin-3 and cell cycle progression. *J. Cell Physiol.* 127:403–409.

Land, H., Parada, L. F., Weinberg, R. A. 1983. Cellular oncogenes and multistep carcinogenesis. *Science* 222:771–778.

Lang, R. A., Metcalf, D., Gough, N. M., Dunn, A. R., Gonda, T. J. 1985. Expression of a hemopoietic growth factor cDNA in a factor-dependent cell line results in autonomous growth and tumorigenicity. *Cell* 43:531–542.

Lee, F., Yokota, T., Otsuka, T., Gemmell, L., Larson, N., Luh, J., Arai, K.-I., Rennick, D. 1985. Isolation of cDNA for a human granulocyte-macrophage colony-stimulating factor by functional expression in mammalian cells. *Proc. Natl. Acad. Sci. USA* 82:4360-4364.

Metcalf, D. 1984. *The Hemopoietic Colony-Stimulating Factors.* Amsterdam: Elsevier.

Metcalf, D. 1986. The molecular biology and functions of the granulocyte-macrophage colony-stimulating factors. *Blood* 67:257–267.

Migliaccio, G., Migliaccio, A. R. 1987. Cloning of human erythroid progenitors (BFU-E) in the absence of fetal bovine serum. *Br. J. Haematol.* 67:129–134.

Moloney, J. B. 1960. Biological studies on a lymphoid leukemia virus extracted from sarcoma S.37. I. Origin and introduction investigations. *J. Natl. Cancer Inst.* 24:933.

Nagata, S., Tsuchiya, M., Asano, S., Kaziro, Y., Yamazaki, T., Yamamoto, O., Hirata, Y., Kubota, N., Oheda, M., Nomura, H., Ono, M. 1986. Molecular cloning and expression of cDNA for human granulocyte colony-stimulating factor. *Nature (London)* 319:415–417.

Neel, B. G., Hayward, W. S., Robinson, H. L., Fang, J., Astrin, S. M. 1981. Avian leukosis virus-induced tumors have common proviral integration sites and synthesize discrete new RNAs: oncogenesis by promoter insertion. *Cell* 23:323–334.

Nishizuka, Y. 1984. The role of protein kinase C in cell surface signal transduction and tumour promotion. *Nature (London)* 308:693–698.

Nowell, P. C., Hungerford, D. A. 1960. Chromosome studies on normal and leukemic human leukocytes. *J. Natl. Cancer Inst.* 25:85.

Patil, S. R., Merrick, S., Lubs, H. A. 1971. Identification of each chromosome with a modified Giemsa stain. *Science* 173:821–822.

Pledger, W. J., Stiles, C. D., Antoniades, H. N., Scher, C. D. 1977. An ordered sequence of events is required before Balb/c-3T3 cells become committed to DNA synthesis. *Proc. Natl. Acad. Sci. USA* 74:4481–4485.

Raskind, W. H., Papayannopoulou, T. H., Hammond, W. P. 1988. T (8;21) with a phenotype of chronic myeloid leukemia. *Am. J. Hematol.* 28:266–269.

Rauscher, R. J. 1962. A virus-induced disease of mice characterized by erythrocytopoiesis and lymphoid leukemia. *J. Natl. Cancer Inst.* 29:515.

Rauscher, R. J. III, Sambucetti, L. C., Curran, T., Distel, R. J., Spiegelman, B. M. 1988. A common DNA binding site for fos protein complexes and transcription factor AP-1. *Cell* 52:471–480.

Reitz, M. S., Gallo, R. C. 1987. Human T-cell leukemia viruses. In: *The Molecular Basis of Blood Diseases*, Stamatoyannopoulous, G., Nienhuis, A. W., Leder, P., Majerus, P. W., eds., pp. 377–406. Philadelphia: Saunders.

Rodbell, M. 1980. The role of hormone receptors and GTP-regulatory proteins in membrane transduction. *Nature (London)* 284:17–22.

Roussel, M. F., Rettenmier, C. W., Sherr, C. J. 1988. Introduction of a human colony-stimulating factor-1 gene into a mouse macrophage cell line induces CSF-1 independent, but not tumorigenicity. *Blood* 71:1218.

Roussel, M. F., Dull, R. J., Rettenmier, C. W., Ralph, P., Ullrich, A., Sherr, C. J. 1987. Transforming potential of the c-FMS proto-oncogene (CSF-1 receptor). *Nature (London)* 325:549.

Rowley, J. D. 1982. Identification of the constant chromosome regions involved in human hematologic malignant disease. *Science* 216:749–751.

Rozengurt, E. 1986. Early signs in the mitogenic response. *Science* 234:161.

Scherr, C. J., Rettenmier, C. W., Sacca, R., Roussel, M. F., Look, A. T., Stanley, E. R. 1985. The *c-fms* proto-oncogene product is related to the receptor for the mononuclear phagocyte growth factor, CSF-1. *Cell* 41:665–676.

Shih, C., Shilo, B. Z., Goldfarb, M. P., et al. 1979. Passage of phenotypes of chemically transformed cells via transfection of DNA and chromatin. *Proc. Natl. Acad. Sci. USA* 76:5714.

Slamon, J. D., de Kernion, J. B., Verma, I. M., Cline, M. J. 1984. Expression of cellular oncogenes in human malignancies. *Science* 224:256–262.

Stam, K., Heisterkamp, N., Grosveld, G., DeKlein, A., Verma, R. S., Coleman, M., Dosik, H., Groffen, J. 1985. Evidence of a new chimeric *bcr/c-abl* mRNA in patients with chronic myelocytic leukemia and the Philadelphia chromosome. *N. Engl. J. Med.* 313:1429–1433.

Stiles, C. D., Capone, G. T., Scher, C. D., Antoniades, H. N., Van Wyle, J. J., Pledger, W. J. 1979. Dual control of cell growth by somatomedins and platelet-derived growth factor. *Proc. Natl. Acad. Sci. USA* 76:1279–1283.

Taub, R., Kirsch, I., Morton, C., et al. 1982. Translocation of the *c-myc* gene into the immunoglobulin heavy chain locus in human Burkitt lymphoma and murine plasmacytoma cells. *Proc. Natl. Acad. Sci. USA* 79:7837–7841.

Temin, H. M., Mizutani, S. 1970. RNA-dependent DNA polymerase in virions of Rous sarcoma virus. *Nature (London)* 226:1211–1213.

Tooze, J., ed. 1973. *The Molecular Biology of Tumor Viruses*, pp. 18–73. Cold Spring Harbor, New York: Cold Spring Harbor Laboratory.

Ullrich, A., Riedel, H., Yarden, Y., Coussesn, L., Gray, A., Dull, T., Schlessinger, J., Waterfield, M. D., Parker, P. J. 1986. Protein kinases in cellular signal transduction: tyrosine kinase growth factor receptors and protein kinase C. *Cold Spring Harbor Symp. Quant. Biol.* 51:713–723.

Varmus, H. 1987. Cellular and viral oncogenes. In: *The Molecular Basis of Blood Diseases*, Stamatoyannopoulos, G., Nienhuis, A. W., Leder, P., Majerus, P. W., eds., pp. 271–346. Philadelphia: Saunders.

Virchow, R. 1860. *Cellular Pathology*. London: John Churchill, original publisher. Birmingham, Alabama: Gryphon Editions, Special Edition, Classics of Medicine Library, 1978.

Waterfield, M. D., Scrace, G. T., Whittle, N., Stroobant, P., Johnsson, A., Wasteson, A., Westermark, B., Heldin, C.-H., Huang, J. S., Denel, T. F. 1983. Platelet-derived growth factor is structurally related to the putative transforming protein p28[515] of simian sarcoma virus. *Nature (London)* 304:35–39.

Weinberg, R. A. 1983. A molecular basis of cancer. *Sci. Am.* 249:126–142.

Weinberg, R. A. 1987. Oncogenes and the mechanisms of carcinogenesis. In: *Scientific American Medicine*, Vol. 2, Rubenstein, E., Federman, D. D., eds., pp. 1–11. New York: Scientific American.

Wintrobe, M. M., Lee, G. R., Boggs, D. R., Bittell, T. C., Foerster, J., Athens, J.W, Lukens, J. N. 1981. In: *Clinical Hematology*, 8th Ed., pp. 1449–1453. Philadelphia: Lea and Febiger.

Wong, G. G., Witek, J. S., Temple, P. A., Wilkens, K. M., Leary, A. C., Luxenberg, D. P., Jones, S. S., Brown, E. L., Kay, R. M., Orr, E. C., Shoemaker, C., Golde, D. W., Kaufman, R. J., Hewick, R. M., Wang, E. A., Clark, S. C. 1985. Human GM-CSF: molecular cloning of the complementary DNA and purification of the natural and recombinant proteins. *Science* 228:3–8.

Yang, Y. C., Ciarletta, A. B., Temple, P. A., Chung, M. P., Kovacic, S., Witek-Gianotti, J. S., Leary, A. C., Kriz, R., Donahue, R. E., Wong, G. G., Clark, S. C. 1986. Human IL-3 (multi-CSF): identification by expression cloning of a novel hematopoietic growth factor related to murine IL-3. *Cell* 47:3–10.

Yunis, J. J. 1983. The chromosomal basis of human neoplasia. *Science* 221:227–236.

PART III.
Observed ELF
Effects

7 Interaction of ELF Electric and Magnetic Fields with Neural and Neuroendocrine Systems

LARRY E. ANDERSON
Biology and Chemistry Department
Pacific Northwest Laboratory
Richland, Washington

CONTENTS

HUMAN STUDIES 140
    Laboratory Studies 141
    Epidemiological Studies 142
ANIMAL STUDIES 143
    Neurochemistry 143
    Neuroanatomy 145
    Neurophysiology 146
    Neuroendocrinology 147
    Behavior 148
SUMMARY AND CONCLUSIONS 150
ACKNOWLEDGMENT 151
REFERENCES 151

Biological effects reported to occur in humans or animals exposed to extremely low frequency (ELF) electromagnetic fields most often appear to be associated with the nervous system. This apparent sensitivity of neural tissue to ELF might be expected because sensitivity to electrical signals is a major feature of the nervous system. Additionally, neural and neuroendocrine systems comprise an essential element of the interaction between animals and their environment. Understanding the interactions between ELF and the primary elements of nervous system function (transmittal of sensory input from external stimuli, central processing of such information, efferent innervation of tissues or organs along with possible hormonal output) may provide much needed insight on possible links between ELF exposure and observed biological consequences. It should also be noted in the broader context of health issues that influences of ELF fields may be mediated by the nervous system indirectly through neuroendocrine or endocrine responses.

Nervous system parameters were seldom measured in early ELF research, although many of the observed effects, especially those involving behavior, are closely tied to nervous system function. Reports from the U.S.S.R. in the late 1960s and early 1970s claimed a variety of exposure-related neurological symptoms, including headaches, fatigue, and increased excitability in switchyard workers (Asanova and Rakov, 1966; Korobkova et al., 1972). Evidence supporting a direct tie between ELF field exposure and these reported, somewhat subjective, functional changes was relatively weak. Nevertheless, the findings did spur increased interest in such research throughout the world. In the succeeding two decades, considerable effort has been mounted to assess exposure to ELF electromagnetic radiation for potential biological consequences. Results of these studies are summarized in numerous sources, including Sheppard and Eisenbud (1977), Phillips and Kaune (1977), National Academy of Sciences (1977), Anderson and Phillips (1985), and Polk and Postow (1987).

Initial studies on ELF exposure that related to nervous system function could be classified in three general areas: (1) assessments of activity or startle-response behavior; (2) evaluations of stress-related hormones (e.g., corticosteroids); and (3) general measurements of central nervous system (CNS) responses [e.g., electroencephalograms (EEG) and interresponse times]. The collected data were often contradictory, resulting in confusing claims of both effects and noneffects from ELF exposure. In an attempt to resolve these inconsistencies and to understand the extent and nature of ELF–tissue interaction, studies have been expanded within the last decade to include a broader range of neurological assessments. This chapter focuses on research that examines the effects of power-frequency fields on neural and neuroendocrine systems.

HUMAN STUDIES

Information on the effects of ELF field exposure on the human is somewhat limited. In surveys of people living or working in the vicinity of high-voltage lines, data have been collected in a small number of epidemiological studies as well as in a few laboratory and clinical studies. Although human data are of primary interest and concern, these must be interpreted with caution because of certain persistent and often present methodological problems, including small sample sizes with limited statistical power, lack of quantitative data on exposure levels, and inappropriate control groups.

Laboratory Studies

With the exception of an examination of some behavioral endpoints, few human studies have addressed the effects of ELF exposure on neural or neuroendocrine systems. The most detailed laboratory experiments using human subjects have probably been those conducted by Hauf in Germany (Hauf, 1974, 1976a,b). In the early 1970s, Hauf conducted extensive clinical evaluations on more than 100 volunteer subjects who had been exposed for brief periods (two 45-min sessions for 3 consecutive days) to 50-Hz fields of 1, 15, or 20 kV/m. No field-related effects were observed in many of the neurological tests, including electroencephalogram (EEG) patterns. A slight diminution in reaction time, as well as an increase in norepinephrine levels, was seen in exposed subjects. These data, and data from a small epidemiological study involving 32 subjects exposed in a 380-kV facility (Bauchinger et al., 1981), are summarized in a semicritical review by Hauf (1982). Again, no changes in behavior or EEG patterns were observed. In a further attempt to accurately define the cause of the slight variations indicated, Hauf and co-workers conducted a study on the effects of injected currents (200 μA) in humans. These currents, calculated to approximate the displacement currents expected from the earlier electric-field exposures, caused no alterations in either reaction times or EEG. The overall results led Hauf to conclude that the few slight effects observed in exposed subjects were probably caused by "unspecified stimulation" (Hauf, 1976b).

A more recent study in which humans were exposed in the laboratory has been conducted by Cabanes in France (Cabanes and Gary, 1981). In this study, no specific neural parameters were measured; rather, the object of the experiment was to determine the threshold for perception of a 50-Hz electric field. Subjects maintained one of three different body positions during exposure to varying electric fields. Thresholds of perception ranged from 0.35 kV/m (in 4% of subjects) to greater than 27 kV/m. With their arms held against the body, 40% of the subjects could not detect 27 kV/m. Deno and Zaffanella (1982) obtained similar results using 60-Hz fields; 5% of exposed subjects were able to detect a 1-kV/m field, and the median field strength value for perception was 7 kV/m. Perceptual threshold variation among individuals may provide a reasonable explanation for the great variability in biological data obtained to date.

In a recent comprehensive, double-blind investigation of the effects of exposure of 12 men to 60-Hz electric (9 kV/m) and magnetic (16 A/m) fields, Graham et al. (1987) evaluated performance measures as well as physiological data (electroencephalogram, electrocardiogram, skin conductance, and respiration). Most neurological tests showed no effects from exposure; however, significant field-related alterations were observed in several components of the auditory and visual evoked response. Although of interest, the results should be viewed with caution because a number of the alterations showed reversals of effect during the days of exposure.

ELF magnetic fields of moderate intensity [10 millitesla (mT) or more] produce a visual phenomenon known as "phosphenes" in humans (Lovsund et al., 1979). This phenomena appears to arise from the induction of electric currents in the retina and is highly frequency dependent, with maximum effects at approximately 20 Hz. However, no perception of the field was observed at 5- to 10-fold lower field intensities (Tucker and Schmitt, 1978). In humans, other magnetic-field effects on the functioning nervous system or on the biochemistry of neural tissue are relatively unexplored, and few effects have been found (Sheppard, 1978, 1983). Friedman et al. (1967) and Medvedev et al. (1976) reported increased reaction time and latency of sensorimotor reactions, respectively. However, Beischer et al. (1973) observed no effect on reaction time in humans. No effect of 50-Hz magnetic field exposure was observed in human EEG patterns at 2- to 7-mT field intensities (Sander et al., 1982).

Epidemiological Studies

Few epidemiological studies containing neurological data analyses have been reported during the past decade. Stoops and Janischewsky (1979) and Knave et al. (1979) described fairly extensive health surveys conducted on electrical workers in which a wide range of biological variables were examined over 5 to 10 years. In neither study were significant effects on nervous system function reported. The Stoops study is one of only a few epidemiological studies in which field characteristics and length of exposures were measured and described. Results in both cases agree with the few other human studies examining neurological parameters in showing no significant effects. Unfortunately, both studies are based on small numbers of subjects (30 and 53 men, respectively).

Of particular interest to the topic discussed in this book are the results of an occupational history study by Lin et al. (1985). Work histories were obtained for nearly 1000 people who died of brain tumors in Maryland (1969–1982). Jobs in which electromagnetic-field exposures were likely to occur were noted more frequently among victims of primary brain tumors (gliomas and astrocytomas) when compared to controls. In the case-control analysis, exposure to ELF fields was ranked according to job category, with the relative occurrence of tumors in each category as follows: definite (2.15), probable (1.95), possible (1.44), or no exposure (1.00). Although the data are limited to presumed rather than measured exposures, the results of this study nevertheless suggest an association between some aspect of occupation (possibly ELF field exposure) and cancer.

In reviewing all available literature concerning neural effects in humans exposed to low-intensity electric and magnetic fields, the data are suggestive but not entirely conclusive of weak behavioral, neurophysiological, or neurochemical effects.

ANIMAL STUDIES

Many areas of biological investigation can be most effectively addressed by employing various nonhuman animal species. Experiments evaluating electromagnetic-field effects have been performed using rodents (mice and rats), birds, dogs, swine, and nonhuman primates. Exposure levels in the ELF range have varied in field strength from a few volts per meter to more than 120 kV/m or from a few microtesla to more than 10 mT.

Studies that examine neurochemical, neuroanatomical, neurophysiological, and neuroendocrinological functions in animals are discussed in the following subsections. In addition, nervous system-related areas of behavior and circadian function are briefly reviewed, particularly as they pertain to specific physiological impacts of ELF exposure on the nervous system. Expanded discussions of pineal gland response to electromagnetic fields as well as circadian function in general are addressed in other chapters.

Neurochemistry

The relationship between the neurotransmitters norepinephrine and epinephrine and the physiological responses of stress/arousal are well established. As research expanded in the investigation of ELF field exposures for potential biological effects, measurement of these neurotransmitters became the primary assessment used to examine the state of the nervous system, partly because of the ease of measuring these chemicals in blood, urine, or brain tissue. Additionally, in early observations, ELF fields were reported to act as mild stressors (Dumansky et al., 1976; Marino et al., 1977). Unfortunately, a number of potential methodological problems in these early studies raised questions as to their validity (Michaelson, 1979).

Catecholamine levels were measured in both urine and blood after exposure of rats to 100 kV/m, 60-Hz fields (Groza et al., 1978). Significant increases in epinephrine levels in both fluids were reported following acute (6-hr to 3-day) exposures, but no changes were observed in norepinephrine or epinephrine levels with longer term (12-day) exposures. Increased norepinephrine levels were also measured in the blood of rats exposed to lower strength 50-Hz fields (50 V/m and 5.3 kV/m; Mose, 1978). A companion paper by Fisher et al. (1978) examined norepinephrine content in brain tissue of rats exposed to 5.3 kV/m for 21 days. After 15 min of exposure, the levels increased rapidly; after 10 days of exposure, however, levels were significantly decreased when compared to those of a control group. Portet and Cabanes (1988) reported no changes in adrenal epinephrine or norepinephrine in 2-month-old rats exposed for 8 hr/day to a 50-kV/m, 50-Hz field.

Reports of changes in other neurotransmitters also exist. Wolpaw et al. (1987) reported decreases in cerebrospinal fluid concentrations of the major metabolites of dopamine and serotonin in macaques exposed to 60-Hz electric and magnetic fields. Researchers in the U.S.S.R. have investigated neurochemical responses to 50-Hz magnetic fields of high field intensity. In rats exposed to 40 mT, Kolodub and Chernysheva (1980) observed altered brain metabolism as indicated by decreased levels of glycogen, creatine phosphate, and glutamine, and by increased DNA content. Mobilization of adrenal catecholamines and corticosteroids was also observed in rats exposed to 20-mT, 50-Hz fields (Sakharova et al., 1981, and Udintsev and Moroz, 1974, respectively).

Some of the differences in results of various laboratories may be caused by species-specific responses. Alternatively, taking circadian fluctuations in levels of neurotransmitters into account may be a critical factor in determining whether or not an effect is observed. Vasquez and co-workers (1988) reported significant changes in diurnal patterns of several biogenic amines when rats were exposed for 4 weeks to 60-Hz electric fields.

Yet another neurochemical parameter, the enzyme acetylcholinesterase (AChE), has been measured in rats exposed to 50-Hz electric fields. Kozyarin (1981) reported that AChE activity in blood was higher than normal, by approximately 25%, in both young and old animals exposed to 15 kV/m for 60 days, 30 min/day. Brain levels of AChE were lower than normal in exposed animals. At the end of exposure, all values returned to normal within 1 month, leading the investigators to conclude that electric fields can cause changes in the functional condition of the CNS, although the changes appear not to be permanent.

Corticosteroids have frequently been assessed in the blood of animals exposed to electric fields. The rapidly changing response of these adrenal steroids to various stimuli, however, tends to introduce great variability into assessments of this physiological endpoint (Michaelson, 1979). Hackman and Graves (1981), for instance, observed an acute, transient increase in corticosterone levels in plasma of mice exposed to 25 or 50 kV/m; levels of the steroid return to normal within a short time. Dumansky et al. (1976) also showed an increase in corticosteroids in rats exposed for 1, 3, or 4 months to 5 kV/m. In contrast, in a study conducted by Marino et al. (1977)

a decrease in serum corticosteroid was reported in animals exposed for 30 days to 15 kV/m.

Results that appear to contradict both Marino's and Dumansky's data were described by Free et al. (1981) and Portet and Cabanes (1988). Rats or rabbits were exposed to 100 kV/m and 50 kV/m, respectively, for 30 or 120 days; exposed and control animals did not differ in corticosterone levels. Studies conducted in Michaelson's laboratory (Quinlan et al., 1985) showed no effect on corticosterone in rats exposed for short periods (1–3 hr) at 100 kV/m. Gann (1976) reported similar results in evaluating adrenal secretion in dogs exposed to 15 kV/m, that is, no alterations from exposure were observed in corticosterone levels.

In general, neurochemical data provide some evidence that exposure to electric and/or magnetic fields in the ELF range causes small changes in nervous system function. A small number of experiments and questionable methodology in some of these studies weaken the conclusions that might be drawn. Nevertheless, the findings are reasonably consistent across studies and support the hypothesis that ELF exposure appears to result in increased arousal in animals and possibly has other impacts on the nervous system, including alterations in circadian rhythms. The data also support the notion that observable effects on neural tissue per se may be of rather short duration.

Neuroanatomy

Neuroanatomical changes in animals after exposure to ELF fields have been reported in relatively few studies. For the most part, these investigations have reported no significant alterations in the morphology of brain tissue from field exposure. Carter and Graves (1975) and Bankoske et al. (1976) studied chicks exposed to 40 kV/m and observed no effects in the morphology of the CNS. Phillips and co-workers (1978) conducted experiments in which rats were exposed to 100 kV/m for 30 days. They also reported no morphological evidence of an electric-field effect in the CNS. Nor did Gona (1987) find any evidence of consistent abnormalities in brain tissue of rats developing during exposure to a combination of electric (100-kV/m) and magnetic (1-mT and 0.1-mT) fields.

A limited number of studies do report effects on CNS morphology. For example, Hansson (1981a,b) reported significant changes in the cell structure of Purkinje's cells of the cerebellum in rabbits exposed to 14 kV/m. Exposed animals showed disintegration of Nissl bodies and endoplasmic reticulum, as well as an abnormal accumulation of lamellar bodies. Reduced numbers of mitochondria, reduced arborization of dendritic branches, and an absence of hypolemmal cisterns were also evident in these cells. These observed changes are in conflict with other reports in different species and should be interpreted with some degree of caution. The animals were exposed outdoors and showed some evidence of health deficits (whether caused by the electric field, other environmental factors, or some combination of these conditions is not clear).

Two rather convincing arguments suggest that these results may be artifactual and not electric field related. First, such dramatic alterations in CNS morphology do not correspond to the obvious lack of significant CNS functional deficits in the thousands of animals exposed to date. Second, in a recent study conducted in France, no ultrastructural changes were seen in the cerebella of young rabbits exposed to 50 kV/m (Portet and Cabanes, 1988).

Reports from a laboratory in the U.S.S.R. also suggest pathomorphological changes in brain tissue of rats exposed to 50-Hz magnetic fields of 20 mT and above (Toroptsev and Soldatova, 1981; Soldatova, 1982). It is difficult to make any conclusive evaluations from the limited data presented in these papers. The possibility of synergistic effects from ELF field exposure and a stressful environment cannot be entirely ruled out as a source of those effects that have been reported.

Neurophysiology

The nervous system is, by its very nature, electrically sensitive, and therefore it has been assumed to be particularly vulnerable to influence by externally applied electric or magnetic fields. To some degree, this assumption has been verified by experimental results, although in the area of neurophysiology there exists a confusing array of studies claiming both effects and noneffects of ELF field exposure.

In a study by Blanchi et al. (1973), significant changes in EEG activity were seen when guinea pigs were exposed for a half-hour to a 100-kV/m, 50-Hz field. Takashima et al. (1979) examined EEG from rabbits (with silver-electrode implants for recording the EEG) exposed at 1–10 MHz modulated at 15 Hz. After 2–3 weeks of exposure, the EEG were abnormal. However, if the electrodes were removed during actual exposure, the EEG returned to normal. The investigators thus concluded that the effect on the EEG was caused by the local fields created by the presence of the electrodes in the cranial cavity. EEG responses in cats exposed to 50-Hz magnetic fields (8 hr/day at 20 mT) showed short-term decreases in power-density spectra (Silney, 1979) but only shortly after the magnetic field was switched on. The activity of spontaneously firing neurons in the brains of anesthetized rats were studied during exposure to 50-, 30-, and 15-Hz electric fields (Blackwell, 1986). Low-strength fields of 100 V/m produced no alterations in neuronal firing rate; however, some synchronization of firing occurred with the period of the exposure waveform at 15 and 30 Hz.

In assessments similar to EEG but somewhat more specific, Jaffe et al. (1983) measured the visual-evoked response (VER) in 114 rats exposed in utero through 20 days postpartum to a 65-kV/m, 60-Hz electric field. No consistent statistically significant effects of exposure were observed. Evoked potentials have also been examined in macaques exposed to combined electric and magnetic fields (Wolpaw et al., 1987). As in Jaffe's studies, the VER, as well as auditory-evoked potentials, gave no indication of change from exposure. An attenuation of the late components of the somatosensory-evoked potentials was observed in exposed animals, changes that may have resulted from alterations in the attention capacity of the animals due to an increase in the number of stimuli to exposed animals.

Two additional neurophysiological studies have provided clear, replicable results. An enhanced neuronal excitability in synaptic junctions was observed in rats chronically exposed to 60-Hz, 100-kV/m fields for 30 days (Jaffe et al., 1980). No other parameters tested showed any differences between exposed and sham-exposed animals. Jaffe and co-workers (1981) also examined a wide range of physiological parameters of the peripheral nervous system and neuromuscular junctions. The only effect of exposure to E fields observed was a slightly faster recovery from fatigue after chronic stimulation in slow-twitch muscle (the soleus).

The evidence from neurophysiological studies indicates a general lack of effects on the peripheral nervous system with the possible exception of an enhancement of neuronal excitability in the autonomic nervous system. In the CNS, the experimental evidence is less clear, with at least some studies indicating perturbations in the electrical activity of the CNS from ELF field exposure.

Neuroendocrinology

The studies that have examined possible effects of ELF field exposure on neuroendocrine function can be divided into two classes: the work on the pineal gland and its associated hormone products, and studies related to corticosteroids of the adrenal gland. Data on the first area are somewhat limited but generally straightforward; the latter effort has been addressed in many studies, and, as indicated in the section on neurochemistry, has been plagued with conflicting results.

The effect of field exposure on the pineal gland and its associated neurohormones is treated in detail in Chapter 8. Briefly, concentrations of neuroendocrine substances in pineal glands were significantly changed when rats were exposed for 30 days to 130, 65, 10, or 1.9 kV/m (Wilson et al., 1981; Reiter et al., 1988). Both melatonin and an associated biosynthetic enzyme, serotonin-N-acetyl transferase (SNAT) showed a dark-phase decrease in exposed animals; levels of 5-methoxytryptophol, another neuroendocrine product of the gland, were increased at night. Tests at intervals from 1 hr to 4 weeks after the start of exposure showed that at least 3 weeks of exposure were required for the effect to reach significance

(Wilson et al., 1986). The position of the pineal gland as a "neuroendocrine transducer" and its integral tie with the circadian cycle make this effect particularly interesting. Several other indications that electric-field effects relate to circadian cycling seem to tie these results together and emphasize the possibility of electric-field influences on such rhythms (Russell and Ehret, 1982; Sulzman and Murrish, 1987). In studies using magnetic-field exposure, nocturnal pineal components in mice and rats have been shown to be sensitive to exposure (Semm, 1983; Welker et al., 1983; Kavaliers et al., 1984). Some evidence points to retinal sensitivity as the possible site of the ELF field–organism interactions that give rise to pineal response to electromagnetic fields (Olcese et al., 1985; Reuss and Olcese, 1986).

Circadian function in squirrel monkeys has also been investigated for effects of ELF fields. Exposing the monkeys to a range of electric-field intensities (2.6, 26, and 39 kV/m) accompanied by a 100-μT magnetic field, Sulzman and Murrish (1987) reported intensity-related effects. No changes in activity or feeding were observed in monkeys exposed to 2.6 kV/m after 2 weeks of exposure. However, 33% of the monkeys exposed to 26 kV/m and 75% of those exposed to 39 kV/m showed significant alterations in their circadian cycles.

Although firm conclusions cannot yet be drawn regarding potential health impacts from ELF effects on circadian or biological rhythms, it appears to be well established that electromagnetic fields alter the circadian timing mechanisms in mammals.

Behavior

There is an extensive literature on the behavioral effects of electric fields. Field intensities in these studies range from less than 1 V/m to more than 100 kV/m. Behavioral studies in several species of animals have provided evidence of perception of ELF fields. Rats probably can detect the presence of electric fields at levels as low as 1 kV/m, although the average threshold for perception appears to be between 4 and 10 kV/m (Stern et al., 1983; Sagan and Stell, 1984). Graves et al. (1978) observed that pigeons detect 60-Hz electric fields at 32 kV/m, and Kaune et al. (1978) reported detection limits of fields in pigs to be approximately 30 to 35 kV/m. The work of Graham et al. (1987) indicated that human volunteers could perceive a 9-kV/m, 60-Hz field in certain postures.

Differential responses to a gradient of voltage levels has been demonstrated in rats by other investigators. Hjeresen et al. (1980) observed avoidance behavior to 60-Hz fields greater than 75 kV/m. Twenty-four-hour tests showed field aversion at 75 kV/m and higher and a preference for fields of 25 and 50 kV/m. Exposed rats were also more active during the first hour of exposure, concurring with the results of Rosenberg et al. (1981). In a similar study with swine, Hjeresen et al. (1982) showed that pigs spent most of their time out of the field at 30 kV/m. The basis for these preference/aversion behaviors has not been identified. However, a change in environmental factors, such as relative humidity, can cause pronounced changes in perception levels (Weigel and Lundstrom, 1987).

Most investigations on ELF effects on behavior have been conducted using rodents as experimental subjects exposed to a wide range of field intensities. In such studies, many of which present somewhat contradictory results, motor activity is the most commonly measured behavioral endpoint. Smith et al. (1979) examined exploratory activity in rats exposed to a 60-Hz, 25-kV/m field and found no effect of exposure. Babovich and Kozyarin (1979), however, observed changes in unconditioned reflexes produced by daily exposure to 50-Hz, 7-kV/m fields. These effects had disappeared after 1 month of exposure. Hilmer and Tembrock (1970) reported increased activity at night in rats exposed to 50 Hz and 50–70 kV/m. These results are in contrast to those of Bawin et al. (1979), who reported decreased motor activity at night in rats exposed to 1 kV/m.

Some of these activity changes may be produced by some type of peripheral stimulation in the animals during exposure to the field, particularly during the initial exposures. Behavioral evidence, in this regard, supports the neuroendocrinological data. Rosenberg et al. (1981) reported that the general activity level in mice increased during the inactive phase of their circadian cycle when initially exposed to a 60-Hz, 100-kV/m field. This arousal response, however, was quickly extinguished with repeated exposures. It was further demonstrated that field strengths less than 50 kV/m seldom produced reliable arousal responses (Rosenberg et al., 1983). Activity changes have also been observed in animals exposed to ELF magnetic fields of low intensity (Persinger, 1969; Persinger and Foster, 1970; Smith and Justensen, 1977). In contrast, studies conducted at higher field intensities have shown no evidence of magnetic-field-associated effects on animal behavior (Davis et al., 1984; Creim et al., 1985).

Nonhuman primates provide an excellent model for generalizing the experimental findings of ELF electric-field exposure to man. Several such studies were conducted in the early 1970s and showed essentially no behavioral effects in animals exposed at very low field strengths (7–100 V/m; National Academy of Sciences, 1977). The few experiments conducted since that review include behavioral studies in the baboon and a recently initiated study using the pigtailed macaque. In the former study, baboons showed sensitivity to a 60-Hz field at 30 kV/m with minor behavioral alterations (Rogers et al., 1987). Exposed animals showed increased startle/arousal responses that quickly adapted out with continued exposure. The authors concluded that none of the behavioral effects seen in baboons were permanent or harmful.

In the experimental studies that have been conducted to determine if ELF fields cause behavioral alterations, remarkably few robust effects have been demonstrated (Lovely, 1988). Effects that have been observed, usually arousal or activity responses, probably result from detection and possible perception of the electric field by the animal. In those cases, fields seem to function as do other low-level novel stimuli, and adaptation occurs quickly with no apparent cumulative behavioral effects. For animals exposed to ELF magnetic fields the basis for observed animal responses to fields of low intensity is less clear.

SUMMARY AND CONCLUSIONS

Over the past three decades, a multitude of studies have been initiated to determine the extent to which electrical environments containing power-frequency electric or magnetic fields pose a health hazard to living organisms (particularly humans). In general, the biological effects reported in many of these studies have not confirmed any pathological effects, even after prolonged exposures to high-strength (100-kV/m) fields or high-intensity magnetic (10-mT) fields. However, much work remains to be accomplished before the observed effects and their biological consequences are clearly understood.

Areas in which effects have been demonstrated appear to be primarily associated with the nervous system, including altered neuronal excitability, altered circadian levels of pineal hormones, and indications of transient arousal responses. Additionally, in several instances in which unconfirmed or controversial data exist, observed effects may or may not be real; for example, changes in serum catecholamines or corticosteroids, morphology of brain tissue, and changes in electroencephalographic waveforms. It is not yet known whether these and other putative effects are caused by a direct interaction of the electromagnetic fields with tissue or an indirect interaction, that is, a physiological response from detection and/or sensory stimulation by the field. The nature of the physical mechanisms involved in field-induced effects is obscure, and such knowledge is one of the urgent goals of current research.

Experimental results have provided evidence for various neurological effects of ELF field exposure in specific species. The extrapolation of specific effects that occurred under controlled laboratory conditions to a general assessment of the health risk for a human population exposed to electric and/or magnetic fields is very tenuous. At least four considerations are essential to implement such an extrapolation with validity: (1) the relationship of specific laboratory conditions to the real-world environment; (2) the relevance of effects in laboratory animals to other species, particularly humans; (3) dosimetric considerations, including scaling between species; and (4) an evaluation of the biological consequences of observed effects.

Many of the experiments reviewed in this chapter were designed to study the effects of electric fields under laboratory conditions; a few, to study magnetic fields under those conditions. Because most experiments were performed at frequencies of 50 or 60 Hz, field strengths and intensities usually corresponded to those characteristic of power lines, but factors other than the electric field may have affected experimental results. Such factors, for example, ozone, ions, spark discharge, audible noise, vibration, etc., can produce biological effects and must be recognized and controlled to determine whether the electric or magnetic field is actually the agent responsible for the observed effects. Further, any extrapolation of effects from one species to another depends on the mechanism(s) by which fields exert influence on biological systems. This requires a knowledge of the biological structures and functions involved, as well as dosimetric scaling of exposure from the test animal to another species. We hope that the rapidly expanding knowledge of ELF field interactions with neural tissues and systems will hasten this understanding.

Perhaps most difficult to answer is the question of when the occurrence of a "biological effect" constitutes a health hazard. There are, as yet, no well-documented examples of ELF field-related neuropathological conditions. Generally, many of the biological effects reported thus far are quite subtle, and differences between exposed and unexposed subjects may be masked by normal biological variations. Careful examination of the data, however, emphasizes our limited understanding of biological interactions with electromagnetic fields, particularly about the long-term nature of many health factors, such as the time required for carcinogenic processes to develop. Questions concerning the potential health implications of ELF electromagnetic-field exposure, particularly in the area of neural and neuroendocrine effects and carcinogenesis, have yet to be fully clarified. As is discussed in other chapters in this book, it is entirely possible that ELF effects on rhythms, particularly those mediated by the neural and neuroendocrine systems, could play an important role in bioeffects and health-related questions.

ACKNOWLEDGMENT

Work was supported by the U.S. Department of Energy, Office of Energy Storage and Distribution, under contract DE–AC06–76RLO 1830.

REFERENCES

Anderson, L. E., Phillips, R. D. 1985. Biological effects of electric fields: An overview. In: *Biological Effects and Dosimetry of Nonionizing Radiation: Static and ELF Electromagnetic Fields*, Grandolfo, M., Michaelson, S. M., Rindi, A., eds., pp. 345–378. New York: Plenum.

Asanova, T. P., Rakov, A. I. 1966. The state of health of persons working in electric fields of outdoor 400 and 600 kV switchyards. In: *Hygiene of Labor Professional Diseases 5*, Spec. Pub. No. 10, Knickerbocker, G., trans. Piscataway, New Jersey: IEEE Power Engineering Society.

Babovich, R. D., Kozyarin, I. P. 1979. Effects of low frequency electrical fields (50 Hz) on the body. *Gig. I. Sanit.* 1:11–15.

Bankoske, J. W., McKee, G. W., Graves, H. B. 1976. *Ecological Influence of Electric Fields*. EPRI EA-178, Project 129, Interim Rep. 2. Palo Alto, California: Electric Power Research Institute.

Bauchinger, M., Hauf, R., Schmid, E., Dreps, J. 1981. Analysis of structural chromosome changes and SCE after occupational long-term exposure to electric and magnetic fields from 380 kV systems. *Radiat. Environ. Biophys.* 19:235–238.

Bawin, S. M., Sabbot, I., Bystrom, B., Sagan, P. M., Adey, W. R. 1979. Effects of 60-Hz environmental electric fields on the central nervous system of laboratory rats. In: *Proceedings of the International Union of Radio Science*, Spring Meeting, June 18–22, 1979, Seattle, Washington.

Beischer, D. E., Grissett, J. D., Mitchell, R. R. 1973. *Exposure of Man to Magnetic Fields Alternating at Extremely Low Frequency*. NAMRL 1155 (NTIS No. AD754058). Pensacola, Florida: Naval Aerospace Medical Research Laboratory.

Blackwell, R. P. 1986. Effects of extremely-low-frequency electric fields on neuronal activity in rat brain. *Bioelectromagnetics* 7:425–434.

Blanchi, D., Cedrini, L., Cepia, F., Meda, E., Re, G. 1973. Exposure of mammalians to strong 50-Hz electric fields: (2) Effect on heart's and brain's electrical activity. *Arch. Fisiol.* 70:33.

Cabanes, J., Gary, C. 1981. Direct perception of the electric field. In: *International Conference on Large High Tension Electric Systems*, pp. 22–81. Paris: CIGRE.

Carter, J. H., Graves, G. H. 1975. *Effects of High Intensity AC Electric Fields on the Electroencephalogram and Electrocardiogram of Domestic Chicks: Literature Review and Experimental Results*. University Park: Pennsylvania State University.

Creim, J. A., Lovely, R. H., Kaune, W. T., Miller, M. C., Anderson, L. E. 1985. 60-Hz magnetic fields: Do rats avoid exposure? In: *Abstracts of the Seventh Annual Meeting of the Bioelectromagnetics Society*, June 16–20, San Francisco, California, p. 58. Frederick, Maryland: Bioelectromagnetics Society.

Davis, H. P., Mizumori, S. J. Y., Allen, H., Rosenzweig, M. R., Bennett, E. L., Tenforde, T. S. 1984. Behavioral studies with mice exposed to DC and 60-Hz magnetic fields. *Bioelectromagnetics* 5:147–164.

Deno, D., Zaffanella, L. 1982. Electrostatic effects of overhead transmission lines and stations. In: *Transmission Line Reference Book 345 kV and Above*. 2nd Ed. Palo Alto, California: Electric Power Research Institute.

Dumansky, Y. D., Popovich, V. M., Prokhvatilo, E. V. 1976. Hygiene assessment of an electromagnetic field created by high-voltage power transmission lines. *Gig. I. Sanit.* 8:19–23.

Fisher, G., Udermann, H., Knapp, E. 1978. Ubt das netzfrequente wechselfeld zentrale wirkungen aus? *Zentralbl. Bakteriol. Abt. 1 Orig. B Hyg. Krankenhaushyg. Betriebshyg. Praev. Med.* 166:381–385.

Free, M. J., Kaune, W. T., Phillips, R. D., Cheng, H. C. 1981. Endocrinological effects of strong 60-Hz electric fields on rats. *Bioelectromagnetics* 2:105–121.

Friedman, H., Becker, R. O., Bachman, C. H. 1967. Effects of magnetic fields on reaction time performance. *Nature (London)* 213:949–950.

Gann, D. J. 1976. *Final Report*. Electric Power Research Institute RP 98-02. Baltimore, Maryland: Johns Hopkins University.

Gona, A. G. 1987. *Effects of 60-Hz Electric and Magnetic Fields on the Developing Rat Brain. Final Report*. Albany, New York: Health Research, Inc.

Graham, C., Cohen, H. D., Cooke, M. R., Phelps, J. W., Gerkovich, M. M., Fotopoulos, S. S. 1987. A doubleblind evaluation of 60-Hz field effects on human performance, physiology, and subjective state. In: *Interaction of Biological Systems with Static and ELF Electric and Magnetic Fields*, Anderson, L. E., Weigel, R. J., Kelman, B. J., eds., pp. 471–486, Proceedings of the 23rd Annual Hanford Life Sciences Symposium, Richland, Washington. CONF 841041. Springfield, Virginia: National Technical Information Service.

Graves, H. B., Carter, J. H., Kellmel, D., Cooper, L. 1978. Perceptibility and electrophysiological response of small birds to intense 60-Hz electric fields. *IEEE Trans. Power Appar. Sys. PAS* 97:1070–1073.

Groza, P., Carmacia, R., Bubuiann, E. 1978. Blood and urinary catecholamine variations under the action of a high voltage electric field. *Physiologie* 15:139–144.

Hackman, R. M., Graves, H. B. 1981. Corticosterone levels in mice exposed to high intensity electric fields. *Behav. Neural Biol.* 32:201–213.

Hansson, H. A. 1981a. Lamellar bodies in Purkinje nerve cells experimentally induced by electric field. *Brain Res.* 216:187–191.

Hansson, H. A. 1981b. Purkinje nerve cell changes caused by electric fields: Ultrastructual studies on long term effects on rabbits. *Med. Biol.* 59:103–110.

Hauf, R. 1974. Effect of 50 Hz alternating fields on man. *Elektrotech. Z.* B26:318–320.

Hauf, R. 1976a. Effect of electromagnetic fields on human beings (Einfluss elektromagnetisches Felder auf den menschen). *Elektrotech. Z.* B28:181–183.

Hauf, R. 1976b. Influence of 50 Hz alternating electric and magnetic fields on human beings. In: *Recherches sur les Effets Biologique des Champs Electrique et Magnétique, Revue Generale de l'Electricité*, Special Issue, pp. 31–49 (July 1976).

Hauf, R. 1982. Electric and magnetic fields at power frequencies, with particular reference to 50 and 60 Hz. In: *WHO Regional Publication, European Series, No. 10, Nonionizing Radiation Protection, Vol. VIII*, pp. 175–198. Copenhagen, Denmark: World Health Organization.

Hilmer, H., Tembrock, G. 1970. Untersuchungen zur lokomotorischen aktivität weisser ratten unter dem Einfluss von 50-Hz hochspannungs-wechsel Feldern. *Biol. Zentralbl.* 89:1.

Hjeresen, D. L., Kaune, W. T., Decker, J. R., Phillips, R. D. 1980. Effects of 60-Hz electric fields on avoidance behavior and activity of rats. *Bioelectromagnetics* 1:299–312.

Hjeresen, D. L., Miller, W. C., Kaune, W. T., Phillips, R. D. 1982. A behavioral response of swine to 60-Hz electric field. *Bioelectromagnetics* 2:443–451.

Jaffe, R. A., Laszewski, B. L., Carr, D. B. 1981. Chronic exposure to a 60-Hz electric field: Effects on neuromuscular function in the rat. *Bioelectromagnetics* 2:227–239.

Jaffe, R. A., Laszewski, B. L., Carr, D. B., Phillips, R. D. 1980. Chronic exposure to a 60-Hz electric field: effects on synaptic transmission and peripheral nerve function in the rat. *Bioelectromagnetics* 1:131–147.

Jaffe, R. A., Lopresti, C. A., Carr, D. B., Phillips, R. D. 1983. Perinatal exposure to 60-Hz electric fields: effects on the development of the visual evoked response in the rat. *Bioelectromagnetics* 4:327–340.

Kaune, W. T., Phillips, R. D., Hjeresen, D. L., Richardson, R. L., Beamer, J. L. 1978. A method for the exposure of miniature swine to vertical 60-Hz electric fields. *IEEE Trans Biomed. Eng. BME* 25(3):276–283.

Kavaliers, M., Ossenkopp, K. P., Hirst, M. 1984. Magnetic fields abolish the enhanced nocturnal analgesic response to morphine in mice. *Physiol. Behav.* 32:261–264.

Knave, B., Gamberale, F., Berstrom, S., Birke, E., Iregren, A., Kolmodin-Hedman, B., Wennberg, A. 1979. Long-term exposure to electric fields. A cross-sectional epidemiologic investigation of occupationally exposed industrial workers in high-voltage substations. *Electra* 65:41–54.

Kolodub, F. A., Chernysheva, O. N. 1980. Special features of carbohydrate-energy and nitrogen metabolism in the rat brain under the influence of magnetic fields of commercial frequency. *Ukr. Biokhim. Zh.* 3:299–303.

Korobkova, V. A., Morozov, Y. A., Stolarov, M. S., Yakub, Y. A. 1972. Influence of the electric field in 500 and 750-kV switchyards on maintenance staff and means for its protection. In: *International Conference on Large High Tension Electric Systems*, 28 August 1972, p. 23–06. Paris: CIGRE.

Kozyarin, I. P. 1981. Effects of low frequency (50 Hz) electric fields on animals of different ages. *Gig. I. Sanit.* 8:18–19.

Lin, R. S., Dischinger, P. C., Conde, J., Farrell, K. P. 1985. Occupational exposure to electromagnetic fields and the occurrence of brain tumors. An analysis of possible associations. *J. Occup. Med.* 27:413–419.

Lovely, R. H. 1988. Recent studies in the behavioral toxicology of ELF electric and magnetic fields. In: *Electromagnetic Fields and Neurobehavioral Function*, O'Connor, M. E., Lovely, R. H., eds., pp. 327–348. New York: Alan R. Liss.

Lovsund, P., Oberg, P. A., Nilsson, S. E. G. 1979. Influence on vision of extremely low frequency electromagnetic fields. *Ophthalmology* 57:812–821.

Marino, A. A., Berger, T. J., Austin, B. P., Becker, R. O., Hart, F. X. 1977. In vivo bioelectrochemical changes associated with exposure to extremely low frequency electric fields. *Physiol. Chem. Phys.* 9:433–441.

Medvedev, M. A., Urazaev, A. M., Kulakov, I. U. A. 1976. Effect of a constant and low frequency magnetic field on the behavioral and autonomic responses of the human operator. *Zh. Vyssh. Nervn. Deyat. Im. I. P. Pavlova* 26:1131–1136. (Transl. 1978; *Sov. Neurol. Psychiatry* 11:3–13.)

Michaelson, S. M. 1979. Analysis of studies related to biologic effects and health implications of exposure to power frequencies. *Environ. Prof.* 1:217–232.

Mose, J. R. 1978. Problems of housing quality. *Zentralbl. Bakteriol. Abt. 1 Orig. B Hyg. Krankenhaushyg. Betriebshyg. Praev. Med.* 166:292–304.

National Academy of Sciences. 1977. *Biological Effects of Electric and Magnetic Fields Associated with Proposed Project Seafarer*. Report of Committee on Biosphere Effects of Extremely-Low-Frequency Radiation, Division of Medical Sciences, Assembly of Life Science. Washington, D.C.: National Academy of Sciences.

Olcese, J., Reuss, S., Vollrath, L. 1985. Evidence for the involvement of the visual system in mediating magnetic field effects on pineal melatonin synthesis in the rat. *Brain Res.* 333:382–384.

Persinger, M. A. 1969. Open-field behavior in rats exposed prenatally to a low intensity-low frequency, rotating magnetic field. *Dev. Psychobiol.* 2:168–171.

Persinger, M. A., Foster, W. S., IV. 1970. ELF Rotating magnetic fields: Prenatal exposures and adult behavior. *Arch. Meteorol. Geophys. Biol. B* 18:363–369.

Phillips, R. D., Kaune, W. T. 1977. *Biological Effects of Static and Low-Frequency Electromagnetic Fields: An Overview of United States Literature*. EPRI Special Rep. EA-490-SR. Palo Alto, California: Electric Power Research Institute.

Phillips, R. D., Chandon, J. H., Free, M. J., et al. 1978. *Biological Effects of 60-Hz Electric Fields on Small Laboratory Animals*. Annual Report HCP/T1830-3. Washington, DC: U.S. Department of Energy, Division of Electrical Energy Systems.

Polk, C., Postow, E. 1987. *CRC Handbook of Biological Effects of Electromagnetic Fields*. Boca Raton, Florida: CRC Press.

Portet, R. T., Cabanes, J. 1988. Development of young rats and rabbits exposed to a strong electric field. *Bioelectromagnetics* 9:95–104.

Quinlan, W. J., Petrondas, D., Lebda, N., Pettit, S., Michaelson, S. M. 1985. Neuro-endocrine parameters in the rat exposed to 60-Hz electric fields. *Bioelectro-magnetics* 6:381–389.

Reiter, R. J., Anderson, L. E., Buschbom, R. L., Wilson, B. W. 1988. Reduction of the nocturnal rise in pineal melatonin levels in rats exposed to 60-Hz electric fields in utero and for 23 days after birth. *Life Sci.* 42:2203–2206.

Reuss, S., Olcese, J. 1986. Magnetic field effects on the rat pineal gland: role of retinal activation by light. *Neurosci. Lett.* 64:97–101.

Rogers, W. R., Feldstone, C. S., Gibson, E. G., Polonis, J. J., Smith, H. D., Cory, W. E. 1987. Effects of high-intensity, 60-Hz electric fields on operant and social behavior of nonhuman primates. In: *Interaction of Biological Systems with Static and ELF Electric and Magnetic Fields*, Anderson, L. E., Weigel, R. J., Kelman, B. J., eds., pp. 365–378, Proceedings of the 23rd Annual Hanford Life Sciences Symposium, Richland, Washington. CONF 841041. Springfield, Virginia: National Technical Information Service.

Rosenberg, R. S., Duffy, P. H., Sacher, G. A. 1981. Effects of intermittent 60-Hz high-voltage electric fields on metabolism, activity and temperature in mice. *Bioelectromagnetics* 2:291–303.

Rosenberg, R. S., Duffy, P. H., Sacher, G. A., Ehret, C. F. 1983. Relationship between field strength and arousal in mice exposed to 60-Hz electric field. *Bioelectromagnetics* 4:181–191.

Russell, J. J., Ehret, C. F. 1982. The effect of electric field exposure on rat circa-dian and infradian rhythms. In: *Proceedings of the Fourth Annual Meeting of the Bioelectromagnetics Society*, June-July, 1982, Los Angeles, California. Frederick, Maryland: Bioelectromagnetics Society.

Sagan, P. M., Stell, M. E. 1984. Behavioral study of mechanisms for detection of electric fields. In: *Abstracts of the Sixth Annual Meeting of the Bioelectromagnetics Society*, July 1984, Atlanta, Georgia. Frederick, Maryland: Bioelectromagnetics Society.

Sakharova, S. A., Ryzhov, A. I., Udintser, N. A. 1981. Mechanism of the sympathoadrenal system's response to the one time action of a variable magnetic field. *Kosm. Biol. Aviakomischeskaya. Med.* 15:52.

Sander, R., Brinkmann, J., Kuhne, B. 1982. Laboratory studies on animals and human beings exposed to 50-Hz electric and magnetic fields. In: *International Congress on Large High Voltage Electrical Systems*, Paris, France, Paper 36-01. Paris: CIGRE.

Semm, P. 1983. Neurobiological investigations on the magnetic sensitivity of the pineal gland in rodents and pigeons. *Comp. Biochem. Physiol. A.* 76:683–689.

Sheppard, A. R. 1978. Magnetic field interactions in man and other mammals: An overview. In: *Proceedings of the Biomagnetic Effects Workshop*, Tenforde, T., ed., pp. 28–31. New York: Plenum.

Sheppard, A. R. 1983. *Biological Effects of High Voltage AC Transmission Lines*. A report to Montana Department of Natural Resources and Conservation, Helena, Montana.

Sheppard, A. R., Eisenbud, M. 1977. *Biological Effects of Electric and Magnetic Fields on Extremely Low Frequency*. New York: New York University Press.

Silney, J. 1979. Effects of electric fields on the human organism. In: *Institut zur Enforschung electrischer unfalle*, p. 39. Cologne: Medizinisch – Technischer Berichte.

Smith, R. F., Justesen, D. R. 1977. Effects of a 60-Hz magnetic field on activity levels of mice. *Radio Sci.* 12(6S):279–286.

Smith, M. T., D'Andrea, J. A., Gandhi, O. P. 1979. Behavioral effects of strong 60-Hz electric fields in rats. *J. Microwave Power* 14:223–228.

Soldatova, L. P. 1982. Sequence of pathomorphological reactions to the effect of alternating magnetic fields. *Arkh. Anat. Gistol. Embriol.* 83:12–15.

Stern, S., Laties, V. G., Stancampiano, C. V., Cox, C., de Lorge, J. O. 1983. Behavioral detection of 60-Hz electric fields by rats. *Bioelectromagnetics* 4:215–247.

Stoops, G. J., Janischewsky, W. 1979. *Epidemiological Study of Workers Maintaining HV Equipment and Transmission Lines in Ontario*. Canadian Electrical Association Research Report, Montreal, Quebec, Canada.

Sulzman, F. M., Murrish, D. E. 1987. *Effects of Electromagnetic Fields on Primate Circadian Rhythms*. Final Report to the New York State Power Lines Project. Albany, New York: New York State Department of Health.

Takashima, S., Onaral, B., Schwan, H. P. 1979. Effects of modulated RF energy on the EEG of mammalian brains. Effects of acute and chronic irradiations. *Radiat. Environ. Biophys.* 16:15–27.

Toroptsev, I. V., Soldatova, L. P. 1981. Pathomorphological reaction of cerebrocortical neural elements to alternating magnetic field. *Arkh. Patol.* 43:33–36.

Tucker, R. D., Schmitt, O. H. 1978. Tests for human perception of 60-Hz moderate strength magnetic fields. *IEEE Trans. Biomed. Eng. BME* 25:509–518.

Udintsev, N. A., Moroz, V. V. 1974. Response of the pituitary-adrenal system to the action of a variable magnetic field. *Byull. Eksp. Biol. Med.* 77:51–53.

Vasquez, B. J., Anderson, L. E., Lowery, C. I., Adey, W. R. 1988. Diurnal patterns in brain biogenic amines of rats exposed to 60-Hz electric fields. *Bioelectromagnetics* 9:229–236.

Weigel, R. J., Lundstrom, D. L. 1987. Effect of relative humidity on the movement of rat vibrissae in a 60-Hz electric field. *Bioelectromagnetics* 8:107–110.

Welker, H. A., Semm, P., Willig, R. P., Commentz, J. C., Wiltschko, W., Vollrath, R. 1983. Effects of an artificial magnetic field on serotonin-*N*-acetyltransferase activity and melatonin content of rat pineal. *Exp. Brain Res.* 50:426–432.

Wilson, B. W., Chess, E. K., Anderson, L. E. 1986. 60-Hz electric field effects on pineal melatonin rhythms: time course for onset and recovery. *Bioelectromagnetics* 7:239–242.

Wilson, B. W., Anderson, L. E., Hilton, D. I., Phillips, R. D. 1981. Chronic exposure to 60-Hz electric fields: Effects on pineal function in the rat. *Bioelectromagnetics* 2:371–380.

Wolpaw, J. R., Seegal, R. F., Dowman, R. I., Satya-Murti, S. 1987. *Chronic Effects of 60 Hz Electric and Magnetic Fields on Primate Central Nervous System Function.* Final Report to the New York State Power Lines Project, New York State. Albany, New York: New York Department of Health.

8 ELF Electromagnetic-Field Effects on the Pineal Gland

BARY W. WILSON
LARRY E. ANDERSON
Pacific Northwest Laboratory
Richland, Washington

CONTENTS

In this chapter, we review data, primarily from chronobiological and neuroendocrine studies, that suggest that ELF electric and magnetic fields can affect pineal function. We then propose ways in which alterations or deficits in the normal function of the pineal gland might be implicated in the etiology of certain cancers. To consider these data in context, we begin with a brief historical introduction regarding the various physiological roles ascribed to the pineal organ.

BACKGROUND

Anatomical description of the pineal gland probably predates that of any other endocrine organ. From the writings of Galen in the second century B. C., it is clear that he recognized the pineal. In the seventeenth century, Descartes postulated, possibly because of its central location in the brain, that the pineal organ was the "seat of the soul." He also put forward the idea that psychic phenomena and thoughts have their origin and come to fruition in the pineal.

In 1879, Koelliker felt justified in stating, on the basis of microscopic examination of both calcified and uncalcified glands, that the pineal was "undoubtedly an insignificant organ." (It was later noted, however, that in the rat the blood-flow-to-tissue-weight ratio of the pineal was exceeded only by that of the kidney.) In 1898 Heubner noted that pineal tumors in man were sometimes associated with precocious puberty. Wurtman et al. (1968a) later argued that precocious puberty followed destruction of pineal tissue, whereas tumors resulting in hyperfunction of the gland were associated with delayed puberty.

Early in the twentieth century, pineal extracts attracted interest because they could reversibly blanch the skin of amphibia (McCord and Allen, 1917). In 1958 Lerner isolated an active compound from approximately 250,000 bovine pineal glands, and showed it to be the most potent frog-skin-lightening agent known, active at concentrations of 10^{-13} g/ml (Lerner et al., 1958, 1959). This compound was called melatonin and, in an early application of organic mass spectrometry, was shown to be N-acetyl-5-methoxytryptamine.

In 1965, Wurtman and Axelrod set forth a hypothesis that has continued to be the basis for successful experiments in pineal function and physiology; namely, that the pineal organ, like the adrenal medulla, functions as a "neuroendocrine transducer," converting neuronal input to hormonal output (Wurtman and Axelrod, 1965).

PINEAL GLAND PHYSIOLOGY

Biosynthesis of Melatonin

Metabolic pathways for the biosynthesis of melatonin from serotonin are shown in Fig. 1. All enzymes involved in this pathway are found in the mammalian pineal gland. The enzyme that adds the methyl group to the 5-hydroxy group of the indole nucleus (Axelrod and Weissbach, 1960), namely hydroxyindole-O-methyltransferase (HIOMT), was thought to be virtually exclusive to the pineal (Axelrod et al., 1961). More recently, however, there have been reports of its occurrence in other structures. For example, it has been detected in the retina (Cardinali and Rosner, 1971), which is not surprising considering the evolution of the pineal, and in the Harderian gland (Vlahakes and Wurtman, 1972), an endocrine organ associated with the nictitating membrane in rodents. Harderian glands are not present in humans; the principal site of HIOMT activity in the human is the pineal, and the bulk of circulating melatonin is derived from that organ.

Calcification of the pineal gland increases with age, but enzyme studies of pineals from humans up to 70 years of age show that they retain their capacity to synthesize melatonin (Wurtman et al., 1964). Histological studies from puberty to old age have shown that the proportion of the gland occupied by parenchymal cells is not decreased (Tapp and Huxley, 1972). More recent studies indicate that there are age-related changes in melatonin synthesis by the pineal (Waldhauser and Steger, 1986). Melatonin levels appear to decrease around the time of puberty, and become progressively less with age. The reduction in melatonin synthesis at puberty presents a consistent picture of pineal function with regard to sexual maturation, considering the antigonadotropic action of melatonin.

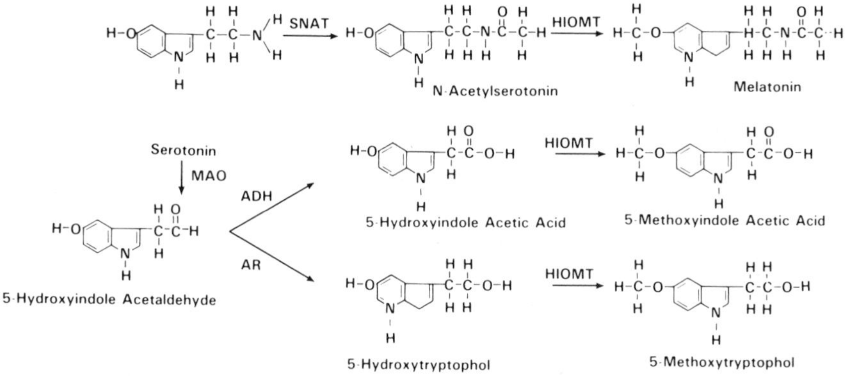

Fig. 1. Biosynthetic pathways for melatonin and other serotonin metabolites. Abbreviations: AR, aryl reductase; AH, aryl hydrogenase; MAO, monoamine oxidase; SNAT, serotonin N-acetyltransferase.

Other Pineal-Specific Indoles

As noted, the enzyme HIOMT, which is necessary for the synthesis of melatonin, is localized mainly in the pineal gland. This enzyme can participate in the synthesis of other 5-methoxy indoles: 5-methoxytryptamine (O-methyl serotonin), and 5-methoxytryptophol, for example, are physiologically active compounds that exhibit circadian rhythms in the pineal (McIsaac et al., 1965; Koslow, 1976; Wilson et al., 1978). Figure 1 depicts a fairly complete scheme for the biosynthesis of serotonin metabolites.

There is strong evidence to suggest that 5-methoxytryptophol is an antigonadotropic compound, its primary action being directed against follicle-stimulating hormone (FSH) rather than luteinizing hormone (LH) (Fraschini and Martini, 1970). In human plasma, 5-methoxytryptamine is quickly converted to its major urinary metabolite, 5-methoxyindole acetic acid (Wilson and Snedden, 1979). Alterations in 5-methoxytryptamine content have been measured in the cerebrospinal fluid (CSF) and in the brain tissue of depressive suicides (Shaw et al., 1967). McIsaac (1961) has put forward the hypothesis that certain 5-methoxy indoles may be implicated in behavioral changes. For example, he suggests that the 5-methoxylation of the indole nucleus is an important factor in the enhanced psychomimetic activity of the various hallucinogenic drugs, such as the ergot alkaloids, that are structurally related to serotonin.

Melatonin Measurement

Before the work of Wilson et al. (1977) and Lewy and Markey (1978), there was no reliable, sensitive assay for melatonin. Indeed, its existence in the human circulation was not unambiguously demonstrated until 1976 (Smith et al., 1976), although it had been previously inferred from results of bioassays (Pelham et al., 1973). Consequently, possible physiological functions of melatonin had been investigated only by indirect means, such as analysis of pineal enzyme activity or observations after pinealectomy and the exogenous administration of the hormone. Early mass spectrometric assays have largely been abandoned in favor of specific radioimmunoassays that have become commercially available (Arendt, 1978; Arendt et al., 1985). Pineal function can now be assessed by measurement of melatonin or its metabolites in the blood, urine, and CSF as well as in the pineal itself.

A particularly important addition to the analytical tools available for study of pineal function has been the development of a radioimmunoassay for 6-hydroxy melatonin sulfate (6-OHMS), a stable melatonin metabolite excreted in the urine. This assay allows noninvasive determination of pineal function. Levels of this metabolite have been compared with those of plasma melatonin (Bojkowski and Arendt, 1988) and exhibit the expected circadian rhythm. Morning void urines, which represent accumulation of the metabolite in the kidneys overnight, contain nominally 5- to 10-fold higher concentrations than do urines collected during the daytime.

Chronobiology

Some of the earliest observations suggested that melatonin production had a circadian variation, high at night and low during the day (Axelrod et al., 1965). It was further shown that this rhythm directly correlated with the light:dark conditions of the environment: constant darkness leading to continuously elevated melatonin production (as evidenced by increased HIOMT activity) and the reverse in constant light (Wurtman et al., 1963; Klein and Weller, 1971). This probably relates to the curious evolution of the pineal. In fish, amphibia, and birds it contains two parts, one of which is a sensory end-organ with discrete photoreceptor cells responding directly to light and having neuronal connections to the central nervous system (CNS). Because it lies near the surface of the skull, the pineal gland has been logically termed the "third eye." In mammals the pineal has evolved significantly, losing the photoreceptor cells but retaining specific endocrine cells of the pineal, the pinealocytes. This evolutionary change does not prevent the gland from responding to light, since it receives indirect sensory input from the retina (Moore et al., 1968).

Thus, serotonin and its catabolic metabolites, including *N*-acetyl serotonin, melatonin, 5-methoxytryptamine, and 5-methoxytryptophol, exhibit strong circadian rhythms in most reptiles, birds, and mammals that have been studied. A number of ultradian rhythms superimposed on pineal gland function have also been reported. These include an approximately monthly rhythm in women that is synchronized to the menstrual period (Arendt et al., 1985; Hariharasubramanian et al., 1985), and an annual rhythm in plasma melatonin levels that may be especially pronounced at higher latitudes (Dubbels and Khoory, 1986). In studies on the annual melatonin rhythms in man, Bojkowski and Arendt (1988) have shown that the total 24-hr excretion of 6-0HMS does not vary significantly with the time of year. The timing, or acrophase, of the nighttime peak does vary significantly however (by approximately 2 hr) at 52°N latitude (Fig. 2). It is also important to note that exposure to light of sufficient intensity during the dark cycle leads to an immediate drop in circulating melatonin content in several species, including man (Lewy et al., 1982).

Neuroendocrine Function

The pineal organ is described as a "neuroendocrine transducer" (Wurtman and Axelrod, 1965) because it serves as a locus for the conversion of a neural input concerning the environment to a hormonal output, primarily melatonin. Melatonin acts as a neuroendocrine mediator for the biological clock located in the suprachiasmatic nucleus, and produces the primary hormonal signal for synchronizing endocrine rhythms within the body (Cardinali, 1974). In mammals, it is clear that one important target of melatonin is the CNS (Anton-Tay, 1974). Innervation of the pineal is

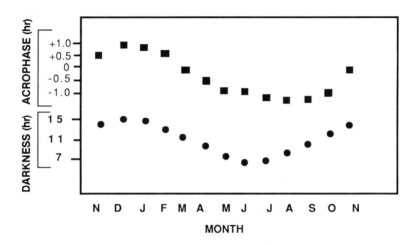

Fig. 2.   Change in acrophase for humans in relation to the duration of the daily dark period. (Adapted from Bojkowski and Arendt, 1988.)

mainly from fibers of the superior cervical ganglion (Moore et al., 1968). However, fibers originating in the hypothalamus and in the optic system regions of the brain also innervate the gland (Moller and Mikkleson, 1988). The neuroendocrine function of the pineal as related to stress and reproduction is discussed in Chapter 5 by Reiter (this volume).

It is now recognized that pineal function in the endocrine system is inhibitory for most other endocrine glands (Wurtman and Cardinali, 1974; Nir, 1978). Pinealectomy in rats results in endocrine changes that include increases in circulating levels of gonadal steroids as well as FSH and prolactin (Kamberi et al., 1971). Pineal ablation also causes measurable functional changes in the hypothalamus, pituitary, and gonads, as well as in the thyroid and parathyroid glands (Nir, 1978). Although some of these effects may be secondary, there is evidence for direct interaction of the pineal gland with the pituitary (Bindoni and Raffaele, 1968) and the gonads (Reiter, 1981).

Research into the effects of melatonin on the brain and behavior has shown that exogenous melatonin readily enters the brain; uptake from CSF is even greater than from blood (Wurtman et al., 1964; Anton-Tay and Wurtman, 1969). Melatonin has the effect of inhibiting protein synthesis in brain slices (Orsi et al., 1973), and pinealectomy in neonatal rats is reported to significantly decrease myelin formation and brain development (Relkin et al., 1973; Relkin and Schreck, 1975). Zisapel et al. (1988) have identified melatonin receptors in the hypothalamus and optic system regions of the brain. These receptor populations exhibit a clear circadian rhythm that appears to be largely light:dark driven.

Bowers et al. (1984) demonstrated the pineal gland's sensitivity to neuronal input during experiments wherein the intercarotid nerve, which together with the superior cervical ganglion innervates the pineal gland, was either sectioned, crushed, or frozen. Each procedure was effective in inhibiting pineal synthesis of *N*-acetyltransferase (NAT) and melatonin. On recovery from the freezing treatment, the intercarotid nerve was able to conduct nerve impulses, and an artificial 5-Hz signal was effective in eliciting pineal NAT response. However, spontaneous pineal rhythms were not observed at any time after treatment. Similarly, Kneisley et al. (1978) observed disruptions in the rhythms of excreted melatonin in human patients with cervical spinal cord transection. Cervical ganglion lesions also resulted in changes in electroencephalograph patterns (Adey et al., 1968). These investigators also noted sleep disruptions in quadriplegic patients.

Melatonin can induce sleep in both man (Anton-Tay, 1974; Cramer et al., 1974) and a variety of animals (Barchus et al., 1967; Hishikawa et al., 1969). Specific effects of melatonin on rapid eye movement (REM) sleep have been reported in man (Mouret et al., 1974). Rapidly fluctuating photoperiods, a presumed pineal stimulant, also affect REM (Rechtschaffen et al., 1969; Chamblin and Drew, 1971). Studies employing computer analysis of continuous EEG recordings found no significant difference between melatonin-induced sleep in man and natural sleep.

Only recently has there been a recognition of the variety in neurotransmitter receptor sites present in the pineal organ. Along with the well-established β-adrenergic receptors that are primarily responsible for stimulation of melatonin synthesis, the mammalian pineal contains α-adrenergic receptors and receptors for gamma amino butyric acid (GABA), as well as serotonergic, dopaminergic, glutamatergic, and muscarinic cholinergic receptors (Ebadi and Govitrapong, 1986).

Depression and Affective Illness

Pineal gland dysfunction has been implicated as a possible factor in affective illness. This is owing partially to the alterations in normal biological rhythms, including the sleep/wake cycle, that are often symptomatic of affective disorders. Some 30 years ago, Altschule (1957) and Eldred et al. (1961) showed that bovine pineal extracts were effective in lessening the severity of schizophrenic episodes in institutionalized patients. The importance of their findings seemed to diminish with the development and introduction of antipsychotic drugs such as haloperidol, although they were essentially confirmed by the work of Bigelow (1974).

In the 1970s some workers suggested ways in which the pineal may play a role in the etiology of schizophrenia. Their ideas were based largely on the fact that slight changes in the metabolism of melatonin could give rise to potential hallucinogens (McIsaac, 1961; Greiner, 1970). Interest in the pineal as a possible mediator of schizophrenia has diminished in the 1980s; however, the pineal is now thought to be involved in the etiology of depressive illness, although its exact role remains unclear.

Lewy et al. (1982) and others have shown that extending the daily photoperiod by exposure to bright light can be effective in reducing symptoms in patients suffering from seasonal affective disorder syndrome (SADS). SADS patients become depressed during the winter months, but do not show such symptoms during the summer. In a study of 28 patients suffering from major depression, Brown et al. (1987) reported that decreased nocturnal melatonin levels in plasma correlated with depressed mood and reality disturbance, as determined by the Hamilton depression rating. Other studies, including those of Mendlewicz et al. (1980), Branchey et al. (1982), Beck-Friis et al. (1985), and Claustrat et al. (1984), have associated disruptions in melatonin circadian rhythmicity with "major depression." Specific mechanisms by which pineal dysfunction affects mood have not been worked out. On a systemic level, both reduction and phase-shifting of the circadian peak in melatonin concentration have been suggested as factors, although current views favor changes in the acrophase (timing of the melatonin peak) as the most relevant to these mood changes. [For a review of possible ELF field effects on mood and depression, see Wilson (1988).]

Papke et al. (1986) have generated evidence that melatonin is a neuromodulator. Neurons from rat superior cervical ganglia were cocultured with pinealocytes and exhibited action potentials that were longer and of higher amplitude than potentials from neurons grown in the presence of nonneuronal cells. In these same experiments, melatonin caused an increase in action potential duration that was not associated with increases in amplitude. Increasing levels of circulating steroids resulting from reduced melatonin secretion could be a factor in ELF-induced mood changes. Su et al. (1988) have identified specific gamma receptors in the brain and showed that certain gonadal steroids, including estradiol, inhibit gamma-receptor binding. The authors speculated that these sites may be involved in mediation of steroid-induced mood changes. Alterations in mood and affect may follow directly from the loss of melatonin as a neuromodulator or from changes in circulating steroid levels resulting from the loss of melatonin as a gonadal inhibitor.

An interesting consequence of the ELF field-induced reduction in NAT activity, in experiments carried out in our laboratory, was the compensatory increase in the levels of 5-methoxytryptophol (5-MTOL), which is an apparent effect of increased substrate availability in the face of reduced NAT capacity to metabolize serotonin. We hypothesized in this situation that serotonin was being shunted down the monoamine oxidase (MAO) pathway (see Fig. 1). Murphy et al. (1986) conducted experiments to measure the effect of MAO inhibitor treatment on melatonin levels in depressive patients. They found a pineal response complementary to that observed in the case of ELF exposure; when the MAO pathway was shut down, more serotonin was shunted down the NAT pathway, resulting in an approximate threefold increase in melatonin levels. This increase in melatonin was believed to contribute to the antidepressive effects of the MOA inhibitor treatment.

In reviewing the physiology of melatonin thus far, a range of disparate activities have been surveyed. These can be integrated if one considers that the primary targets of melatonin are certain centers of the brain, where melatonin modifies the serotonin-containing neurons (as in the hypothalamus), to affect pituitary secretion, or acts in the CNS, to affect behavior.

PINEAL FUNCTION AND CANCER

Three fundamental assumptions are important to the discussion presented in this section:

1. Maintenance of environmentally mediated rhythms including intradian, circadian, and ultradian biological rhythms, particularly in endocrine hormone and neurotransmitter synthesis and release, is vital to adaptation and homeostasis in animals (see Moore-Ede et al., 1985).
2. Control and feedback mechanisms exist among the nervous, endocrine, and immune systems and allow neural mediation of certain endocrine and immune system functions (see Cotman et al., 1987). Mechanisms of this kind involving pineal gland function have been demonstrated in some of the work reviewed here, and they may be especially important with respect to ELF electric- and magnetic-field effects.
3. ELF electromagnetic fields interact with the nervous systems in a variety of animals, including humans (see O'Conner and Lovely, 1988). Of particular importance is evidence, obtained in several laboratory experiments, that pineal circadian rhythms in serotonin, melatonin, and other tryptophan metabolites are altered or obliterated by appropriate exposure to ELF electric fields (Wilson et al., 1981, 1986). More recent work with healthy human subjects suggests that ELF electric- or magnetic-field exposure may alter pineal function in certain individuals (Wilson et al., 1988).

Pineal function might be linked to the etiology of cancer in at least three fundamental ways:

First, melatonin itself is oncostatic and appears to be a humoral factor that inhibits the proliferation of certain cancer cells.

Second, melatonin enhances certain facets of the immune response, again possibly helping to protect against the development of cancers.

Third, melatonin functions as an inhibitor of the hypothalamic–pituitary–gonadal axis. As such, it may reduce the availability of hormones that are required for the growth of certain hormone-dependent breast, ovarian, and prostate cancers.

Oncostatic Properties of Melatonin

Oncostatic properties of pineal extracts were first demonstrated in 1940 by Nakatani et al. (1940), who observed that an unknown component from bovine pineal gland could inhibit growth of tumor cells in culture. Several subsequent studies confirmed that pinealectomy led to increased growth and proliferation of several cancer cell lines (Rodin, 1963; Das Gupta and Terz, 1967). By the early 1970s, it was clear that melatonin was a pineal factor that could reduce the growth and metastasis of several cancers (Bostelmann et al., 1971; El-Domeiri and Das Gupta, 1973).

Proper pineal function appears protective against cancer (Blask et al., 1988). Blask and Hill (1986) demonstrated that physiological concentrations of melatonin were effective in inhibiting the growth of breast tumor cell lines in vitro, while subphysiological and superphysiological concentrations had no significant effect. [See Regelson and Pierpaoli (1987) for a review of melatonin as an endogenous antitumor factor.]

Melatonin and Immune Function

Melatonin appears to have a stimulatory effect on immune function in the whole animal. Maestroni et al. (1986), for example, showed that pharmacological inhibition of pineal function in mice results in depressed immune response to sheep red blood cells. Appropriately timed injection of melatonin restored normal immune function as determined by this assay. Evening injections of melatonin antagonized the suppression of antibody production resulting from adminstration of the stress hormone corticosterone in mice (Pierpaoli and Maestroni, 1987). These workers have suggested that the endogenous opioid system may mediate the immunoregulatory effects of melatonin. This view was supported by experiments in which the opioid antagonist naltrexone completely abolished the immunoenhancing effects of melatonin.

Melatonin as an Endocrine Inhibitor

A third effect of pineal hypofunction is the loss of gonadal inhibition that is normally exerted by melatonin. Thus, Cohen et al. (1978) proposed that reduced pineal melatonin secretion may be a factor in breast cancer etiology. Women with breast cancer showed reduced urinary melatonin levels (Bartsch et al., 1981). Tamarkin et al. (1982) and Danforth et al. (1982) extended these findings, noting altered melatonin secretion in patients with estrogen receptor- (ER-) positive breast cancer.

Androgenic prostate cancer is analogous to ER-positive breast cancers in that appropriate steroidal hormones are required for continued growth of the tumor. Buzzell et al. (1988) showed inhibition of growth in rat prostatic adenocarcinoma with melatonin. These findings have been confirmed by Philo and Berkowitz (1988) working with the same Dunning tumor cell line. Bartsch et al. (1988) reported that men with cancer of the prostate had lower nocturnal melatonin levels than men who did not have the disease. A pilot study by Leone and colleagues (Leone and Skene, 1989) has confirmed findings of altered melatonin secretion in prostate cancer as compared to controls with benign hypertrophy of the prostate. These investigators also reported alterations in 6-OHMS secretion associated with ovarian cancer in women.

Functional pinealectomy, as defined by Shah et al. (1984), can be achieved by exposing young virgin rats to constant light, thus inhibiting the release of melatonin. This approach has been used in studies on tumor formation and promotion in mammary epithelial cells. Tamarkin et al. (1981) and Shah et al. (1984) have shown a direct inhibitory effect of melatonin on the incidence and latency of dimethylbenzanthracene- (DMBA-) induced breast cancer in rats.

Noting these findings, Stevens (1987) suggested that reductions in melatonin concentration arising from ELF field exposure in rats may result in increased circulating estrogen levels. This may stimulate mammary tissue proliferation and thus increase breast cancer risk. To test this hypothesis, experiments analogous in design to those of Shah et al. (1984) were conducted in our laboratory; ELF electric fields were used instead of light to reduce pineal melatonin output. This work has shown that DMBA-induced rats exposed to 60-Hz electric fields since birth had more tumors per tumor-bearing animal than sham-exposed controls (Leung et al., 1988a). In the course of these studies, Leung found that mean pineal melatonin tended to be lower in ELF electric-field-exposed animals compared to controls. This is consistent with the findings of Tamarkin and others, who identified reduced melatonin and increased prolactin as risk factors in DMBA-induced mammary cancer in rats.

ELF ELECTRIC- AND MAGNETIC-FIELD EFFECTS IN HUMANS AND ANIMALS

Self-reported human symptoms related to ELF exposure are most often associated with subjective-state changes, including headaches, and often involve feelings of fatigue and general malaise (Asanova and Rakov, 1966;

Gavalas-Medici and Magadaleno, 1976; Shandala et al., 1984). In other studies on populations exposed to 50- to 60-Hz fields, increased excitability and irritability were reported. In many cases, a state of hyperactivity that interfered with normal physiological and mental function was observed and attributed to ELF field exposure (Sheppard and Eisenbud, 1977).

In experiments with 60-Hz electric fields, Graham et al. (1984) reported minor changes in evoked response measures subsequent to field exposure. Magnetic (ELF) field exposure can affect perceived flicker fusion frequency and low-light visual acuity in humans (Krause et al., 1985). Wever (1968) reported that 50-Hz field exposure arrested the lengthening of the human circadian cycle that normally occurs in the absence of time cues.

It is clear from studies on laboratory animals that the nervous system is affected by ELF exposure. Hjeresen et al. (1980) reported that electric-field exposure increased avoidance activity, and Jaffe et al. (1980) showed that exposed rats had faster recovery from slow-twitch muscle fatigue than did controls. Ehret et al. (1980) reported that exposed mice entered a state of torpor at electric-field strengths at or above about 75 kV/m. Ehret and Groh (see Chapter 4, this volume) have argued that an important effect of ELF exposure is loss of normal timing in biological circadian rhythms. Such "circadian dyschronism" appears to be a primary consequence of ELF exposure.

Both electric and magnetic fields can alter pineal gland function. Welker et al. (1983) showed that artificial magnetic fields could inhibit NAT activity in the rat pineal. Subsequently, a number of reports have confirmed that low-strength magnetic fields can alter pineal function both in vitro (Semm, 1983) and in vivo (see review by Olcese et al., 1988). It should be noted, however, that most of the work to date on magnetic-field effects has been carried out using low-intensity static fields. Laboratory data on the effects of higher strength AC magnetic fields are still being gathered and assessed, although early data from female rats indicate no alteration.

We have shown that chronic (21-day) exposure to ELF electric fields between approximately 2 and 40 kV/m can abolish the normal nocturnal peak in pineal melatonin in male rats first exposed to these fields at approximately 55 days of age (Wilson et al., 1981, 1983) (Fig. 3). This response appeared to have a threshold between 0.2 and 2 kV/m. At field strengths above approximately 2 kV/m there was no apparent dose response, suggesting the electric fields gave rise to an "all-or-none" effect. In the animals first exposed at 55 days of age, normal melatonin rhythms were reestablished within 3 days after cessation of ELF field exposure at 39 kV/m (Wilson et al., 1986) (Fig. 3C).

Rats exposed from conception until 23 days after birth at field strengths of either 10, 65, or 130 kV/m showed a decrease in the amplitude of the nocturnal melatonin peak, as well as an approximate 1.4-hr delay in the occurrence of peak melatonin concentration, compared to controls (Reiter

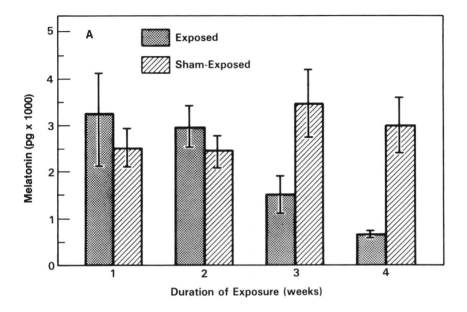

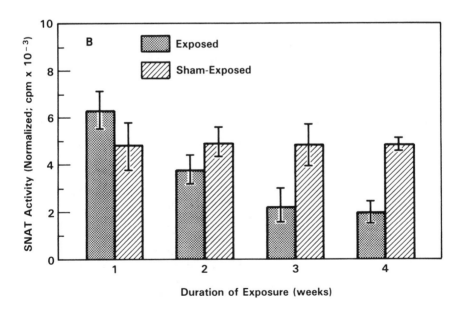

Fig. 3. **(A)** Mean nighttime pineal melatonin levels for rats exposed to 39-kV/m electric fields. Exposed animals were significantly different from controls in third week. **(B)** *N*-Acetyl serotonin activity in animals shown in A. (Figure continued on next page.)

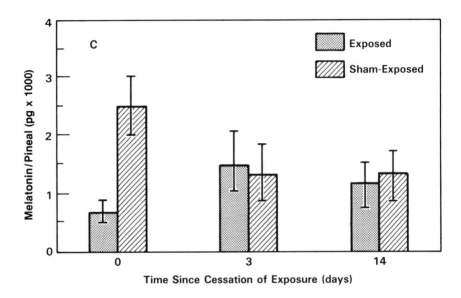

Fig. 3. (continued) **(C)** Recovery of melatonin rhythms in animals at cessation of exposure.

et al., 1988). Figure 4 shows a comparison of pineal melatonin concentration at several time points during the circadian cycle for animals exposed in utero through weaning at 23 days of age versus unexposed control animals.

Further work was done with animals first exposed at 55 days of age to determine what changes in pineal metabolism were associated with the observed reduction in melatonin. Activity of N-acetyltransferase, the rate-limiting enzyme for the production of melatonin, was shown to be depressed after about 3 weeks of exposure (see Fig. 3B). The day:night rhythm in serotonin was abolished, resulting in higher-than-normal nighttime levels of pineal serotonin. Consistent with the reduced NAT activity were decreased levels of its immediate metabolite, N-acetyl serotonin. There also appeared to be a compensatory rise in pineal 5-methoxytryptophol. Higher 5-MTOL levels were assumed to be the result of increased availability of substrate serotonin, which was metabolized via the monamine oxidase pathway (see Fig. 1). Although each of these findings is consistent with the induction of a biochemical lesion in NAT synthesis, the effect has not been studied thoroughly enough to give an indication of what changes in neuronal function may be involved.

Several factors appear to influence rat sensitivity to electric fields. There appear to be marked strain differences, for example. Males appear

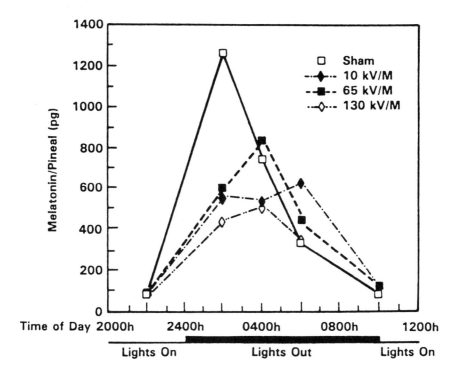

Fig. 4. Effect of ELF electric-field exposure at 10, 65, and 130 kV/m on rats exposed in utero. Note apparent shift in acrophase of exposed animals relative to controls. (Adapted from Reiter et al., 1988.)

more sensitive to magnetic fields than do females, and nonpigmented strains appear more senstive than do pigmented strains. There are a number of studies on rats that have shown no difference in the status of ELF field-exposed animals compared to controls using paradigms that may be anticipated, on the basis of the foregoing discussion, to show an effect. Of most interest among these are the studies of Morris and Phillips (1982), who have consistently shown no difference between ELF field-exposed and control animals in cellular or whole-animal immune response to several antigens.

Evidence for ELF Field Effects on Human Pineal Function

We have presented evidence from various experiments that suggest a possible contributory role for the pineal in the etiology of several disorders, including cancer. It has also been demonstrated, in this laboratory and others, that chronic exposure to ELF fields can affect pineal gland function. The work of Wever (1968) on the effects of 10-Hz electric fields on human circadian rhythms constitutes indirect evidence for an interaction between the pineal gland and ELF fields. Although it is not our intention to argue that ELF field exposure contributes to increased cancer risk, we have conducted experiments to determine if a direct effect on pineal gland function in humans could be demonstrated as a result of exposure to electric and magnetic fields normally encountered in a residential environment.

Domestic electric blanket use occurs at night when the pineal is most active, and represents a highly periodic exposure to ELF fields. Use of these blankets, however, does not require alteration in the normal lifestyle or daily routine of the subjects. Electric and magnetic fields associated with the blankets are readily measurable, especially in comparison to other domestic sources of ELF fields. We therefore designed and executed a study to determine if domestic ELF electric- and magnetic-field exposure, as represented by the use of standard or modified electric blankets, could affect human pineal gland function. Preliminary findings from this study were reported by Wilson et al. (1988) and are only summarized here.

Our primary measure for determination of effect on pineal gland function was to monitor, by radioimmune assay, the nighttime excretion of urinary 6-OHMS. Two types of blankets were used in the study: conventional (snap safety switch), and continuous polymer wire (CPW). The latter had a shorter duty (on-off) cycle and exhibited an approximately 50% higher magnetic field, compared to the former.

Because of the wide variation in 6-0HMS excretion among individuals (Bojkowski and Arendt, 1988), the study was designed so that the volunteers served as their own controls. When the data were considered by individuals, seven subjects from the CPW groups exhibited statistically significant differences in their mean 6-OHMS excretion between at least two of the last three exposure periods. One individual, for example, showed a response pattern typical of that for the overall CPW user population; that is, no change in 6-OHMS excretion between preexposure and exposure period 2 (week 1–5), followed by a significant decrease in exposure period 3 (week 6–10), and a rebound to higher values at the cessation of exposure (Fig. 5). Figure 6 plots raw data of the nighttime melatonin excretion values for study subject 2045. Note the rapid increase in 6-OHMS excretion immediately after cessation of exposure. No statistically significant changes in 6-OHMS excretion occurred in the conventional blanket users among any of the exposure periods on either an individual or a group basis.

Those specific electric- or magnetic-field characteristics of CPW blankets that were responsible for the apparent changes in 6-OHMS excretion have not been determined. There was little difference in the magnitudes of the electric fields generated by the conventional AC as compared

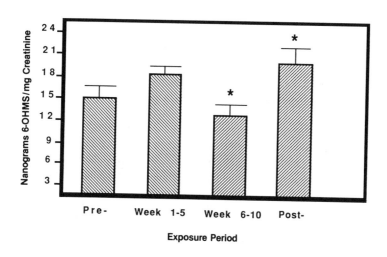

Fig. 5. Overnight excretion of 6-hydroxy melatonin sulfate in female volunteer (#2055) before, during, and after use of a continuous polymer wire (CPW) electric blanket. Values with asterisk (*) are significantly different from those for previous exposure period.

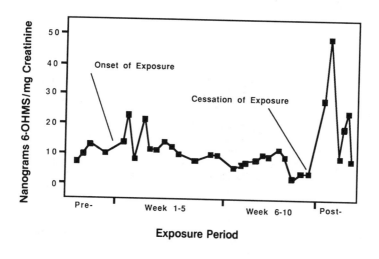

Fig. 6. Overnight excretion of 6-hydroxy melatonin sulfate for female volunteer #2045. Note substantial increase in urinary 6-OHMS after cessation of exposure.

to the CPW AC blankets. Magnetic-field strength, however, was as much as 50% higher in the CPW blankets. Graham et al. (1988) observed an enhancement of electric-field effect on humans when the field was rapidly switched on and off; CPW blankets switched on and off at approximately twice the rate of the conventional units.

Electric blanket use may constitute a special case in residential exposure. Both electric and magnetic fields under electric blankets are among the highest encountered in the home. Exposure generally occurs during the night when the pineal gland is active. Electric blanket use is periodic and may represent 50 to 60 changes in field exposure state per hour. Figure 7 is a graph of magnetic-field exposure during a 24-hr period, including time spent under an electric blanket from approximately 10:00 P.M. to 7:00 A.M., and serves as a graphic example of the contribution of these appliances to daily magnetic-field exposure. The fact that the conventional blanket users showed no response to their blankets in this study might indicate that reduction of fields or changes in duty cycle associated with these blankets may reduce or eliminate their effect on pineal function.

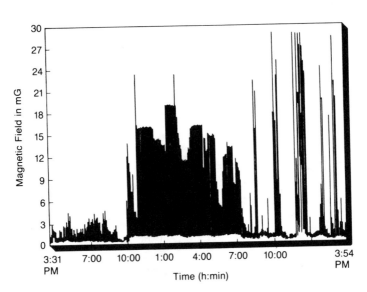

Fig. 7. Magnetic-field exposure for an 8-year-old child shows a large increase in field level from electric blanket use at night. "Spikes" during daytime are mostly from exposures received traveling to and from school.

[Reprinted with permission from Sussman, S. 1987. EMDEX (electric and magnetic field exposure) system. *Tech. Brief RP799–16*. Palo Alto, California: Electric Power Research Institute.]

POSSIBLE MECHANISMS

It now appears very likely that ELF electric and magnetic fields can alter pineal gland function by decreasing melatonin synthesis and release. On the level of the intact organism, there are a number of possible mechanisms for the pineal function response to ELF electromagnetic fields. Several of these have been discussed previously (Wilson et al., 1981). Although NAT enzyme activity is an apparent site of action for the fields, it is likely that more careful measurements will show that the requisite (β-adrenergic) neuronal input for stimulation of NAT activity is altered or missing in exposed animals.

It is possible that the alteration in pineal gland function observed in rats, as well as in certain humans, is a manifestation of stress. Animals placed in electric fields of approximately 39 kV/m or more show clear indications of stress resulting from such exposure. Leung et al. (1988b), for example, have shown that the occurrence and severity of a porphyrin-containing exudate from the Harderian gland of rats correlates with the level of field exposure. Occurrence of the exudate, termed chromodacryorrhea, is an indicator of stress in several rodents, including rats and hamsters. Leung has also reported that ELF exposure at 39 kV/m reduces or eliminates the change in circulating prolactin levels that normally occurs in response to the stress associated with restraining the animal. Thus, the ability of the ELF field-exposed animal to respond appropriately to an additional stressor is impaired.

Considering the location of the pineal gland, both in rats and in humans, it seems unlikely that ELF electromagnetic fields act directly to produce the functional changes that have been observed. However, it should be noted that cultured pinealocytes (i.e., outside the skull) do appear to respond to changes in the local magnetic field (Welker et al., 1983). In pigeons and rodents, magnetic fields are detected by receptors in or near the eyes. Responses to changes in magnetic field can be detected by changes in melatonin content in both the retina and pineal glands of these animals (Olcese et al., 1988). Thus, the possibility of subconscious retinal sensing of ELF fields in humans cannot be ruled out until experiments analogous to those described by Olcese et al. (1988) are performed in humans.

Another possible mechanism is one that we term ganglionic coupling, wherein the ELF field interacts with the nervous system directly to alter aspects of the natural electrical information that is transmitted between neurons. The work of Bowers et al. (1984), cited earlier, suggests that very subtle changes in neural signaling may affect pineal function.

Possible Consequences of ELF-Induced Alterations in Pineal Gland Function

In a report examining possible ELF field effects from the Sanguine/ Seafarer project (see Chapter 1, this volume), the author stated that ELF exposure appeared to act as an agent of stress on the organism and seemed to lower the organism's resistance. This summary of ELF effect from some two decades ago would still seem to be appropriate in describing those findings of the most recent studies which show effects.

In the discussion that follows, we speculate on the ways in which observed pineal effects may be manifested as stress. ELF electric and magnetic fields may be thought of as agents or stimuli from the environment that call for or elicit some adaptive response on the part of the organism. There are almost certainly thresholds for the response, which may be in terms of field strength, flux density, orientation, waveform, rate of change, polarization, or other characteristics. Which, if any, of these characteristics call for an adaptive response may be dependent on the neurological, immune, and endocrine state of the exposed organism or individual. The general response pattern of the CPW blanket volunteers described earlier, and particularly that of the subject shown in Fig. 6, can be used to illustrate how this adaptation process may proceed.

At initial presentation of the field, the subject showed a slight increase in melatonin metabolite excretion. We view this as an initial response on the part of the pineal to a detectable change in the environment that may be termed a stressor. This initial increase was then extinguished as exposure to the stimulation or stress represented by nightly exposure to the field continued. For several individuals in the study, but not for the population in general, 6-OHMS excretion was lower after 4 to 5 weeks of exposure than during preexposure baseline. It appears that the longer term exposure antagonized melatonin production. If allowed to continue, such suppression of melatonin may put the exposured individual at a disadvantage in responding to further stimuli requiring adaption. When the exposure ceased, 6-OHMS levels increased significantly, indicating the removal of the antagonistic "agent." Whether or not this adaptive response constitutes any significant challenge in homeostasis for humans is unclear. It can be imagined, however, that the presence of the field may act as a stimulus or additional stressor agent for individuals with compromised neurological or immune function. Its effect may be manifested in insufficient immune system performance or by changes in endocrine status or mood and affect. These changes may only occur or be observable in certain individuals. Any untoward effects resulting from ELF exposure are likely to result from long-term rather than acute exposure.

CONCLUSIONS

It remains to be determined whether ELF field effects on pineal gland function represent a health risk to humans. The pineal gland and melatonin affect a variety of physiological functions, and it may, therefore, be coincidental that the three main areas of concern that have emerged from epidemiological studies (emotional depression, miscarriages, and cancer) appear to be associated with changes in physiological function that can be directly affected by melatonin. Indeed, removal of the pineal gland in human adults, although a rare occurrence, has not been reported to result in significant adverse affects to the patient. Nonetheless, it should be considered that pineal rhythms are generally strong and reproducible on an individual basis. They are conserved in most of the animals studied, even in the face of severe tryptophan restriction. It is therefore unlikely that they are of no importance to the organism.

Because of melatonin's activity relative to the etiology of cancer, the pineal organ has been fairly termed an "oncostatic gland." Evidence from several sources suggests that ELF electric-field exposure may affect pineal function. Perturbations in the melatonin rhythm have been reported in a number of disease states. These include colonic cancers, prostate adenocarcinomas, and ovarian cancers as well as ER-positive breast cancers. Hence, data such as those indicating increased susceptibility to chemically induced mammary tumors demonstrated in animals with compromised pineal function, and the increased number of tumors per tumor-bearing animal after exposure to ELF electric fields, suggest that further work in the area is needed.

It is now important to determine the relative importance of the electric and magnetic components of the fields in altering pineal function. The effects of changes in variables such as exposure duration, magnitude, waveforms, and relative phase of the fields should be studied. Initial data suggest that these effects are manifest in some individuals and not in others. Further, the physiological significance of these effects is not well understood. Research is also required to determine what factors, including possible genetic and environmental influences, may predispose certain individuals to exhibit an endocrine response to ELF field exposure.

Galen is credited with the observation, in the second century B. C., that melancholy women seemed at greater risk of cancer than women with sanguine dispositions. Galen also left us with the first recorded description of the pineal gland. We do not know if Galen associated pineal function with the occurrence of cancer. Nevertheless, two millenia after he recorded them, his observations would appear to be relevant to a central question regarding the possible health effects of ELF field exposure.

ACKNOWLEDGMENTS

Preparation of this manuscript was supported by the U.S. Department of Energy (DOE) Office of Energy Conservation and Storage. Work on electric-field effects in small animals is supported by U.S. DOE contract DE/AC06–76RLO 1830 with Pacific Northwest Laboratory. The Electric

Power Research Institute also supported portions of the work carried out at Battelle, Pacific Northwest Laboratories and reviewed in this chapter.

REFERENCES

Adey, W. R., Bors, E., Porter, R. W. 1968. EEG sleep patterns after high cervical lesions in man. *Arch. Neurol.* 19:377–383.

Altschule, M. D. 1957. Some effects of aqueous extracts of acetone-dried beef pineal substance in chronic schizophrenia. *N. Engl. J. Med.* 257:919–924.

Anton-Tay, F. 1974. Melatonin – effects on brain function. *Adv. Biochem. Psychopharmacol.* 11:315–324.

Anton-Tay, F., Wurtman, R. J. 1969. Regional uptake of [³H]melatonin in endocrine and nervous tissue and the effects of constant light exposure. *J. Pharmacol. Exp. Ther.* 143:314–318.

Arendt, J. 1978. Melatonin assays in body fluids. *J. Neural Transm. (Suppl.)* 13:265–278.

Arendt, J., Bojkowski, E., Franey, C., Wright, J., Marks, V. 1985. Immunoassay of 6-hydroxy melatonin sulphate in human plasma and urine: abolition of the urinary 24-hr rhythms with atenolol. *J. Clin. Endocrinol. Metab.* 60:1166–1173.

Asanova, T. P., Rakov, A. I. 1966. The state of health of persons working in electric field of outdoor 400- and 500-kV switch yards. In: *Hygiene of Labor and Professional Diseases*, Vol. 5, Knickerbocker, G., trans. IEEE Power Engineering Society Spec. Pub. 10, 1975. Piscataway, New Jersey: Institute of Electrical and Electronics Engineers.

Axelrod, J., Weissbach, W. 1960. Enzymatic O-methylation of N-acetylserotonin to melatonin. *Science* 131:1312.

Axelrod, J., Wurtman, R. J., Snyder, S. H. 1965. Control of hydroxy indole-o-methyl transferase activity in the rat pineal gland by environmental lighting. *J. Biol. Chem.* 240:949–955.

Axelrod, J., McLean, O. D., Albers, R. W., Weissbach, H. 1961. In: *Regional Neurochemistry*, Kety, S. S., Elkes, J., eds., pp. 307–311. Oxford: Pergamon Press.

Barchus, J., DaCosta, F., Spector, S. 1967. Acute pharmacology of melatonin. *Nature (London)* 214:919–920.

Bartsch, C., Bartsch, G., Jain, A. K., Laumas, K. R., Wetterberg, L. 1981. Urinary melatonin levels in human breast cancer patients. *J. Neural Transm.* 52:281.

Bartsch, C., Bartsch, H., Fluchter, S. H., Attanasio, Das Gupta, D. 1988. Evidence for a modulation of melatonin secretion in men with benign and malignant tumors of the prostate: relationship with pituitary hormones. *J. Pineal Res.* 2:121.

Beck-Friis, J., Kjellman, B. F., Aperia, B., Unden, F., Von Rosen, D., Ljunggren, J.-G., Wetterberg, L. 1985. Serum melatonin in relation to clinical variables in patients with major depressive disorder and a hypothesis of a low melatonin syndrome. *Acta Psychiatr. Scand.* 71(4):319–330.

Bigelow, L. B. 1974. Effects of aqueous pineal extract in chronic schizophrenia. *Biol. Psychol.* 8:5–15.

Bindoni, M., Raffaele, R. 1968. Mitotic activity of the adenohypophysis of rats after pinealectomy. *J. Endocrinol.* 41:451–452.

Blask, D., Hill, S. 1986. Effects of melatonin on cancer: studies on MCF-7 human breast cancer cells in culture. *J. Neural Transm. (Suppl.)* 21:433–449.

Blask, D. E., Hill, S. M., Pelletier, D. B. 1988. Oncostatic signaling by the pineal gland and melatonin in the control of breast cancer. In: *The Pineal Gland and Cancer*, Das Gupta, D., Attanasio, A., Reiter, R. J., eds., pp. 195–206. London: Brain Research Promotion.

Bojkowski, C. J., Arendt, J. 1988. Annual changes in 6-sulfatoxymelatonin excretion in man. *Acta Endocrinol. Copenh.* 117:470-476.

Bostelmann, W., Gocke, H., Ernst, B., Tesmann, D. 1971. Der Einfluss einer Melatonin Behandlung auf des wachstum des Walker Carcinosarcoma der Ratte. *Z. Allg. Pathol.* 114:289.

Bowers, C. W., Baldwin, C., Zigmond, R. E. 1984. Sympathetic reinnervation of the pineal gland after postganglionic nerve lesion does not restore normal pineal function. *J. Neurosci.* 4:2010-2015.

Branchey, L., Weinberg, U., Branchey, M., Linkowski, P., Mendlewicz, J. 1982. Simultaneous study of 24-hour patterns of melatonin and cortisol secretion in depressed patients. *Neuropsychobiology* 8(5):225–232.

Brown, R. P., Kocsis, J. H., Caroff, S., Amsterdam, J., Winokur, A., Stokes, P., Frazer, A. 1987. Depressed mood and reality disturbance correlate with decreased nocturnal melatonin in depressed patients. *Acta Psychiatr. Scand.* 76:272-275.

Buzzell, G. R., Amerongon, H. M., Toma, J. G. 1988. Melatonin and the growth of Dunning R3327 rat prostatic adenocarcinoma. In: *The Pineal Gland and Cancer*, Das Gupta, T. K., Attanasio, A., Reiter, R. J., eds. London: Brain Research Promotion.

Cardinali, D. P. 1974. Melatonin and the endocrine role of the pineal gland. In: *Current Topics Experimental Endocrinology, Vol. 2*, James, V. H. T., Martini, L., eds., pp. 107–128. New York: Academic Press.

Cardinali, D. P., Rosner, J. J. 1971. Retinal localization of HIOMT in the rat. *Endocrinology* 89:301–310.

Chamblin, M., Drew, W. G. 1971. The effects of lights-off stimulation on the circadian distribution of REM sleep in the cat. *Commun. Behav. Biol.* 6:11.

Claustrat, B., Chazot, G., Brun, J., Jordan, D., Sassolas, G. 1984. A chronobiological study of melatonin and cortisol secretion in depressed subjects. Plasma melatonin, a biochemical marker in major depression. *Biol. Psychiatry* 19(8):1212–1228.

Cohen, M., Lippman, M., Chabner, B. 1978. Role of pineal gland in etiology and treatment of breast cancer. *Lancet* 2:814–816.

Cotman, C. W., Brinton, R. E., Galaburda, A., McEwen, B., Schneider, D. M., eds. 1987. *The Neuro-Immune-Endocrine Connection*. New York: Raven Press.

Cramer, H., Rudolf, H., Consbruch, U., Kendel, K. 1974. On the effects of melatonin in sleep and behavior in man. *Adv. Biochem. Psychopharmacol.* 11:187–191.

Danforth, D., Tamarkin, L., Chabner, B., Demoss, E., Lichter, A., Lippman, M. 1982. Altered diurnal secretory pattern of melatonin in patients with estrogen receptor-positive breast cancer. *Clin. Res.* 30(2):532A.

Das Gupta, T. K., Terz, J. 1967. Influence of the pineal gland on growth and spread of melanoma in the hamster. *Cancer Res.* 27:1306.

Dubbels, R., Khoory, R. 1986. Circannual changes in melatonin excretion in an Antarctic station. In: *Melatonin in Humans, J. Neural Transm. (Suppl.)* 21:483–484, Wurtman, R. J., Waldhauser, F., eds. New York: Springer-Verlag.

Ebadi, M., Govitrapong, P. 1986. Neural pathways and neurotransmitters affecting melatonin synthesis. *J. Neural Transm. (Suppl.)* 21:125–155.

Ehret, C. F., Rosenberg, R. S., Sacher, G. A., Duffy, P. H., Groh, K. R., Russell, R. J. 1980. *Biomedical Effects Associated with Energy Transmission Systems: Effects of 60-Hz Electric Fields and Circadian and Ultradian Physiological and Behavioral Function in Small Rodents: Annual Report.* Washington, D.C.: U.S. Department of Energy, Division of Electric Energy Systems.

El-Domeiri, A. A. H., Das Gupta, T. K. 1973. Reversal by melatonin of the effect of pinealectomy on tumor growth. *Cancer Res.* 33:2280.

Elred, S. H., Bell, N. W., Sherman, L. J. 1961. A pilot study comparing the effects of pineal extract and a placebo in patients with chronic schizophrenia. *N. Engl. J. Med.* 263:1330-1332.

Fraschini, F., Martini, L. 1970. Rhythmic phenomena and pineal principles. In: *The Hypothalamus*, Martini, L., Fraschini, F., Motta, M. eds., pp. 529–549. New York: Academic Press.

Gavalas-Medici, R. J., Magadaleno, S. R. 1976. *An Evaluation of Possible Effects of 45- and 60-Hz and 75-Hz Electric Fields in Neurophysiology and Behavior of Monkeys. Phase I: Continuous Wave.* ONT Tech. Rep. ADA-008404/6ST. Springfield, Virginia: National Technical Information Service.

Graham, C., Cohen, H. D., Cook, M. R., Gerkovich, M. M., Phelps, J. W., Riffle, D. W. 1988. Effects of intermittent exposure to 60-Hz fields on human cardiac activity. In: *Proceedings of the 10th Annual Bioelectromagnetics Society Meeting*, Stamford, Connecticut, June 1988. Frederick, Maryland: Bioelectromagnetics Society.

Graham, C., Cohen, H. D., Cook, M. R., Phelps, J. W., Gerkovich, M. M., Fotopoulos, S. S. 1984. A double blind evaluation of 60-Hz field effects on human performance physiology, and subjective state. In: *Interaction of Biological Systems with Static and ELF Electric and Magnetic Fields.* CONF 841041. Springfield, Virginia: National Technical Information Service.

Greiner, A. C. 1970. Schizophrenia and the pineal gland. *Postgrad. Med.* Aug:111-118.

Hariharasubramanian, N., Nair, N. P. V., Pilapil, C. 1985. Circadian rhythm of plasma melatonin and cortisol during the menstrual cycle in the pineal gland: endocrine aspects. In: *Advances in the Biosciences, Vol. 53*, Brown, G. M., Wainwright, S. D., eds. Oxford: Pergamon Press.

Heubner, O. 1898. Tumor der glandula pinealis. *Dtsch. Med. Wochenschr.* 24: 214-215.

Hishikawa, Y., Cramer, H., Kuhlo, W. 1969. Natural and melatonin-induced sleep in young chickens–a behavioral and electrographic study. *Exp. Brain Res.* 7:84-95.

Hjeresen, D. L., Kaune, W. T., Decker, J. R., Phillips, R. D. 1980. Effects of 60-Hz electric fields on avoidance behavior and activity in rats. *Bioelectromagnetics* 1:299-312.

Jaffe, R. A., Laszewski, B. W., Carr, D. B., Phillips, R. D. 1980. Chronic exposure to 60-Hz electric field: effects on synaptic transmission and peripheral nerve functions in the rat. *Bioelectromagnetics* 1:113-148.

Kamberi, I. A., Mical, R. S., Porter, J. C. 1971. Effects of melatonin and serotonin on the release of FSH and prolactin. *Endocrinology* 88:1288-1293.

Klein, D. C., Weller, J. L. 1971. Rapid light-induced decrease in pineal serotonin N-acetyltransferase activity. *Science* 177:532-533.

Kneisley, L. W., Maskowitz, M. A., Lynch, H. J. 1978. Cervical spinal lesions disrupt the rhythm in human melatonin secretion. *J. Neural Transm. (Suppl.)* 13:311–323.

Koslow, S. H. 1976. The biochemical and behavioral profile of 5-methoxytryptamine. In: *Trace Amines and the Brain*, Usden, E., Sandler, M., eds., pp. 103–130. Oxford: Pergamon Press.

Krause, K. G., Cremer-Bartels, Mitoskas, G. 1985. Effects of low magnetic field on human and avian retina. In: *The Pineal Gland: Endocrine Aspects*, Brown, G. M., Wainwright, S. D., eds. Oxford: Pergamon Press.

Leone, A. M., Skene, D. 1989. Melatonin rhythms and tumor development: potential influence of electric and magnetic fields. In: *Proceedings, Electromagnetic Fields and Circadian Rhythmicity*, Campbell, S. S., Moore, M. C., eds., Boston, September 2–3, 1989. Institute of Circadian Physiology. Cambridge: Harvard University Press (in press).

Lerner, A. B., Case, J. D., Heinzelman, R. V. 1959. Structure of melatonin. *J. Am. Chem. Soc.* 81:6085.

Lerner, A. B., Case, J. D., Takahashi, Y., Lee, T. H., Mori, W. 1958. Isolation of melatonin, the pineal gland factor that lightens melanocytes. *J. Am. Chem. Soc.* 80:2587.

Leung, F. C., Rommereim, D. N., Miller, R. A., Anderson, L. E. 1988a. Experimental observations in rats exposed to 60-Hz electric fields. In: *Proceedings of the 10th Annual Bioelectromagnetics Society Meeting*, Stamford, Connecticut, June 1988. Frederick, Maryland: Bioelectromagnetics Society.

Leung, F. C., Rommereim, D. N., Stevens, R. G., Wilson, B. W., Buschbom, R. L., Anderson, L. E. 1988b. Effects of electric fields on rat mammary tumor development induced by 7,12-dimethylbenz(a)anthracene (DMBA). In: *Proceedings of the 10th Annual Bioelectromagnetics Society Meeting*, Stamford, Connecticut, June 1988. Frederick, Maryland: Bioelectromagnetics Society.

Lewy, A. J., Markey, S. P. 1978. Analysis of melatonin in human plasma by gas chromatography negative chemical ionization mass spectrometry. *Science* 201:741–742.

Lewy, A. J., Kern, H. A., Rosenthal, N. E., Wehr, T. A. 1982. Bright artificial light suppresses melatonin secretion in humans. *Science* 210:1267–1269.

Maestroni, G. J. M., Conti, A., Pierpaoli, W. 1986. Role of the pineal gland in immunity circadian synthesis and release of melatonin modulates the antibody response and antagonizer immunosuppressive effect of corticosterone. *J. Neuroimmunol.* 13:19–30.

McCord, M. P., Allen, F. P. 1917. Evidence associating pineal gland function with alteration in pigmentation. *J. Exp. Zool.* 23:207–224.

McIsaac, W. M. 1961. A biochemical concept of mental disease. *Postgrad. Med.* Aug:111–118.

McIsaac, W. M., Farrell, G., Taborsky, R. G., Taylor, A. N. 1965. Indole compounds: isolation from pineal tissue. *Science* 145:102–103.

Mendlewicz, J., Branchey, L., Weinberg, U., Branchey, M., Linkowski, P., Weitzman, E. D. 1980. The 24-hour patterns of plasma melatonin in depressed patients before and after treatment. *Commun. Psychopharmacol.* 1:49–56.

Moller, M., Mikklesen, J. D. 1988. The innervation of the mammalian pineal gland complex studied by retro- and antegrade neuronal tracing as well as by immunohistochemistry. *Chin. J. Physiol. Sci.* 4:257.

Moore, R. Y., Heller, A., Bhatnager, R. K., Wurtman, R. J., Axelrod, J. 1968. Central control of the pineal gland: visual pathways. *Arch. Neurol.* 18:208–218.

Moore-Ede, M. C., Sulzman, F. M., Fuler, C. A. 1985. *The Clocks That Time Us: Physiology of the Circadian Timing System.* Cambridge: Harvard University Press.

Morris, J. E., Phillips, D. D. 1982. Effects of 60-Hz electric fields on specific humoral and cellular components of the immune system. *Bioelectromagnetics* 3:341–347.

Mouret, J., Coindet, J., Chouvet, G. 1974. Effet de la pinealectomie sur les etats et rhythmes de sommeil du rat male. *Brain Res.* 81:97–105.

Murphy, D. L., Garrick, N. A., Tamarkin, L., Taylor, P. L., Markey, S. D. 1986. Effects of antidepressants and other psychotropic drugs on melatonin release and pineal gland function. *J. Neural Transm. (Suppl.)* 21:291–310.

Nakatani, M., Ohara, Y., Katagiri, E., Nakano, K. 1940. Studien über die zirbellosen weiblichen weissen Ratten (original in Japanese). *Nippon Butsuri Gakkaishi* 30:323–236.

Nir, I. 1978. Non-reproductive systems and the pineal gland. *J. Neural Transm. (Suppl.)* 13:225–244.

O'Connor, M. E., Lovely, R. H., eds. 1988. *Electromagnetic Fields and Neurobehavioral Function. Progress in Clinical and Biological Research, Vol. 257.* New York: Alan R. Liss.

Olcese, J., Reuss, S., Semm, P. 1988. Geomagnetic field detection in rodents. *Life Sci.* 42:605–613.

Orsi, L., Denari, J. H., Nagle, C. A., Cardinali, D. P., Rossner, R. J. 1973. Effects of melatonin on the synthesis of proteins by the rat hypothalamus, hypophysis, and pineal organ. *J. Endocrinol.* 58:131–132.

Papke, R. L., Podleski, T. R., Oswald, R. E. 1986. Effects of pineal factors on the action potentials of sympathetic neurons. *Cell Mol. Biol.* 6:381–395.

Pelham, R. W., Vaughan, G. M., Sandock, K. L., Vaughan, M. K. 1973. Twenty-four-hour cycle of a melatonin-like substance in the plasma of human males. *J. Clin. Endocrinol. Metab.* 37:341–344.

Pierpaoli, W., Maestroni, G. J. M. 1987. Melatonin: A principal neuroimmunoregulatory and antistress hormone: its antiaging effects. *Immunol. Lett.* 16:355–362.

Philo, R., Berkowitz, A. S. 1988. Inhibition of Dunning growth by melatonin. *J. Urol.* 139:1099–1102.

Rechtschaffen, A., Dates, R., Tobias, M., Whitehead, W. E. 1969. The effects of lights-off stimulation on the distribution of paradoxical sleep in the rat. *Commun. Behav. Biol. A* 3:93–99.

Regelson, W., Pierpaoli, W. 1987. Melatonin, a rediscovered antitumor hormone. Its relation to surface receptors in sex steroid metabolism. Immunologic response and chronobiologic factors in tumor growth and therapy. *Cancer Invest.* 5(4):379–385.

Reiter, R. J., ed. 1981. *The Pineal Gland, Vols. I, II, III.* Boca Raton, Florida: CRC Press.

Reiter, R. J., Anderson, L. E., Buschbom, R. L., Wilson, B. W. 1988. Reduction of the nocturnal rise in pineal melatonin levels in rats exposed to 60-Hz electric fields in utero and for 23 days after birth. *Life Sci.* 42:2203–2206.

Relkin, R., Schreck, L. 1975. Effects of pinealectomy on rat brain myelin. *Proc. Soc. Exp. Biol. Med.* 148:337–338.

Relkin, R., Fok, W. Y., Schreck, L. 1973. Pinealectomy and brain myelination. *Endocrinology* 92:1427–1428.

Rodin, A. E. 1963. The growth and spread of wallar 256 carcinoma in pinealectomized rats. *Cancer Res.* 23:1545.

Semm, P. 1983. Neurobiological investigations on the magnetic sensitivity of the pineal gland in rodents and pigeons. *Comp. Biochem. Physiol.* 76:683–689.

Shah, P. N., Mhatre, M. C., Kothari, L. S. 1984. Effect of melatonin on mammary carcinogenesis in intact and pinealectomized rats in varying photoperiods. *Cancer Res.* 44:3403–3407.

Shandala, M. G., Rudnev, M. I., Varetsky, V. V., Lysina, G. I., Vasilevsky, N. 1984. National literature survey on the effects of electromagnetic fields (0–300 GHz) on the nervous system. In: *Proceedings of the U. S./U. S. S. R. Workshop on Nervous System Effects of Electromagnetic Waves, Vol. 1, U. S. and U. S. S. R. Literature Overviews.* Washington, DC: National Institute of Environmental Health Sciences.

Shaw, F., Camps, F., Eccleston, E. 1967. 5-Hydroxytryptamine in the hindbrain of depressive suicides. *Br. J. Psychiatry* 113:1407–1411.

Sheppard, A. R., Eisenbud, M. 1977. *Biological Effects of Electric and Magnetic Fields of Extremely Low Frequency.* New York: University Press. (See also references cited therein.)

Smith, I., Mullen, P. E., Silman, R. E., Snedden W., and Wilson, B. W. 1976. Absolute identification of melatonin in human plasma and cerebrospinal fluid. *Nature (London)* 260:718–719.

Stevens, R. G. 1987. Electric power use and breast cancer, a hypothesis. *Am. J. Epidemiol.* 125:556–561.

Su, T.-S., London, E. D., Jaffe, J. H. 1988. Steroid binding at J receptors suggests a link between endocrine, nervous, and immune systems. *Science* 240:219–231.

Sussman, S. 1987. EMDEX (electric and magnetic field exposure) system. *Tech. Brief RP799–16.* Palo Alto, California: Electric Power Research Institute.

Tamarkin, L. C., Roselle, D., Reichart, C., Lippman, M., Chabner, B. 1981. Melatonin inhibition and pinealectomy enhancement of 7,12-dimethylbenz(a)-anthracene-induced mammary tumors in the rat. *Cancer Res.* 41:4432.

Tamarkin, L., Danforth, D., Lichter, A., Demoss, R., Cohen, M., Chabner, B., Lippman, M. 1982. Decreased nocturnal plasma melatonin peak in patients with estrogen receptor-positive breast cancer. *Science* 216:1003–1005.

Tapp, E., Huxley, M. 1972. The histological appearance of the human pineal gland from puberty to old age. *J. Pathol.* 103:137–144.

Vlahakes, G., Wurtman, R. J. 1972. A Mf^{++}-dependent hydroxy indole-o-methyltransferase in rat Harderian gland. *Biochim. Biophys. Acta* 61:194–198.

Waldhauser, F., Steger, H. 1986. Changes in melatonin secretion with age and pubescence. *J. Neural Transm.* 21:183–197.

Welker, H. A., Semm, P., Willig, R. P., Commentz, J. C., Wiltschko, W., Vollrath, L. 1983. Effects of an artificial magnetic field on serotonin N-acetyltransferase and melatonin content of the rat pineal gland. *Exp. Brain Res.* 50:426–432.

Wever, R. 1968. Einfluss schawcher electromagnetischer Felder auf die circadiane Periodik des Menschen. *Naturwissenschaften* 55:29–32.

Wilson, B. W. 1988. Chronic exposure to ELF fields may induce depression. *Bioelectromagnetics* 9:195–205.

Wilson, B. W., Snedden, W. 1979. Capillary column GCMS detection of 5-methoxy-tryptamine in human plasma using selected ion monitoring. J. Neurochem. 33:939–941.

Wilson, B. W., Chess, E. K., Anderson, L. E. 1986. 60-Hz electric field effects on pineal melatonin rhythms: time course for onset and recovery. Bioelectromagnetics 7:239–242.

Wilson, B. W., Lynch, H. J., Ozaki, Y. 1978. 5-Methoxytryptophol in rat serum and pineal: detection, quantitation, and evidence for daily rhythmicity. Life Sci. 23:1019–1024.

Wilson, B. W., Anderson, L. E., Hilton, D. I., Phillips, R. D. 1981. Chronic exposure to 60-Hz fields: effects on pineal function in the rat. Bioelectromagnetics 2:371–380.

Wilson, B. W., Anderson, L. E., Hilton, D. I., Phillips, R. D. 1983. Chronic exposure to 60-Hz fields: effects on pineal function in the rat (erratum). Bioelectromagnetics 4:293.

Wilson, B. W., Wright, C. W., Morris, J. A., Stevens, R. G., Anderson, L. E. 1988. Effects of electric blanket use on human pineal gland function: a preliminary report. In: Proceedings of the DOE/EPRI Contractor's Review, Phoenix, Arizona, October 30-November 3, 1988. Washington, D.C.: U.S. Dept. of Energy, Office of Energy Storage and Distribution.

Wilson, B. W., Snedden, W., Muller, P. G., Sillman, R. E., Smith, I., Landon, J. 1977. A gas chromatography-mass spectrometry method for the quantitative analysis of melatonin in plasma and cerebrospinal fluid. Anal. Biochem. 81:283–291.

Wurtman, R. J., Axelrod, J. 1965. The pineal gland. Sci. Am. 213:50-60.

Wurtman, R. J., Cardinali, D. P. 1974. The pineal organ. In: Textbook of Endocrinology, Williams, R. H., ed. Philadelphia: Saunders.

Wurtman, R. J., Axelrod, J., Barchas, J. D. 1964. Age and enzyme activity in the human pineal. J. Clin. Endocrinol. Metab. 24:299.

Wurtman, R. J., Axelrod, J., Kelly, D. E. 1968a. The Pineal. New York: Academic Press.

Wurtman, R. J., Axelrod, J., Anton-Tay, F. 1968b. Inhibition of the metabolism of [³H]melatonin by phenothiazines. J. Pharmacol. Exp. Ther. 161:367–372.

Wurtman, R. J., Axelrod, J., Phillips, C. S. 1963. Melatonin synthesis in the pineal gland controlled by light. Science 142:1071–1073.

Zisapel, N., Laudon, M., Nir, I. 1988. Melatonin receptors in discrete brain regions of mature and aged male rats: age-associated decrease in receptor density and in circadian rhythmicity. Chin. J. Physiol. Sci. 4:292.

9 ELF Effects on Calcium Homeostasis

CARL F. BLACKMAN
Health Effects Research Laboratory
U.S. Environmental Protection Agency
Research Triangle Park, North Carolina

CONTENTS

This chapter is presented from the perspective of a radiation biologist examining the interaction of electric and magnetic fields with biological tissues. The endpoint that we have used is the release of radiolabeled calcium ions from brain tissue, in vitro. Although calcium ions have many critical physiological roles, no attempt has been made to identify which particular roles are affected by exposure to ELF electric and magnetic fields. Rather, the purpose is to define those electromagnetic parameters that are important for the interaction process. Specifically, the intensity and frequency interactions, as well as other conditions that directly affect biological responses to those primary factors, are discussed. A generalized concept is offered as a point from which to view this field-induced phenomenon, for it is through such knowledge that appropriate parameters can be identified for testing. It is left for others to describe particular calcium-ion homeostatic processes that may be affected by the exposure, and, from that knowledge, to speculate on their potential physiological significance.

MATERIALS AND METHODS

Biological Preparation

The preparation used (see Blackman et al., 1982, 1985a) is one that was pioneered by Bawin et al. (1975) more than 15 years ago. The entire forebrain is removed from 1- to 7-day-old chickens, separated at the midline, placed into a physiological salt solution containing radioactive calcium, and incubated for 0.5 hr at 37°C. The tissue is removed from the solution and rinsed so that all the loosely associated calcium is removed. The tissue is then put into an identical salt solution lacking the radioactive calcium label and treated for 20 min either to a particular combination of frequency and intensity of electric and magnetic fields or to sham exposure. At the end of the 20-min treatment, the solution surrounding the brain is sampled for the amount of radioactive calcium that has come off the brain tissue. Four brain tissues are treated at one time to either an exposure or a sham. This process is repeated, alternating between exposure and sham, so that there are usually six to eight sets of exposed/sham pairs in a given test. Appropriate statistical analyses are then performed to evaluate the effect of the exposure (Blackman et al., 1985a).

This tissue preparation is not designed to elicit detailed biological information. The tissue is electrically dead when it is removed from the animal; within 1.5 min after removal, no electrical signal from it can be detected, and when electrically shocked there is no response that would indicate an electrically active tissue (Dr. William Howell, unpublished data). On the other hand, the oxygen utilization capabilities of the tissue were tested 60 min after the tissue had been removed from the animal, at the end of the normal treatment period (Blackman et al., 1981); the brain tissue was still able to metabolize oxygen quite well (Fig. 1). The zero minute in the figure corresponds to the time when the brain is placed into the chamber housing the oxygen electrode. A comparable curve, obtained 1.5 min after the brain was removed from the animal, showed similar kinetics. These data demonstrate that the brain tissue is actively metabolizing at the end of our treatment process, so it is not dead in the metabolic sense but is apparently dead as a functioning organ.

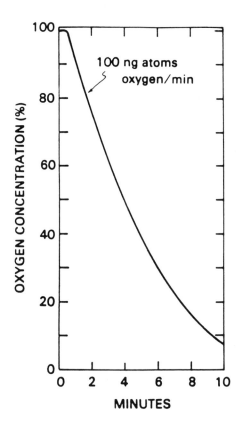

Fig. 1. Oxygen utilization by chick brain tissue, in vitro.
(Reprinted with permission from Blackman et al. 1981. Calcium ion efflux induction in brain tissue by radiofrequency radiation. In: *Biological Effects of Nonionizing Radiation*, Illinger, K.H., ed., ACS Symp. Ser. 157:299–314. Washington, DC: American Chemical Society.)

Exposure System

The exposure system used in these experiments is a Crawford cell (Fig. 2), a specially designed metal box; the outside is grounded and a metal plate in the center of the box is energized to produce an electric field and a magnetic field in the chamber between the plate and the exterior walls (Blackman et al., 1982). The brain tissue samples are in test tubes placed in the middle of the chamber. The exposure chamber is maintained at 37°C during the exposure period.

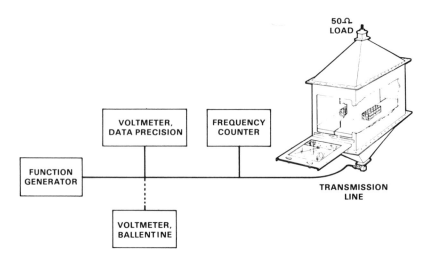

Fig. 2. ELF electromagnetic field exposure system.
(Reprinted with permission from Blackman et al. 1982. Effects of ELF fields on calcium-ion efflux from brain tissue. *Radiat. Res.* 92:510–520.)

EARLY EXPERIMENTS

The release of radioactively labeled calcium ions from the brain tissue was examined over a series of intensities using amplitude-modulated (AM) radio-frequency fields. Some unusual intensity responses were observed that were difficult to interpret because both a radio-frequency and a low-frequency component were present (Blackman et al., 1979, 1980a, 1980b, 1981; Joines and Blackman, 1980, 1981; Joines et al., 1980). We decided to use low-frequency fields directly, without the radio-frequency component, to investigate the intensity response of the brain samples (Blackman et al., 1982). Figure 3 shows the response of the brain tissue to different intensities of 16-Hz electric and magnetic fields. The abscissa is the electric-field intensity in volts peak-to-peak per meter in air (Vpp/m); the corresponding magnetic-field flux density can be approximated from the relationship, 1.6 nT/Vpp/m. The ordinate is a normalized value of the amount of calcium released from the tissue. The interrupted line is the response of tissue exposed to the field, and the solid line is the response of tissues that are sham exposed. There were null results at intensities of 1 and 2 Vpp/m, but a trend begins at 3 Vpp/m and an effect is statistically significant at 5, 6.5, and 7.5 Vpp/m. Null results begin again at 10 Vpp/m and continue through 20 and 30 Vpp/m.

To this point the dose–rate response is not unusual. However, as the dose rate is increased further, the response is unusual. A statistically significant result reappears at 35 Vpp/m and continues through 40, 45,

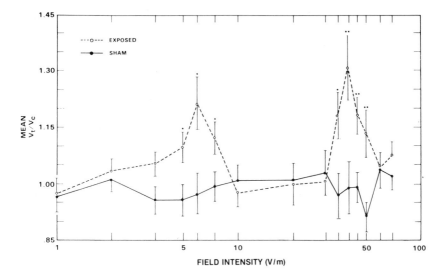

Fig. 3. Dose dependence of calcium ion release induced by 16-Hz fields. Electric-field intensities are in volts peak-to-peak per meter in air; there are corresponding magnetic fields (see text).

(Figure reprinted with permission from Blackman et al. 1982. Effects of ELF fields on calcium-ion efflux from brain tissue. *Radiat. Res.* 92:510–520.)

and 50 Vpp/m. At 55, 60, and 70 Vpp/m, there are null results again. This is a very unusual finding, analogous to the intensity responses found with radio-frequency fields using two different carrier frequencies, 147 MHz and 50 MHz, both amplitude modulated to 16 Hz (Blackman et al., 1981). Other investigators have also reported multiple-intensity responses, that is, intensity regions or windows of electromagnetic fields that produce tissue responses, separated by electromagnetic-field intensities that produce null results (Sheppard et al., 1979; Dutta et al., 1984). This response is substantially different from that expected for a toxic chemical, in that increasing amounts of the chemical would be expected to produce an effect whose magnitude either continues upward, or plateaus, and perhaps eventually falls off. The intensity response that is obtained with electric and magnetic fields is discussed further.

FREQUENCY RESPONSES

The response of the brain tissue preparation to different frequencies of electric and magnetic fields is equally puzzling. The difference in calcium release between exposed and the sham tissues at various combinations of frequency and intensity is shown in Fig. 4 (Blackman et al., 1985a). If there is no statistically significant difference between exposure and sham, that is, p is greater than .05, the result is considered null (open symbols). For those frequencies at which there is a statistically significant difference between the exposed and the sham, a closed symbol is used.

In Fig. 4, the ordinate scale on the left is in volts peak-to-peak per meter (Vpp/m), whereas the scale on the right is in volts root-mean-square per meter (V/m), a more traditional measure. It should be noted that there is a corresponding magnetic field at a flux density that can be approximated from the relationship, 4.59 nT per V/m. The 16-Hz data from Fig. 3 are also displayed in Fig. 4. At the intensities tested, null results occurred at 1, 30, and 42 Hz. The positive effects at 45 Hz span the same intensity range as the 16-Hz effects, whereas there was a null effect at 48 Hz. Because the same intensity range was effective for both 16- and 45-Hz fields, we thought the fundamental frequency was probably at one-third the 45-Hz response, or at 15 Hz instead of 16 Hz. An intensity in the middle of the effective range was selected and tested at 15 Hz; indeed, it was effective. This result led to the testing of a frequency series using that intensity (42.5

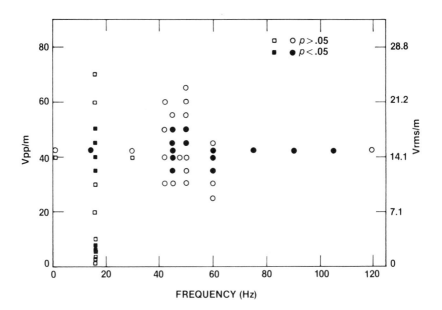

Fig. 4. Dose and frequency dependence of calcium ion release. Symbols: null, *open circle*; statistically significant, *closed circle*. Data taken from Blackman et al. (1982) are denoted by *squares*. There are corresponding magnetic fields (see text).

(Reprinted with permission from Blackman et al. 1985a. Effects of ELF (1–120 Hz) and modulated (50-Hz) RF fields on the efflux of calcium ions from brain tissue, *in vitro. Bioelectromagnetics* 6:1–11.)

Vpp/m = 15 V/m; approximately 69 nT) in 15-Hz increments; null results were found at 1, 30, and 120 Hz, and effects at 45, 60, 75, 90, and 105 Hz. Because there was an effect at 60 Hz, the fundamental frequency for the transmission of electric power in the United States, we decided to examine the neighboring intensities more carefully. We found that there is an intensity region or window in which an effect occurs and outside of which null results occur, similar to the responses at 16 and 45 Hz. Subsequently, 50-Hz fields were found to produce a similar response profile. It should be noted that electric fields of the order of 15 V/m are commonly found in houses (see Chapter 3).

Higher frequencies were tested to determine if a pattern would develop in the frequency profile (Fig. 5). In this figure, results are shown from tests every 15 Hz from 1 to 510 Hz (Blackman et al., 1988b). The dark bars indicate an effect and the open bars a null result. The data from the previous figure to 120 Hz are included. The response at 135 Hz was strongly positive. In contrast, the result at 165 Hz was null; this frequency was examined twice because we expected the pattern of effects at every odd multiple of 15 Hz. A second test again demonstrated a null result at 165 Hz. We did find positive effects at 180 Hz and at 405 Hz that were replicated 4 or 5 months later. Each of these tests takes a week to perform, and thus these repeated tests show the reproducibility of the results over time. However, no pattern was readily apparent on examination of the response profile. Instead of looking for significant variations among differences in mean values alone, we decided that an analysis of the results that included the variance for each mean value would make more complete use of the data. This exploratory reanalysis of the data was for hypothesis generation only.

A readily available parameter for this reanalysis, which includes the variance as well as the mean, is the p value obtained in the previous statistical analysis. The data are plotted in Fig. 5 using p versus frequency up to 510 Hz (Fig. 6). The data cluster with $p < .01$ occurring at 15, 45, 75, 105, and 135 Hz, that is, successive odd multiples of 15 Hz. The next odd multiple, 165 Hz, however, gives a null result but the effect reappears at the next odd multiple and repeats four more times, at 195 through 315 Hz. Hence, there are two ranges of five specific frequencies separated by constant intervals that give effects at $p < .01$. We are not yet certain of the significance of these observations, but this grouping could be the "frequency signature" of one reaction site. In addition, there are effects at 60 and 90 Hz, with a possible trend at 30 Hz. The first odd multiple of 60 Hz would be 180, where there is an effect. The second odd multiple of 60 Hz is 300 Hz, where there is a null result. These latter responses could constitute a second reaction site. The positive effect at 405 Hz is an isolated result that could constitute a third reaction site, although these suggestions are primarily speculative [see Blackman et al. (1988b) for a discussion of the rationale for these groupings]. Some additional comments on the implications of these groupings follow.

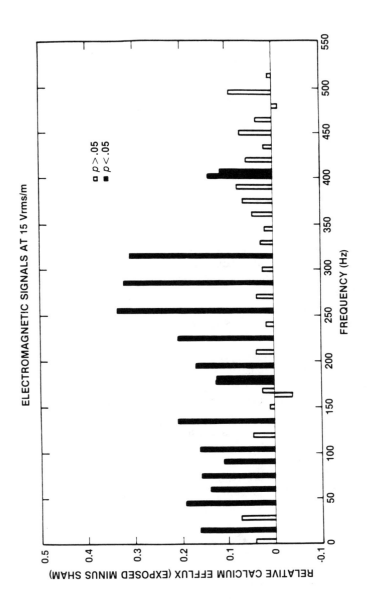

Fig. 5. Extended frequency dependence of calcium ion release (*closed bars*, effect; *open bars*, null). (Reprinted with permission from Blackman et al. 1988b. Influence of electromagnetic fields on the efflux of calcium ions from brain tissue in vitro: a three-model analysis consistent with the frequency response up to 510 Hz. *Bioelectromagnetics* 9(3):215–227.)

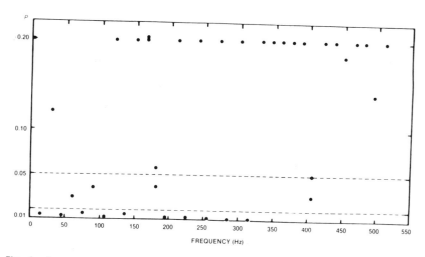

Fig. 6. Frequency dependence of calcium ion release displayed as *p* values. (Reprinted with permission from Blackman et al. 1988b. Influence of electromagnetic fields on the efflux of calcium ions from brain tissue in vitro: a three-model analysis consistent with the frequency response up to 510 Hz. *Bioelectromagnetics* 9(3):215–227.)

Are results reported in the literature that demonstrate a frequency response in another tissue similar to those we observed? Lyle et al. (1983) reported the effects of electromagnetic radiation on the cytotoxic activity of lymphocytes, the ability of the cells to engulf and kill target cells. Lymphocytes are, of course, nonexcitable cells, whereas brain tissue is excitable. It might be concluded that there are various types of reaction sites in excitable tissue and fewer types in nonexcitable tissue. Lyle's data, obtained using a 450-MHz carrier wave amplitude modulated at various frequencies, are shown in Fig. 7. Recently, Dutta et al. (1989) have shown that a 147-MHz carrier wave, amplitude modulated in two frequency regions around 16 Hz and 57.5 Hz, can cause the enhanced release of calcium ions from neuroblastoma cells in culture. These peaks in cytotoxic response at 60 Hz and in the release of calcium ions at 16 and 57.5 Hz appear to be consistent with some of our data.

On the basis of these response groupings, it is possible that there are three separate transduction mechanisms revealed in the frequency response (details follow). Alternative hypotheses have been proposed for some frequencies by other investigators (see Chapters 10 and 11).

INFLUENCE OF THE LOCAL GEOMAGNETIC FIELD

One of our concerns was the distinction between effects caused by the magnetic field and those caused by the electric field. In the Crawford cell

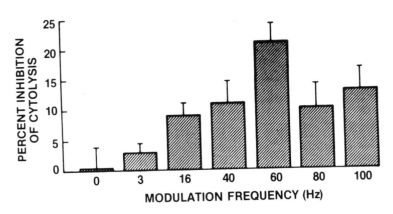

Fig. 7. Inhibition of cytolysis by exposure to amplitude-modulated 450-MHz electromagnetic fields.
(Reprinted with permission from Lyle et al. 1983. Suppression of T-lymphocyte cytotoxicity following exposure to sinusoidally amplitude-modulated fields. *Bioelectromagnetics* 4:281–292.)

(see Fig. 2), an oscillating electric component is established between the center plate and the grounded outer shell. Because there is a 50-Ω load terminating the cell, the current flow in the cell produces an oscillating magnetic component circulating around the center plate. This magnetic component is orthogonal to the electric component. Both oscillating components are in the horizontal plane. The local geomagnetic field (LGF), that is, the field created by the earth but distorted by building structures such as the steel used in construction, is present in our exposure system. This LGF, a DC field, is orthogonal to the horizontal plane containing the oscillating electric and magnetic components. To test whether the LGF had any influence on our results, we placed a pair of Helmholtz coils around our exposure cell and energized them to alter the intensity of the LGF at the location of our samples. We tested whether the normal pattern of an effect at 15 Hz and a null result at 30 Hz could be altered by changing the LGF. The motivation for this line of reasoning came from a consideration of magnetic resonance, as measured in both electron paramagnetic resonance (EPR) and nuclear magnetic resonance (NMR) spectroscopy, which demonstrates a highly efficient interaction between a molecular component and the combination of an oscillating and a DC magnetic field. The DC field could reduce the degeneracy in the energy levels of magnetic spin states, causing states that are only slightly different in energy. We decided to test the hypothesis that the LGF could be involved in the response of brain tissue to electromagnetic fields.

Figure 8 shows the results of this test (Blackman et al., 1985b). The abscissa is the density of the LGF in terms of multiples of the earth's magnetic field density in our Crawford cell, that is, multiples of 0.38 gauss

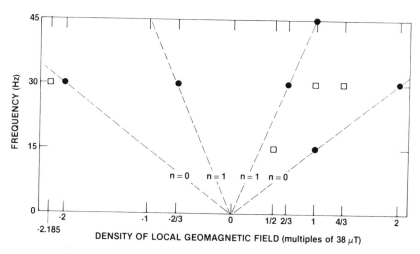

Fig. 8. Influence of the local geomagnetic field on calcium ion release.
(Reprinted with permission from Blackman et al. 1985b. A role for the magnetic field in the radiation-induced efflux of calcium ions from brain tissue, *in vitro. Bioelectromagnetics* 6:327–337.)

(G) (38 µT). Note that LGF measurements need to be made in the Crawford cell at the exact location in the laboratory at which it is used. If the cell is moved to another part of the laboratory the density or the angle of the LGF could be different. The result of exposure at 15 Hz under normal LGF conditions (solid circle) shows that there was a statistically significant difference between exposed and sham treatments. Exposure with the coils energized to reduce the LGF to one-half the normal value produced a null result, that is, the positive effect produced by 15-Hz fields disappeared. The test was repeated and again the effect disappeared. This exposure condition demonstrated a way to cause the positive effect produced by 15-Hz fields to disappear.

However, observing a null result in 15-Hz fields was not so unusual; some investigators have not been able to reproduce the original effects with brain tissue for one reason or another (see following discussion of possible causes of the "Cheshire cat" phenomenon). Further testing was performed to examined the influence of the LGF on exposures with 30-Hz fields, which had produced a null result both for us and for Bawin. When the local geomagnetic field was altered to two-thirds the normal value, an effect appeared. Further, there was no effect when the LGF was boosted to the four-thirds value. Thus, the net DC field density was important, not only the field applied with the Helmholtz coils. Other LGF densities were then examined to determine their influence on the response of the brain tissue to 30-Hz fields (see Fig. 8). These results demonstrated that the local geomagnetic field can determine which frequencies are effective.

The lines in Fig. 8, which helped to identify which LGF densities to test, were drawn from a relationship that a collaborator, Dr. James Rabinowitz, found in a solid-state physics textbook (Kittel, 1976). The relationship is based on a cyclotron resonance model in which a charged entity is accelerated through a DC magnetic field and is acted on by a Lorentz force. This model provided us with a guide for choosing LGF densities to examine at different frequencies; the resonant frequencies are proportional to the density of the local magnetic field times an index, $2i + 1$, where i varies from 0 to 2. The remarkable result is that this model worked within the limited conditions we tested. Both Rabinowitz and I believe that if a Lorentz force mechanism is involved in the results we have obtained with brain tissue, it has to be more sophisticated than a simple cyclotron resonance model of an ion in an ion channel.

Nevertheless, the simple cyclotron resonance model did serve to guide other investigators to successful experiments. Thomas and Liboff heard our report at the Bioelectromagnetic Society Annual Meeting in July 1984. Using two animal behavioral tests to look for 60-Hz magnetic field effects in exposed rodents, they had obtained only null results. When they returned to their laboratory, they adjusted the DC magnetic field to be consistent with this type of model, and in the first four of five animals examined they obtained significant effects in one of the behavioral tests (Thomas et al., 1986). This behavioral test required time estimation by the animals; the animals had to wait 18 sec after a signal before pressing a lever to get a food pellet. Following exposure under appropriately adjusted conditions, the behavioral response of the animals was changed; the animals no longer waited the appropriate length of time between the stimulus and their response, but started responding much sooner. This result was consistent with results obtained by Gavalas-Medici and Day-Magdaleno (1976), showing ELF electric-field effects on the timing response of monkeys (shortening the interresponse interval in some animals), and by Hamer (1968), showing a similar response in man. Thus, the Thomas report demonstrated that our results obtained with in vitro preparations, including the influence of the LGF, could be extrapolated to the in vivo situation.

Subsequently, we focused on effects that were highly significant, that is, $p < .01$. We had shown that brain tissue exposed to 15 Hz is responsive to values of the local geomagnetic field. If the LGF density was shifted to appropriate values, the 30-Hz fields also caused a response. The animal behavioral experiments of Thomas showed that 60 Hz can also be effective (it is possible that the underlying mechanism in these behavioral experiments may be entirely different from that operative in the in vitro brain tissue preparation). We then investigated what would happen if we canceled the local geomagnetic field.

We selected the 315-Hz exposure condition, which was strongly positive (see Fig. 6). Remember that the LGF was perpendicular to the horizontal plane containing the oscillating electric and magnetic components. When the LGF was reduced to zero, the 315-Hz field produced no effect. This result was statistically different from that obtained with the LGF present at 38 µT. Then we maintained a zero-DC field orthogonal to the horizontal plane, that is, keeping the natural LGF at zero, and introduced a new DC field in the horizontal plane with a second set of Helmholtz coils. In one experiment, the new DC magnetic field was aligned parallel to the electric-field component, and in a second experiment, it was parallel to the magnetic-field component. The results showed that a DC field parallel to the electric component, and coincidentally perpendicular to the magnetic component (as in the situation with the natural LGF), restored the ability of the 315-Hz electromagnetic field to alter the efflux of calcium from the brain tissue. However, when the DC field was aligned parallel to the magnetic component, the effect could not be restored in either of two attempts.

These results identify two issues: (1) in our experiments, it is the oscillating magnetic component that is effective for interaction with the DC magnetic field; and (2) the results are consistent with a magnetic resonance phenomenon in which the maximum coupling of the fields with the sample exists when the two magnetic components are orthogonal. It should be noted that in our experiments the density of the AC magnetic component, 69 nT, was approximately one-thousandth of the density of the DC magnetic component, 38 µT (Blackman et al., 1985b, 1988b). Our result with the orientations of the AC and DC magnetic fields appears to disagree with the result obtained with diatom motility in which the two fields had to be parallel for maximal effect (Smith et al., 1987). This disagreement between the two experiments may indicate that a different mechanism is operative in experiments with the diatoms or that the approximately equal densities of the DC and AC components used in the diatom experiments (parallel to each other and each approximately 50 µT) are not easily interpreted in terms of a simple Lorentz-force model.

INFLUENCE OF THE ELECTRIC FIELD DURING DEVELOPMENT

In spite of evidence for the paramount role of the magnetic field in the results, we decided to test whether the oscillating electric field had any influence. The specific question was whether the electric-field environment present during development in incubating eggs could affect the frequency response of the brain tissues of chickens once they are hatched. As can be seen in Fig. 4, the brain tissue responds to different intensities at 50 Hz compared to 60 Hz. However, all the results obtained so far were from chickens incubated in electrically powered incubators at a local university, that is, 60-Hz fields were present during incubation. We decided to incubate the eggs ourselves in either a 50- or a 60-Hz electric-field environment and test the brain tissues to determine if the frequency of exposure during incubation altered the response of the brain tissue after hatching. Eggs were exposed to either 50- or 60-Hz sinusoidal electric fields at a nominal 10 V/m during the 21-day incubation period, and the brain tissues from the hatched chickens were assayed within 1.5 days after hatching for alterations in calcium efflux with either 50- or 60-Hz electric and magnetic fields at 15.9 V/m and 73 nT (Blackman et al., 1988a).

We analyzed two hatches of eggs, each incubation and test requiring 1 month, to have enough brain tissue in each exposure/test category that the statistical tests would be adequately powerful; this introduced an additional complication in the analysis. The full-model analysis of variance included egg exposure, brain exposure, different hatches, and interaction terms between hatches, eggs, and brain exposures. A full analysis of variance on the data showed no statistically significant effect from hatch or from any interaction with hatch. Therefore, the data for the two hatches were collapsed into one, producing a reduced-model analysis of variance that included only egg exposure, brain exposure, and the interaction term. The reduced-model analysis showed that both the two egg exposures and the two brain exposures gave statistically different results, and that there was an interaction between the egg exposures and the brain exposures. No more information could be obtained from the analysis because of this interaction term.

To examine the interaction, Bonferroni-adjusted t tests were performed on each of the six combinations of egg exposure and brain exposure. First, for eggs exposed to 60 Hz (similar to conditions at our supplier), the brain tissue tested at 60 Hz gave a null result. With eggs exposed to 60 Hz, the brain tissue tested at 50 Hz gave a statistically significant alteration in the calcium efflux. This result is consistent with our previous data (see Fig. 4). Thus, we obtained the expected result based on previous observations. In contrast, for eggs exposed to 50 Hz during development, both 50- and 60-Hz tests of the brain tissue showed null results. Thus, exposing the eggs during incubation to 50-Hz electric fields removed the ability of 50-Hz electromagnetic fields to cause change in calcium efflux from the brain tissue. This result was so surprising that the experiment was repeated.

In the second experiment the results of the generalized model were basically the same. There was no interaction with hatching, so the data were collapsed and the model looked only at the effect of egg exposure, brain exposure, and interaction terms. Again the results were the same as the first experiment: the two egg exposures and the two brain exposures gave statistically different results, and there was an interaction between the egg exposures and the brain exposures.

Using the Bonferroni-adjusted t tests of all six combinations, when the eggs were exposed to 60-Hz electric fields during development the 60-Hz test of brain tissue showed a null result while the 50-Hz test showed a positive effect. In contrast, 50-Hz exposure of the eggs during development again produced a null result for both frequencies, replicating the results we had obtained previously.

Could this result directly from the specific placement of the eggs in the exposure system and not directly from the different frequencies of electric fields? The exposure system was composed of three parallel plates, with the center plate grounded and the outer two plates energized at 50 Hz and at 60 Hz (Joines et al., 1986; Blackman et al., 1988a). The eggs exposed to 60-Hz electric fields were always in the bottom of the system and those exposed to 50 Hz in the top. Another exposure was performed in which the position of the two frequencies were reversed to determine if there was a position effect in our exposure system for the eggs. However, exposure of the eggs to 60-Hz electric fields in the new position gave the same result as in previous tests. The full-model analysis of variance was the same, the reduced-model analysis of variance was the same, and the Bonferroni-adjusted t tests gave the same result. Again, exposure of the eggs to 50 Hz removed the ability of the 50-Hz electric and magnetic fields to produce the enhanced efflux.

In summary, our results were reproduced in three independent experiments. The results for eggs exposed to 60-Hz electric fields agreed with published results. Further, this result was not an artifact caused by the position of the eggs in the exposure system. These results indicate that 10-V/m electric fields are biologically active in the developing egg; the induced electric current in the egg, 0.126 μA/m^2, is well below the level normally considered to be biologically effective (see Blackman et al., 1988a). The electric-field intensity to which the eggs were exposed during incubation is commonly found in homes in the United States. Last, these results indicate the ambient electric-field environment in which samples are exposed before testing may contribute to the so-called Cheshire cat phenomenon, that is, the elusiveness of an effect when the tests are repeated under apparently identical conditions. The flux density of the local geomagnetic field may also be important.

POSSIBLE MECHANISMS OF ACTION

Mechanisms involved with the stimulation of calcium release from brain tissue preparations are of considerable scientific interest. A major criterion for any mechanism is that it explain the effects of very low field intensities (Blackman et al., 1988b). In our experiments, the magnetic component is in the nanotesla range, well below the maximum found in homes (5–10 mT) and the electric component is 10 to 15 V/m, a level frequently found in homes (Sheppard and Eisenbud, 1977). Normally one would consider these field intensities innocuous.

We have recently proposed a generalized mechanism that requires three distinct steps: transduction, amplification, and expression (Blackman,

1988). We assume there must be a very efficient transduction site that converts the electromagnetic energy into a chemical change (Blackman, 1988). And yet, because so little energy is available, this chemical change has to be very small even if the coupling is 100% efficient. To produce the type of change we have seen, the field must interact with a biological amplification mechanism, such as a trigger for some more global process, or provide energy for some energy storage mechanism that subsequently causes the changes. This biological amplification concept, which assumes that the system is energetically poised to go through a cooperative transition, is not a new suggestion, having been first proposed for membrane surfaces by Adey (1975).

Further, the effects we have observed do not necessarily result from direct field interaction with calcium ions. The calcium release could be a secondary or tertiary response. This conclusion is drawn from the observation that the neutral sugars, mannitol or sucrose, which are used as monitors of extracellular space, can substitute for calcium ion as a monitor of the changes produced by the fields at a few of the frequencies tested (Blackman et al., 1987). The field-induced changes could easily be on the membrane itself, or some component associated with the membrane, which then causes changes in many processes both exterior to the cell and inside the cell (for example, see Byus et al., 1987, 1988).

How do our data relate to these concepts? The distinctive responses of the brain tissue to various frequencies of electromagnetic fields, together with the dependence of some frequencies on the LGF, are a partial characterization of the transduction processes that efficiently convert electromagnetic energy into chemical change. We have speculated that Lorentz force or magnetic resonance phenomena may be the underlying mechanisms (see Blackman, 1985). For example, the prominent responses at $p < .01$ (two groups of five frequencies in Fig. 6) may constitute the response of unpaired electrons in a transition metal–ligand complex. Further, the 405-Hz result may be caused by the magnetic resonance for ^{13}C, which is 406.8 Hz at our LGF. For either of these hypothetical transduction mechanisms to be the cause of our results, they must be coupled to a molecular ensemble that assumes different configurations, perhaps a planar one and a more bulky one that are close in energy, so that the fields can affect the valence electrons to initiate transitions between the configurations. If these molecular sites are embedded in a large structure, such as that of the plasma membrane poised at a transition, then temporally coherent, field-induced molecular changes at multiple sites could influence the dynamics in the larger membrane structure leading to the observed results.

In addition, there are several ways that fields of specific frequencies could interact with ions via a Lorentz force mechanism to affect their activity at receptor sites on membranes, causing functional changes in membrane components that would lead to alterations in ion fluxes, induction of intracellular components, and hormone-induced responses. The data grouping for the 60-, 90-, and 180-Hz responses may operate through such an alternative mechanism.

The Lorentz force model is also embodied in the cyclotron resonance concept first suggested by Blackman (1985) and by Polk (1986), and developed by Liboff (1985), Chiabrera et al. (1985, 1987), McLeod and Liboff (1987), Durney et al. (1988), and Liboff and McLeod (1988). Although such simple models seem to predict limited portions of the data, they cannot give a full accounting of the results. Nevertheless, the development of such models aids in the design of new experiments to test and refine those models. The report of magnetic-field alterations in diatom mobility, in which the oscillating magnetic component must be parallel to the DC magnetic component to produce a change, is just the opposite of what we have found with calcium efflux from the brain tissue preparation. Thus, there could be other models applicable to particular situations. At present, these possibilities have not yet been developed into a coherent framework.

The unusual intensity responses we have observed with both ELF-modulated radio-frequency (RF) radiation and directly with ELF waves have proven to be a mystery, awaiting explanation for more than 10 years. For both RF and ELF fields, the calcium efflux phenomenon displays at least two intensity "windows" or regions in which responses are elicited by the fields but outside of which null results are seen. At least two windows have been observed for calcium efflux, from chick brain preparations (Blackman et al., 1980b, 1981, 1982) and from human neuroblastoma cells in culture (Dutta et al., 1984). We recently reported six intensity windows for calcium efflux from brain tissue preparation exposed to 50-MHz electromagnetic fields, amplitude modulated at 16 Hz (Blackman et al., 1989) (Fig. 9).

To explain this multiple-intensity response, we have invoked the emerging evidence for dynamic responses of cooperative biological systems (for details, see Blackman et al., 1989). Dynamic processes, including chaos, are recognized in biological systems when the system responses are examined at the macroscopic level. These processes are characterized by dependencies on the driving frequencies and intensities for particular expressions of the macroscopic variables. Thus, the pattern of intensity responses we have observed, and some of the patterns of the frequency responses, may be caused by the dynamic responses of a biological system, particularly one that is sensitized to minor disturbances because it is poised at a chemical instability, perhaps a cooperative or phase transition. One site of such processes is the cell membrane, which is thought to be the principal location for many of the effects (Adey et al., 1982; Blackman and Wilson, 1983; Byus et al., 1987, 1988). This explanation for the frequency and possible intensity dependencies may also apply to the responses observed by Smith et al. (1987) for diatom mobility and by Liboff et al. (1987) for calcium flux in lymphocytes.

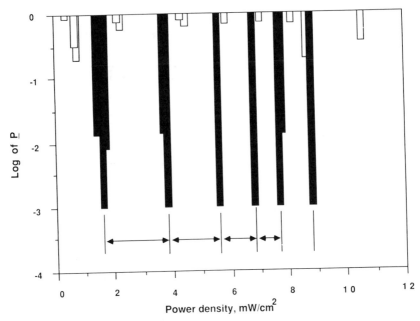

Fig. 9. Power density dependence of calcium ion release displayed as *p* values (*closed bars*, effect; *open bars*, null).
[(Figure reprinted with permission from Blackman et al. 1989. Multiple power density windows and their possible origin. *Bioelectromagnetics* 10(2) (in press).]

Most recently, Goodman and colleagues have reported electromagnetic field-induced effects on the membrane surface properties of cells (Goodman et al., 1986; Marron et al., 1988). Their results show that exposure to the electric component causes an increase in the surface charge density, while exposure to the magnetic component causes an increase in lipophilicity of the cell, that is, the propensity of the cell to partition into the lipid phase rather than the aqueous phase of a two-phase test system. Although we do not know how to generalize that result, it independently demonstrates that membranes are involved in cellular responses to electric and to magnetic fields.

CONCLUSION

Calcium ions play a major role as a second messenger in many signaling mechanisms in cells (see Chapter 14). We have demonstrated in this chapter that under specific combinations of conditions, electric and magnetic fields can influence biological processes as monitored by calcium ion release from tissues. Other examples of field interactions with biological systems can be found in Chapters 10 and 11. Taken together, the evidence overwhelmingly indicates that electric and magnetic fields can alter normal calcium homeostasis and lead to changes in the response of biological systems to their environment. It remains to be clearly demonstrated that these field-induced perturbations force the biological systems beyond the normal physiological range to a level at which the preexposure equilibrium cannot be restored and permanent changes occur that could result in an unhealthy biological response.

ACKNOWLEDGMENTS

I thank Drs. Berman and Phillips for helpful suggestions during the preparation of this manuscript. This document has been reviewed in accordance with the U.S. Environmental Protection Agency policy and approved for publication. Mention of trade names of commercial products does not constitute endorsement or recommendation for use.

REFERENCES

Adey, W. R. 1975. Introduction: effects of electromagnetic radiation on the nervous system. *Ann. N.Y. Acad. Sci.* 247:15–20.

Adey, W. R., Bawin, S. M., Lawrence, A. F. 1982. Effects of weak amplitude-modulated microwave fields on calcium efflux from awake cat cerebral cortex. *Bioelectromagnetics* 3:295–307.

Bawin, S. M., Kaczmarek, L. K., Adey, W. R. 1975. Effects of modulated VHF fields on the central nervous system. *Ann. N.Y. Acad. Sci.* 247:74–81.

Blackman, C. F. 1985. The biological influences of low-frequency sinusoidal electromagnetic signals alone and superimposed on RF carrier waves. In: *Interaction Between Electromagnetic Fields and Cells*, Chiabrera, A., Nicolini, C., Schwan, H. P., eds., pp. 521–535. NATO ASI Series A97. New York: Plenum.

Blackman, C. F. 1988. Stimulation of brain tissue in vitro by extremely low frequency, low intensity, sinusoidal electromagnetic fields. In: *Electromagnetic Waves and Neurobehavioral Function*, Lovely, R. H., O'Connor, M. E., eds., pp. 107–117. New York: Alan R. Liss.

Blackman, C. F., Wilson, B. S. 1983. Distribution of label in studies on the effects on nonionizing radiation on the association of calcium ions with brain tissue. Abstract GJ-24, *5th Annual Meeting of the Bioelectromagnetics Society*, 12–17 July, Boulder, Colorado, p. 28. (Available from Bioelectromagnetics Society, One Bank Street, Suite 307, Gaithersburg, Maryland 20878.)

Blackman, C. F., Benane, S. G., House, D. E. 1987. ELF electromagnetic fields cause enhanced efflux of neutral sugars from brain tissue in vitro. Abstract P-B9, *Ninth Annual Meeting of the Bioelectromagnetics Society*, Portland, Oregon, June 21–25, 1987. (Available from Bioelectromagnetics Society, P. O. Box 3729, Gaithersburg, Maryland 20878.)

Blackman, C. F., Joines, W. T., Elder, J. A. 1981. Calcium ion efflux induction in brain tissue by radiofrequency radiation. In: *Biological Effects of Nonionizing*

Radiation, Illinger, K. H., ed., ACS Symp. Ser. 157:299–314. Washington, DC: American Chemical Society.

Blackman, C. F., Elder, J. A., Weil, C. M., Benane, S. G., Eichinger, D. C., House, D. E. 1979. Induction of calcium-ion efflux from brain tissue by radio-frequency radiation: effects of modulation frequency and field strength. Radio Sci. 14(6S):93–98.

Blackman, C. F., Benane, S. G., Elder, J. A., House, D. E., Lampe, J. A., Faulk, J. M. 1980a. Induction of calcium-ion efflux from brain tissue by radiofrequency radiation: effect of sample number and modulation frequency on the power-density window. Bioelectromagnetics 1:35–43.

Blackman, C. F., Benane, S. G., Joines, W. T., Hollis, M. A., House, D. E. 1980b. Calcium-ion efflux from brain tissue: power-density vs. internal field-intensity dependencies at 50-MHz RF radiation. Bioelectromagnetics 1:277–283.

Blackman, C. F., Benane, S. G., Kinney, L. S., House, D. E., Joines, W. T. 1982. Effects of ELF fields on calcium-ion efflux from brain tissue, in vitro. Radiat. Res. 92:510-520.

Blackman, C. F., Benane, S. G., House, D. E., Joines, W. T. 1985a. Effects of ELF (1–120 Hz) and modulated (50 Hz) RF fields on the efflux of calcium ions from brain tissue, in vitro. Bioelectromagnetics 6:1–11.

Blackman, C. F., Benane, S. G., Rabinowitz, J. R., House, D. E., Joines, W. T. 1985b. A role for the magnetic field in the radiation-induced efflux of calcium ions from brain tissue, in vitro. Bioelectromagnetics 6:327–337.

Blackman, C. F., Benane, S. G., House, D. E., Joines, W. T., Spiegel, R. J. 1988a. Effect of ambient levels of power-line-frequency electric fields on a developing vertebrate. Bioelectromagnetics 9(2):129–140.

Blackman, C. F., Benane, S. G., Elliott, D. J., House, D. E., Pollock, M. M. 1988b. Influence of electromagnetic fields on the efflux of calcium ions from brain tissue in vitro: a three-model analysis consistent with the frequency response up to 510 Hz. Bioelectromagnetics 9(3):215–227.

Blackman, C. F., Kinney, L. S., House, D. E., Joines, W. T. 1989. Multiple power density windows and their possible origin. Bioelectromagnetics 10(2): 115–128.

Byus, C. V., Pieper, S., Adey, W. R. 1987. The effects of low-energy 60-Hz environmental electromagnetic fields upon the growth-related enzyme ornithine decarboxylase. Carcinogenesis 8:1385–1389.

Byus, C. V., Kartun, K., Pieper, S., Adey, W. R. 1988. Increased ornithine decarboxylase activity in cultured cells exposed to low-energy modulated microwave fields and phorbol ester tumor promoters. Cancer Res. 48:4222–4226.

Chiabrera, A., Bianco, B., Viviani, R. 1987. Effect of DC and AC magnetic fields on ligand binding. Abstract H6, Ninth Annual Meeting of the Bioelectromagnetics Society, Portland, Oregon, June 21–25, 1987. (Available from Bioelectromagnetics Society, P. O. Box 3729, Gaithersburg, Maryland 20878.)

Chiabrera, A., Bianco, B., Caratozzolo, F., Giannetti, G., Grattarola, M., Viviani, R. 1985. Electric and magnetic field effects on ligand bonding to the cell membrane. In: Interactions Between Electromagnetic Fields and Cells, Chiabrera, A., Nicolini, C., Schwan, H. P., eds., pp. 253–280. NATO ASI Series A97. New York: Plenum.

Durney, C. H., Rushforth, C. K., Anderson, A. A. 1988. Resonant AC-DC magnetic fields: calculated biological response. *Bioelectromagnetics* 9:315–336.

Dutta, S. K., Ghosh, B., Blackman, C. F. 1989. Radiofrequency radiation-induced calcium ion-efflux enhancement from human and other neuroblastoma cells in culture. *Bioelectromagnetics* 10(2): 197–202.

Dutta, S. K., Subramoniam, A., Ghosh, B., Parshad, R. 1984. Microwave radiation-induced calcium ion efflux from human neuroblastoma cells in culture. *Bioelectromagnetics* 5:71–78.

Gavalas-Medici, R., Day-Magdaleno, S. R. 1976. Extremely low frequency weak electric fields affect schedule-controlled behavior of monkeys. *Nature (London)* 261:256–258.

Goodman, E. M., Sharpe, P. T., Greenebaum, B., Marron, M. T. 1986. Pulsed magnetic fields alter the cell surface. *FEBS Lett.* 199(2):275–278.

Hamer, J. 1968. Effects of low-level, low-frequency electric fields on human reaction time. *Commun. Behav. Biol.* A 2(5):217–222.

Joines, W. T., Blackman, C. F. 1980. Power density, field intensity, and carrier frequency determinants of calcium-ion efflux from brain tissue. *Bioelectromagnetics* 1:271–275.

Joines, W. T., Blackman, C. F. 1981. Equalizing the electric field intensity within chick brain immersed in buffer solution at different carrier frequencies. *Bioelectromagnetics* 2:411–413.

Joines, W. T., Blackman, C. F., Hollis, M. A. 1980. Broadening of the RF power-density window for calcium-ion efflux from brain tissue. *IEEE Trans. Biomed. Eng.* BME 28:568–573.

Joines, W. T., Blackman, C. F., Spiegel, R. J. 1986. Specific absorption rate in electrically-coupled biological samples between metal plates. *Bioelectromagnetics* 7:163–176.

Kittel, C. 1976. *Introduction to Solid State Physics.* New York: Wiley.

Liboff, A. R. 1985. Cyclotron resonance in membrane transport. In: *Interactions Between Electromagnetic Fields and Cells*, Chiabrera, A., Nicolini, C., Schwan, H. P., eds., pp. 281–296. NATO ASI Series A97. New York: Plenum.

Liboff, A. R., McLeod, B. R. 1988. Kinetics of channelized membrane ions in magnetic fields. *Bioelectromagnetics* 9(1):39–51.

Liboff, A. R., Smith, S. D., McLeod, B. R. 1987. Experimental evidence for ion cyclotron resonance mediation of membrane transport. In: *Mechanistic Approaches to Interactions of Electric and Electromagnetic Fields with Living Systems*, Blank, M., Findl, E., eds., pp. 109–132. New York: Plenum.

Lyle, D. B., Shechter, P., Adey, W. R., Lundak, R. L. 1983. Suppression of T-lymphocyte cytotoxicity following exposure to sinusoidally amplitude-modulated fields. *Bioelectromagnetics* 4:281–292.

Marron, M. T., Goodman, E. M., Sharpe, P. T., Greenebaum, B. 1988. Low frequency electric and magnetic fields have different effects on the cell surface. *FEBS Lett.* 230:13–16.

McLeod, B. R., Liboff, A. 1987. Cyclotron resonance in cell membranes: the theory of the mechanism. In: *Mechanistic Approaches to Interactions of Electric and Electromagnetic Fields with Living Systems*, Blank, M., Findl, E., eds., pp. 109–132. New York: Plenum.

Polk, C. 1986. Physical mechanisms by which low-frequency magnetic fields can affect the distribution of counterions on cylindrical biological cell surfaces. *J. Biol. Phys.* 14:3–8.

Sheppard, A. R., Eisenbud, M. 1977. *Biological Effects of Electric and Magnetic Fields of Extremely Low Frequency.* New York: New York University Press.

Sheppard, A. R., Bawin, S. M., Adey, W. R. 1979. Models of long-range order in cerebral macromolecules: effects of sub-ELF and of modulated VHF and UHF fields. *Radio Sci.* 14(6S):141–145.

Smith, S. D., McLeod, B. R., Liboff, A. R., Cooksey, K. 1987. Calcium cyclotron resonance and diatom mobility. *Bioelectromagnetics* 8(3):215–227.

Thomas, J. R., Schrot, J., Liboff, A. R. 1986. Low-intensity magnetic fields alter operant behavior in rats. *Bioelectromagnetics* 7:349–357.

PART IV.
Possible
Mechanisms

10 Electromagnetic Fields, Cell Membrane Amplification, and Cancer Promotion

W. ROSS ADEY
Pettis Memorial Veterans Administration Medical Center
and Loma Linda University School of Medicine
Loma Linda, California

CONTENTS

(continued)

CONTENTS (continued)

Historically, evolving concepts of cellular organization have emphasized a limiting membrane defining the physical boundaries of a living cell. The membrane may enclose a bacterium, or a single-celled protozoan leading an independent existence, or a cell in the tissues of a higher organism. From these earlier concepts of the cell membrane as merely an enclosure, emphasis now focuses on its role as a window through which the cell, as a unitary biological element, senses and responds to its chemical and electrical environment.

Narrow gutters separate cells in tissue, forming the extracellular space (ECS). These gutters form pathways for the flow of intrinsic electrochemical currents generated by cells, as well as for currents that are the induced components of environmental electromagnetic (EM) fields. They are also avenues for intercellular metabolic exchange. They thus form a route by which cells may "whisper together" in an often faint and private language (Young, 1951). There are inward and outward chemical and electrical signals through cell membranes. By the use of weak EM fields to manipulate these signals, a connected picture is now emerging of the sequence and energetics of major events that couple stimuli to the cell interior by signals that arise when hormone, neurotransmitter and antibody molecules bind to their cell-surface receptor sites. Outward signals to neighboring cells can also be manipulated by imposed EM fields.

There is evidence that this intercellular communication plays an essential role in regulation of cell growth. Functional isolation of a cell from its neighbors by the separate or joint actions of EM fields and chemical cancer promoters, both acting at cell membranes, may lead to unregulated growth with tumor formation. Imposed EM fields have thus become an important tool in pursuit of this new model of *epigenetic carcinogenesis* (Newmark, 1987; Slaga and Butterworth, 1987; Trosko, 1987; Yamasaki, 1987), which defines a key role for cell membranes and intercellular communication in tumor formation, rather than the traditional emphasis on damage to DNA in cell nuclei (Adey, 1987a, 1988a,b,c).

Studies of the structural organization of cell membranes have revealed numerous strands of protein (*intramembranous particles*, IMP) inserted into the thin double layer of fat molecules that form the plasma membrane. IMPs span the membrane from inside to outside with external protrusions into the fluid surrounding the cell; their outer tips are terminal glycoprotein strands that sense electric fields and form receptor sites for chemical-stimulating molecules. The IMPs have functional contacts inside the cell with key elements of the cell machinery, including enzymes and the numerous fine tubes of the cytoskeleton. IMPs "float" in the sea of lipid molecules of the plasma membrane, leading to the generally accepted *fluid mosaic* model of the cell membrane (Singer and Nicolson, 1972). Thus, these intramembranous strands form signaling pathways by which external stimuli are sensed and conveyed to the cell interior (Fig. 1).

Functional studies of electrical and chemical intercellular communication have focused on specialized regions of contact between adjacent cell membranes. These regions form *gap-junctions* that couple cells electrically and chemically.

Initial stimuli associated with weak EM oscillations and with binding of humoral molecules at their membrane receptor sites elicit a highly *cooperative modification of calcium binding* to glycoproteins along the membrane surface (Adey, 1988a,b). This longitudinal spread is consistent with the direction of flow of extracellular currents associated with physiological activity and from imposed EM fields. It is also consistent with spreading activation of a membrane-related calcium-dependent enzyme, protein kinase C, that is also a specific receptor for cancer-promoting phorbol esters (Nishizuka, 1983).

This cooperative modification of surface calcium binding is an *amplifying stage*, with evidence from concurrent manipulation of these initial events by imposed EM fields that there is a far greater increase or decrease in calcium efflux than is accounted for in the energy of the field or in the events of receptor–ligand binding (Bawin et al., 1975; Bawin and Adey, 1976; Lin-Liu and Adey, 1982). There is further striking evidence for the nonequilibrium character of this modification in calcium binding in its occurrence in quite narrow frequency and amplitude windows (Bawin et al., 1978a; Adey, 1981a,b; Blackman et al., 1979, 1982, 1985a,b).

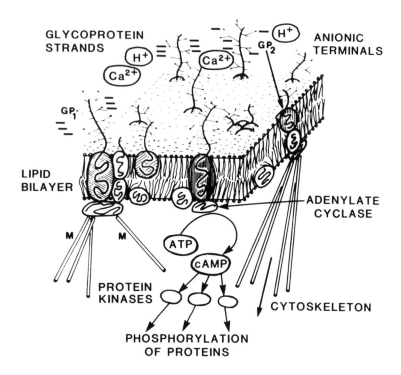

Fig. 1. Fluid mosaic model of cell membranes offers structural basis for tissue interactions with EM fields. Intramembranous particles (IMP) in lipid bilayer have external protruding glycoprotein strands, negatively charged on their amino sugar terminals that form receptor sites for antibodies, neurotransmitters, and hormones; they attract Ca ions. Stimulating molecules and EM fields alter surface calcium binding in Stage 1 of transmembrane signal coupling. In Stage 2, transmembrane signals pass along IMPs that act as coupling proteins to the interior. Stage 3 modulates intracellular enzyme activity. (Modified from Singer and Nicolson, 1972.)

NATURAL AND MAN-MADE ELECTROMAGNETIC ENVIRONMENTS AND THEIR BIOLOGICAL SIGNIFICANCE

Electromagnetic radiation with wavelengths longer than the ultraviolet region of the spectrum (photon energies less than about 12 eV) does not possess sufficient energy to cause ionization. Therefore, there has been a persisting view in certain areas of the physical sciences that nonionizing EM fields are incapable of inducing bioeffects other than by heating (Foster and Guy, 1986). This view overlooks the possibility for cooperativity in biomolecular systems and the profoundly important role that cooperativity appears to play in the detection of tissue components of nonionizing EM fields. As revealed by field effects, these interactions are nonlinear and nonequilibrium in character. They are consistent with quantum processes involving long-range interactions between electric charges on cell-surface macromolecules.

Since our studies have shown similar sensitivities in a wide range of tissues and cell types, we conclude that these electrochemical sensitivities may be a general property of all cells. The nature of these interactions is so far removed from the concepts and models that have guided research in ionizing radiation that expertise in the latter area can offer little to the search for underlying mechanisms; equilibrium thermodynamics and the classical models of the statistical mechanics of matter appear equally inappropriate in their applications to most key questions on the biological effects of nonionizing EM fields.

In the workplace, in the home, and in external environments, the use of devices and systems employing EM energy has increased exponentially over the past half-century. These create an almost infinite variety of EM fields covering a very broad spectrum extending from extremely low frequencies below 100 Hz to millimeter microwaves and the far-infrared region. These artificial fields exceed in intensity the natural background levels in the same spectral region by many orders of magnitude. The proliferation of diverse sources has dramatically increased the potential for public exposure to virtually all this broad spectrum.

Typical suburban environments inside and outside the home are complex (Table I). In addition to an obvious peak at powerline frequencies, there is often a peak in the amplitude-modulated (AM) broadcast band. Frequency-modulated (FM) broadcast stations are the main contributors in the very high frequency (VHF) spectrum. Electric fields directly under high-voltage power lines (220–345 kV) are typically 5–15 kV/m. Electric

TABLE I.
Typical Environmental Field Levels

Electric blanket	200 V/m
High-voltage power line	10,000 V/m, DC or 60 Hz
Hair dryer	30 G
Microwave oven	5 mW/cm^2 (at door) (130 V/m)
Handy-Talkie	5 mW/cm^2 (at head) (130 V/m)
U. S. city suburbia (RF—primarily FM broadcast)	1–4 µW/cm^2 (2–4 V/m)

blankets expose the user to about 200 V/m. Permitted leakage at the door of a microwave oven is 130 V/m (5 mW/cm^2) and a handy-talkie or cellular carphone produces about the same field level at the head.

What are the resulting field levels induced in body tissues? This depends in varying degree on the body geometry with respect to field orientation and frequency (Adey, 1981a). For extremely low frequency (ELF) fields, a simple capacitance model of coupling to electric fields suffices. Coupling efficiency is therefore low at these low frequencies. Thus, a 10-kV/m, 60-Hz field would produce tissue gradients of about 0.1 mV/cm. On the other hand, radio-frequency (RF) fields couple much more efficiently, particularly as dimensions of the whole body, or portions of the body such as the head or limbs, approach a significant fraction of a wavelength at the imposed field frequency. For example, an RF field in the frequency range 100–300 MHz at an incident energy of 1.0 mW/cm^2 (61 V/m) will induce tissue gradients of the order of 10–100 mV/cm in the human body, a gradient impossible to achieve by exposure to fields at powerline frequencies because breakdown in air, which acts as an insulator, will occur long before an environmental ELF field reaches the required level for such induced tissue gradients.

The orientation of the head with respect to the earth's geomagnetic field has been shown to influence activity of the pineal gland (Semm, 1983). In pigeons, guinea pigs and rats, firing rates of about 20% of pineal cells respond to changes in direction and intensity of the earth's magnetic field. The pineal peptide hormone melatonin exercises a strong regulation on circadian rhythms. Nocturnal inversion of the horizontal component of the geomagnetic field decreases synthesis and secretion of melatonin and the activity of its synthesizing enzymes (Welker et al., 1983). Efflux of calcium from chick brain tissue in response to ELF fields is also sensitive to concomitant levels of static magnetic fields. Halving the local geomagnetic field with a Helmholtz coil renders a previously effective 15-Hz field ineffective; and doubling the geomagnetic field causes an ineffective 30-Hz signal to become effective (Blackman et al., 1985b).

Bioelectrical Sensitivities to Intrinsic and Induced Electromagnetic Fields

As a perspective on the biological significance of these induced fields, a number of studies indicate that ELF fields producing tissue gradients around 10^{-7} V/cm are involved in essential physiological functions in marine vertebrates, birds and mammals including man (Table II) (Adey, 1981a). In vitro studies have also shown sensitivities at these very low intensities for calcium efflux (Bawin and Adey, 1976) and in bone growth (Fitzsimmons et al., 1986).

TABLE II.
Bioelectrical Sensitivities to ELF Fields

Organism	Function	Tissue gradient	Imposed field
Sharks and rays	Navigation and predation	10^{-8} V/cm	DC to 10 Hz
Birds	Navigation	10^{-7} V/cm	0.3 G
Birds	Circadian rhythms	10^{-7} V/cm	10 Hz, 2.5 V/m
Monkeys	Subjective time estimations	10^{-7} V/cm	7 Hz, 10 V/m
Man	Circadian rhythms	10^{-7} V/cm	10 Hz, 2.5 V/m

Comparison with Intrinsic Cell and Tissue Neuroelectrical Gradients

Membrane potential	10^{-5} V/cm
Synaptic potential	10^{-3} V/cm
Electroencephalogram	10^{-1} V/cm

With RF fields that are sinusoidally amplitude modulated at ELF frequencies, induced tissue electric gradients can be substantially higher, and at levels of 10–100 mV/cm (cited previously) are in the same amplitude range as intrinsic oscillations generated biologically, such as in the electroencephalogram (EEG). Induced fields at these higher levels also produce a wide range of biological interactions. These responses include entrainment of brain EEG rhythms at the same frequencies as ELF components of imposed fields, conditioned EEG responses to imposed fields, and modulation of brain and behavioral states; in nonnervous tissues, strong effects on cell membrane functions include modulation of intercellular communication through gap-junction mechanisms, reduction of cell-mediated cytolytic immune responses, and modulation of intracellular enzymes that are molecular markers of signals that arise at cell membranes and are then coupled to the cell interior.

PHYSIOLOGICAL BENCHMARKS IN ELECTRICAL ORGANIZATION OF CELLS AND TISSUES

In most cells in the resting state, there is a steady *membrane potential* of approximately 0.1 V between the inside and the outside of the cell, with the interior of the cell negative with respect to the exterior. The membrane potential exists across the extremely thin lipid plasma membrane, typically about 40 Å thick, a membrane so thin that in consequence there is an enormous electric gradient of 10^5 V/cm across the cell membrane. This large gradient is altered by about 10^3 V/cm in synaptic activation in nerve cells (see Table II). In sharp contrast, physiological electric oscillations in fluid surrounding cells are many orders of magnitude weaker than this natural barrier of the membrane potential. For example, the gradient of an EEG measured across the dimensions of a single cell is a mere 0.1 V/cm in the extracellular fluid, six orders of magnitude less than the electrical barrier of the membrane potential.

It is therefore not surprising that such weak physiological gradients in fluid surrounding cells have been denied a physiological role. Nevertheless, many organisms including man are sensitive to tissue gradients in the range 0.1–100 mV/cm (Adey, 1981a,b). These sensitivities have been confirmed in cell and tissue cultures for many cell types, including lymphocytes (Byus et al., 1984; Lyle et al., 1983, 1988), liver cells (Byus et al., 1987), ovary cells (Byus et al., 1988), bone cells (Luben et al., 1982; Luben and Cain, 1984; Cain and Luben, 1987; Cain et al., 1987), cartilage cells (Hiraki et al., 1987), and nerve cells (Dixey and Rein, 1982). Embryonic bone matrix formation is increased by exposure to even far weaker gradients down to 10^{-7} V/cm (Fitzsimmons et al., 1986). These interactions emphasize the importance of *amplification* in their ultimate effects on intracellular mechanisms discussed subsequently.

These observations have been viewed cautiously by many biologists as beyond the realm of a possible physical reality. As is discussed subsequently, it is necessary to consider them in the context of the role of cooperative processes and associated nonlinear electrodynamics at cell membranes seen with imposed electromagnetic fields (Adey, 1984, 1986, 1988a,b,c; Adey and Lawrence, 1984). These cooperative phenomena are in the realm of nonequilibrium thermodynamics, far removed from traditional equilibrium models of cellular excitation that have focused on depolarization of the membrane potential and on associated massive changes in ionic equilibria across the cell membrane.

The Physiological Dilemma of Biological Responses to Weak Environmental and Intrinsic Electromagnetic Fields

These equilibrium models have also been offered as adequate for an understanding of the first events in cell membrane transductive coupling of electrochemical stimuli at the cell surface. For nervous tissue, it has been generally accepted that the models of Hodgkin and Huxley (1952) appropriately describe both the sequence and the energetics of excitatory events.

The diverse cellular effects of weak imposed electromagnetic fields just discussed here and in Table II, as well as others to be described, indicate that these are inappropriate models. We may note that low-frequency pulsed magnetic fields effective in therapy of ununited fractures (Bassett,

1987) induce tissue gradients of around 3 mV/cm and extracellular current densities around 10^{-6} A/cm^2. They modify enzyme activity in cultured bone cells and their secretion of collagen in response to parathyroid hormone, responses at one-millionth of threshold transmembrane currents predicted by Hodgkin–Huxley models (Luben et al., 1982). These models were originally offered only in the context of a mathematical description of major perturbations in Na$^+$ and K$^+$ equilibria at a certain epoch in the temporal and energetic sequences of excitation in nerve fibers. Their extrapolation by others to address threshold phenomena in cellular systems would appear beyond the scope of the original intent.

Pathways for Electric Current Flow in Tissues

The question of the mechanism of physiological response becomes apparent from a consideration of distribution of current flowing in a tissue composed of cells with a high membrane resistance and bathed in a strongly conducting fluid (Cole, 1940). Typical cell membrane resistances are in the range of 3,000–100,000 Ω/cm^2. Extracellular fluid has a typical specific resistance of only 50 Ω/cm. Thus, although the extracellular space forms only about 10% of the conducting cross section of typical tissue, it is clearly a preferred pathway, carrying at least 90% of any imposed or induced current. The longitudinal current flow at cell surfaces seems to have considerable importance in the action of pericellular fields on calcium binding at cell surfaces (Bawin and Adey, 1976; Bawin et al., 1975, 1978a), and may involve concurrent activation of cascades of calcium-dependent protein kinases (Byus et al., 1984) along the membrane, including some involved in growth regulation and responses to chemical cancer promoters (Kikkawa et al., 1983; Nishizuka, 1983, 1984).

Three Stages in Transmembrane Signaling from Cell Surface to Interior

In summary, intramembranous protein particles (IMP) placed within cell membranes provide an essentially direct *inward* path between the cell surface and intracellular enzymatic systems and organelles. As a functional model, there is a minimal sequence of three steps in this transductive coupling and each is calcium dependent:

1. The first weak electrochemical events associated with binding of stimulating molecules at their receptor sites and with EM fields are sensed by cell-surface glycoproteins.
2. These surface events are amplified and then signaled to the cell interior by transmembrane portions of IMPs.
3. These signals are coupled internally to intracellular enzyme systems, and through the tubes and filaments of the cytoskeleton, to the nucleus and to other organelles.

An *outward* stream of electrical and chemical signals also passes through cell membranes and links adjoining cells, partly through gap-junctions. Both inward and outward signals are sensitive to a broad spectrum of weak EM fields, including amplitude-modulated RF fields.

Cooperative Modification of Calcium Binding by Electromagnetic Fields at Cell Surfaces, with Amplification of Initial Signals

Most studies of EM field effects on tissue Ca^{2+} have used cerebral tissue, including cerebral cortex in awake cats (Adey et al., 1982), isolated chick cerebral hemisphere (Bawin and Adey, 1976; Bawin et al., 1975, 1978a,b; Blackman et al., 1979, 1982, 1985a,b), cultured neurons (Dutta et al., 1984), and cerebral synaptosome fractions (Lin-Liu and Adey, 1982). Use of imposed EM fields to manipulate initial events in molecular binding at cell membrane receptor sites through use of imposed EM fields causes a far greater increase or decrease in Ca^{2+} than is accounted for in the events of receptor–ligand binding (Kaczmarek and Adey, 1973, 1974; Bawin et al., 1975; Bawin and Adey, 1976; Lin-Liu and Adey, 1982), and the non-equilibrium character of this altered binding is attested by its occurrence in narrow frequency and amplitude windows.

"Tuning curves" (*frequency windows*) of altered Ca^{2+} efflux as a function of low frequencies in imposed EM fields were first seen in pioneering studies by Bawin et al. (1975; Bawin and Adey, 1976), either as a response to simple low-frequency fields or with low-frequency amplitude modulation of RF fields. Maximal sensitivities were noted at frequencies around 16 Hz and were less at higher and lower frequencies.

Neither size nor geometry is a primary determinant of these interactions. They have been noted in biota having an enormous range of physical dimensions, ranging from awake cerebral cortex in intact animals to cultured neurons, and finally in isolated terminals of cerebral nerve fibers (synaptosomes) with mean diameters around 0.7 μm. They have been reported with RF fields amplitude modulated at low frequencies; with low-frequency electric fields; with low-frequency EM fields; and with combined low-frequency EM fields and static magnetic fields.

In addition to frequency windows, these studies have revealed *intensity windows* in modification of calcium binding. Blackman et al. noted an intensity window in cerebral tissue, confirmed by Bawin et al. (1978b), for RF fields in the range 0.1–1.0 mW/cm^2 (61 V/m in air) with sinusoidal amplitude modulation at 16 Hz. At cellular dimensions, these RF fields typically produce EEG-level gradients (10–100 mV/cm). Calcium-dependent processes also exhibit intensity windows, as in the activity of the enzyme ornithine decarboxylase (ODC), essential for growth in all eukaryotic cells, in liver cells exposed to electric fields in the range 0.10–10 mV/cm (Byus et al., 1987).

CONCEPTS OF COOPERATIVITY IN BIOMOLECULAR SYSTEMS

Many functional linkages between participating elements in the dynamic patterns found in biomolecular systems are characterized by *cooperativity*, defined as ways in which components of a macromolecule, or a system of macromolecules, act together to switch from one stable state of a molecule to another (see Adey, 1988b, for review). These joint actions frequently involve phase transitions, hysteresis, and avalanche effects in input–output relationships (Wyman, 1948; Wyman and Allen, 1951; Schmitt et al., 1975). There is much evidence that molecular organization in biological systems needed to sense weak stimuli, whether thermal, chemical or electrical, may reside in joint functions of molecular assemblies or their subsets (Katchalsky, 1974; Katchalsky et al., 1974), with dynamic patterns developing in populations of elements as a result of their complex flow patterns. These flow patterns can undergo sudden transitions to new self-sustaining arrangements that will be relatively stable over time. Because these dynamic patterns are initiated by continuous inputs of energy, they are classed as "dissipative" processes. For this reason, *they occur far from equilibrium with respect to at least one important parameter in the system* (Katchalsky and Curran, 1965). Also, two or more quite distinct mechanisms can give rise to the same dynamic pattern (Othmer and Scriven, 1971).

Initial triggers to cooperative processes may be weak and the amplified responses many orders of magnitude larger, as in the sharply nonlinear release of $^{45}Ca^{2+}$ from binding sites in cerebral tissue by added calcium ions (Kaczmarek and Adey, 1973) and by weak EM fields (as previously discussed) (Bawin et al., 1975; Bawin and Adey, 1976); and in a series of calcium-dependent processes at cell membranes that include the large generation of cyclic adenosine monophosphate (cAMP) by glucagon binding to membrane receptors (Rodbell et al., 1974); the generation of cAMP by binding of parathyroid hormone to its membrane receptors and the modulation of this process by weak EM fields (Luben et al., 1982; Cain et al., 1987); the amplification of immune responses in patching and capping at cell membranes (Yahara and Edelman, 1972) and in the modulation of cell-mediated cytotoxicity of lymphocytes by weak EM fields (Lyle et al., 1983, 1988); and in the swimming behavior of bacteria elicited by small concentrations of an attractant (Koshland, 1975).

There are enormous amplifications between initial triggers in these systems and the ensuing responses, raising questions about thresholds and the minimum size of an effective triggering stimulus. In evaluating the extent of cooperativity in biological systems, the focus has usually been on effects of a change in an external parameter on the equilibrium constant of a given reaction (Schwarz, 1975; Blank, 1976). Sharp transitions from one highly stable state to another characteristic of cooperative processes can also be achieved by noncooperative means, but require much larger transition energies and the transitions occur more slowly. Sharp and fast transitions characteristic of many biological systems thus involve cooperative interactions, as in the individually weak forces in a series of hydrogen bonds or in hydrophobic reactions (Engel and Schwarz, 1970; Schwarz, 1975).

However, when compared with tissue electric gradients induced by imposed EM fields cited earlier, the requisite electric gradients found in some experimental molecular transitions are very large. For example, long-lasting conformation changes occur in poly(A)•poly(U) and in ribosomal RNA with pulsed electric fields of 20 kV/cm and with a decay time of 10 μsec (Neumann and Katchalsky, 1972). These sensitivities for nucleic acid chains in pure solution contrast sharply with effects of low-frequency trains of EM pulses on DNA synthesis in cultured cells, where significant effects occur at field intensities in the range 10^{-8} to 10^{-8} tesla (T) (Takahashi et al., 1986). Observed EM field interactions with cells and tissues based on oscillating low-frequency electric gradients between 10^{-7} and 10^{-1} V/cm would involve degrees of cooperativity many orders of magnitude greater than in the example just cited (Adey, 1988d). This discrepancy appears to relate in part to far greater sensitivities of cellular systems to low-frequency oscillating EM fields than to imposed step functions or to DC gradients that have been used in many experiments and models to test levels of cooperativity in biological systems (Blank, 1972).

Cell Membrane Receptor Proteins as Substrates for Inward Signaling and Energy Transfer

Much attention is now focused on the long strands of membrane receptor proteins as models of coupling proteins in studies of the nature of transmembrane signals. Although still incomplete, their classification has identified two major structural types. They share important common features. All have an external amino-terminus, negatively charged and located on a terminal glycoprotein strand; their inner ends are characterized by a carboxyl group. All have a short hydrophobic segment placed within the lipid bilayer (plasma membrane), contrasting sharply with the strings of hydrophilic amino acids that form the much longer external and internal segments of the strand. Their principal differences are in the number of consecutive crossings by which the strand threads itself in and out of the plasma membrane.

In their simplest forms, membrane receptor proteins appear to cross the plasma membrane only once. Those in this category include the human epidermal growth factor (EGF), the nerve growth factor (NGF), and the insulin receptor. In a second group, the strand crosses the membrane seven times. Receptors in this group include M_1- and M_2-muscarinic receptors, alpha- and beta-adrenergic receptors, rhodopsin, the substance K receptor, and probably the parathyroid hormone (PTH) receptor. As with the single-crossing receptors, each amino acid domain lying within the plasma membrane appears to form a hydrophobic region. It is noteworthy that two members in this second category, rhodopsin and the PTH receptor, exhibit sensitivity to electromagnetic fields.

Structure of Receptor Proteins Crossing the Plasma Membrane Once

The entire 1210-amino-acid sequence of the EGF receptor protein has been deduced by Ullrich et al. (1985), with striking findings on the sequences that make up the extracellular, intramembranous and cytoplasmic portions of the chain. Extracellular and intracellular segments are each composed of about 600 *hydrophilic* amino acids. The salient and surprising feature is the extremely short length of the presumed intramembranous segment of 23 amino acids, predominantly *hydrophobic*, and with only a single amino acid with a sidechain capable of hydrogen bonding.

Subsequent studies have shown that the NGF receptor protein also has a strikingly similar segment of 23 hydrophobic amino acids within the membrane (Radeke et al., 1987), suggesting that this configuration plays a fundaental role in processes of transmembrane signaling. This view is strengthened by studies with a chimeric protein constructed of the extracellular portion of the insulin receptor protein joined to the transmembrane and intracellular domains of the EGF receptor protein (Riedel et al., 1986). In this molecule, the EGF receptor kinase domain of the chimeric protein is activated by insulin binding. The authors concluded that insulin receptors and EGF receptors employ closely related or identical mechanisms for signal transduction across the plasma membrane.

Despite the hydrophobic character of the short transmembrane coupling segment, addition of EGF to human epidermal cell cultures causes a twofold increase in cytoplasmic free Ca^{2+} within 30–60 sec (Moolenaar et al., 1986).

Functional Organization of Receptor Proteins Making Seven Plasma Membrane Crossings

In a paper entitled "The Return of the Magnificent Seven," Hanley and Jackson (1987) discussed the work of Masu et al. (1987) in cloning the substance K receptor and revealing a possible arrangement with four extracellular and four intracellular domains connected by seven membrane-spanning hydrophobic segments. This study did not reveal an expected complementarity at the nucleotide level between the peptide and its cognate receptor, and thus the site of ligand binding to the receptor remained unclear.

Later studies by Lefkowitz and his colleagues (Kobilka et al., 1988) have examined chimeric $alpha_2$- and $beta_2$-adrenergic receptors to determine domains involved in effector coupling and ligand binding specificity. They identified seven hydrophobic domains consistent with transmembrane spanning segments (Fig. 2). Their effector coupling is to guanine nucleotide regulatory proteins (G proteins). *The seventh hydrophobic domain (counted away from the external amino terminus) appears to be a major determinant of both agonist and antagonist ligand-binding specificity.* However, the authors caution that it cannot be concluded that this seventh hydrophobic domain forms the ligand-binding pocket, because it may confer ligand-binding specificity by interacting with the domains directly involved in formation of the binding site.

Possible Arrangement of Hydrophobic Domains

These hydrophobic domains may form membrane-spanning alpha-helices, as suggested by electron diffraction studies in bacteriorhodopsin (Henderson

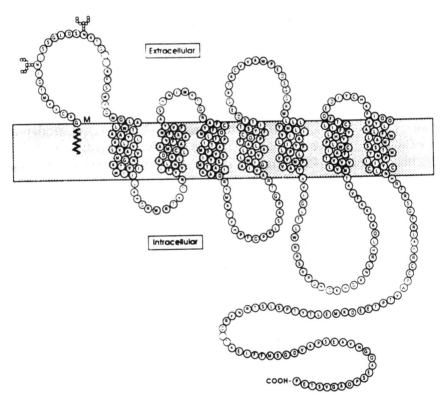

Fig. 2. Model for insertion of substance-K receptor in plasma membrane. Segments spanning lipid bilayer (plasma membrane) are sequence of seven hydrophobic domains. (From Hanley and Jackson, 1987.)

and Unwin, 1975). The less hydrophobic amino acids of these alpha-helices are likely to project toward the interior of the molecule, while the more hydrophobic may form a boundary with the plasma membrane. The alpha-helices lying adjacent to one another are presumed to have evolved in ways that minimize steric and electrostatic repulsive forces between them.

Implications from These Studies About the Essential Nature of the Transmembrane Signal

Ullrich et al. (1985) concluded that a hydrophobic segment as short as 23 amino acids is probably too short to be involved in conformation changes, and that the hydrophobic character of the segment makes unlikely its participation in either ionic or proton movement by coulombic forces. As an alternative, they have suggested that an EGF-induced conformation change in the extracellular segment of the receptor protein strand may be transmitted to the cytoplasmic domain by movement of this short intramembranous segment in and out of the lipid bilayer, or by receptor aggregation. The hydrophobic character of the intramembranous segment and the generally hydrophobic nature of the surrounding plasma membrane suggest that charged molecules and ions within this lipid bilayer are stripped of their hydration shells. Thus, their movements would not be subject to vibration damping by water molecules, although they would be constrained by the extremely high electrostatic gradient between inner and outer leaves of the lipid bilayer (10^5 V/cm). Intrusion of a protein strand into an artificial phospholipid bilayer induces coherent states between charges on the tails of adjoining phospholipid molecules for considerable distances away from the protein strand, and at the same time, they behave more rigidly (McConnell, 1975).

What role may this hydrophobic intramembranous portion of a receptor protein play in transduction of electromagnetic fields? The evidence favors transduction of visible light in these regions of rhodopsin molecules. For ELF magnetic fields, Luben et al. (1982, 1984) concluded that the receptor protein for the parathyroid (PTH) hormone is a probable site of field transduction, based on studies of PTH liganding to its receptor and on differential actions of these ELF fields on collagen synthesis in bone cells stimulated by PTH or by 1,25-dihydroxyvitamin D_3. In continuing studies, Luben and Duong (1989) have incompletely sequenced the PTH receptor in approximately half the molecule, with evidence from a sequence of four consecutive hydrophobic segments that its molecular configuration is consistent with other receptors having seven hydrophobic domains.

Is there evidence that transmembrane movement of ions is mediated by hydrophobic protein domains in this hydrophobic environment? If so, what may be the mechanisms? Addition of epidermal growth factor (EGF) to human epidermal cell cultures causes a two- to fourfold increase in cytoplasmic free Ca^{2+} within 30–60 sec (Moolenaar et al., 1986). This

EGF-induced signal appears to result from Ca^{2+} entry via a voltage-independent protein channel, since it is not accompanied by a change in membrane potential; it is completely dependent on extracellular Ca^{2+} moving into the cell interior. This action is inhibited by cancer-promoting phorbol esters, which have a specific membrane receptor (Ca^{2+}-dependent protein kinase C). EM field interactions with these cancer promoters are discussed subsequently.

Soliton Wave and Cyclotron Resonance Models of Ionic Movements in Hydrophobic Domains at Cell Membranes

In ionic movements through these hydrophobic domains, physical models offer several options. We have hypothesized that this transmembrane signaling may involve nonlinear vibration modes in helical proteins and generation of Davydov-Scott soliton waves (Lawrence and Adey, 1982). Although evidence for soliton waves in DNA and helical proteins is not conclusive and has been criticized on theoretical grounds (Lawrence et al., 1987), there is strong evidence for nonlinear, nonequilibrium processes at critical steps in transmembrane coupling, based on windows of EM field frequency, amplitude, and time of exposure that determine stimulus effectiveness (Adey, 1988b).

Polk (1984) first suggested that free (unhydrated) Ca^{2+} ions in the earth's geomagnetic field would exhibit cyclotron resonances about 10 Hz, and that these cyclotron currents would be as much as five orders of magnitude greater than the Faraday currents if the Ca^{2+} ions exhibited nearest-neighbor coherence. Interactions between the earth's geomagnetic field and a weak low-frequency EM field (40 V/m peak to peak in air; estimated tissue components, 10^{-7} V/cm) modify Ca^{2+} efflux from chick cerebral tissue (Blackman et al., 1985b). For example, halving the local geomagnetic field with a Helmholtz coil rendered ineffective a previously effective 15-Hz field, and doubling the geomagnetic field caused an ineffective 30-Hz signal to become effective.

In the cyclotron resonance model, most of the singly and doubly charged ions of biological interest have gyrofrequencies in a vacuum in the range 10–100 Hz for a mean value of the earth's geomagnetic field of 0.5 gauss (G) (Liboff, 1985). Liboff hypothesizes that imposed EM fields at frequencies close to a given resonance may couple to the corresponding ionic species in such a way as to selectively transfer energy to these ions. He proposes that data from the Blackman experiments previously discussed may relate to cyclotron resonance in singly ionized K^+, with secondary effects on Ca^{2+} efflux.

Intercellular Communication Through Gap-Junctions

Specialized regions of contact between membranes of adjacent cells (Robertson, 1963) serve to couple cells electrically (Furshpan and Furakawa,

1962). High-resolution electron microscopy with lanthanum staining reveals a thin 2- to 3-nm cleft containing protein to which Revel and Karnovsky (1967) assigned the term gap-junction. These gap-junctions are perforated by numerous tiny tubes (connexons) 1.5 nm in diameter that span the entire membrane; when connexons of adjacent cell membranes come into register, the interiors of adjacent cells are effectively in continuity.

These connexons thus provide a physical substrate for ionic coupling and transfer of essential metabolic substances between cells in the process of *metabolic cooperation* (see Fletcher et al., 1987a, for review). Studies by Pitts and coworkers (Pitts and Finbow, 1986) demonstrated that intimately contacting cells could exchange products of metabolism. For example, mutant cells having a defective purine salvage pathway that prevented them from utilizing a specific nucleotide precursor could, nevertheless, incorporate a metabolite of that molecule into their nucleic acid as long as they contacted cells with a functional metabolic pathway.

It remains for future research to establish the exact nature of this metabolic cooperation, since the identity of material in recipient cell nuclei remains unknown, and no precise change has so far been identified in metabolic events as a result of metabolite transfer; but the fact that dansylated amino acids can pass from one cell of a contacting pair to its partner (Johnson and Sheridan, 1971) supports its physiological significance. Cell density-dependent suppression of transformation (Herschmann and Brankow, 1986, 1987) further supports this concept.

Oncogenes may also interrupt intercellular pathways via gap-junctions, possibly by expression of peptides that act at cell membranes as spurious growth factors (Castagna, 1987). In experiments using normal rat kidney cells (NRK) that bear a proviral insert from a temperature-sensitive mutant of Rous sarcoma virus (LA25), Atkinson et al. (1981, 1986) demonstrated an unequivocal relationship between gap-junctions and intercellular communication. They used the temperature sensitivity of the viral gene to show that intercellular communication was fully effective at 39°C, but that both communication channels and identifiable gap-junctions were lost at 33°C, the virus permissive temperature.

LOW-FREQUENCY ELECTROMAGNETIC FIELDS IN CARCINOGENESIS

Disruption of intercellular communication through gap-junctions leads to serious disorders in growth control, including effects on tissue repair and neoplastic transformation (Loewenstein, 1977, 1979, 1981). Controlled cell growth occurs in the presence of gap-junctions, but in their absence, growth may be unregulated. This can be reversed in vitro if cancer cells make contact with normal cells (Newmark, 1987). Chemical cancer promoters, such as phorbol esters which have specific receptors at cell membranes, disrupt intercellular communication through gap-junctions. They are enhanced in this action by athermal microwave fields amplitude modulated at low frequencies (Fletcher et al., 1986, 1987b). As discussed next, disruption of gap-junction communication is now viewed as a prime factor in cancer promotion and tumor formation (Adey, 1987a,b, 1988a; Trosko, 1987; Yamasaki, 1987).

Tumor formation is thought to involve at least two steps, an early step of *initiation* and a later *promotion* effect (Fig. 3). A single agent (a complete carcinogen) may cause both events, or two or more separate events may be necessary, working together in the proper sequence.

EPIGENETIC OR NONGENOTOXIC CARCINOGENESIS

The multistage model of carcinogenesis outlined earlier predicts multiple mechanisms and genes in processes that proceed through multiple defined stages: initiation, promotion, and progression. Development of a fully malignant tumor involves complex interactions between environmental (chemicals, nonionizing and ionizing radiation, viruses) and endogenous (genetic, hormonal) factors (Weinstein, 1988). The overall process can occupy a major fraction of the lifetime of the individual (Foulds, 1954; Slaga et al., 1978; Weinstein et al., 1984).

Transitions between successive stages can be enhanced or inhibited by different types of agents. Therefore, individual stages may involve

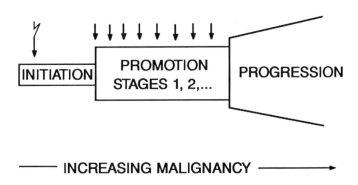

Fig. 3. Model of multistage carcinogenesis from studies in mouse skin. Initiation results from only a single exposure to a carcinogen that appears to damage nuclear DNA. Promotion involves multiple exposures at certain intervals to agents that do not damage DNA directly. Promotion leads to conversion of benign to malignant tumors, with progession increasing degree of malignancy. (After Weinstein, 1988.)

qualitatively different mechanisms at the cellular and genetic levels; in consequence, establishment and maintenance of a malignant tumor may involve multiple cellular genes, and at the same time, multiple types of changes may ensue in genomic structure and function in response to continuing exposure of cells to environmental factors (Weinstein, 1988).

Initiation is generally thought to involve a change in the cell's genetic stores of DNA, but that change is not expressed and a tumor does not result unless one or more promoting agents act repeatedly at a later time. After initiation, the time between exposure to a promoter and appearance of a tumor may be brief. Initiated cells may remain quiescent if they are not stimulated by a promoter, and cancer may never develop if sufficient exposures to promoters do not occur. In a specific context, tobacco proteins are both initiators and promoters. Because of cigarette smoking's promotional attributes, risks of lung cancer decline rapidly after a smoker rejects the habit.

Promotion occurs in previously initiated cells by the action of agents having very weak or no carcinogenetic activity when tested alone, but they markedly enhance tumor yield when applied repeatedly following a low or suboptimal dose of a carcinogen (Slaga et al., 1978; Berenblum, 1982; Kikkawa et al., 1983; Nishizuka, 1984; Berridge, 1987). Skin tumor promoters include croton oil, certain phorbol esters in croton oil, some synthetic phorbol esters, certain *Euphorbia* lattices, extracts of unburned tobacco, tobacco smoke condensate, dinitrobenzene, benzpyrene and benzoyl peroxide.

In contrast to initiating agents, phorbol ester tumor promoters do not bind to DNA but act by binding to membrane-associated receptors. They thus produce their initial effects at the epigenetic level (Weinstein, 1988). The most widely used phorbol ester in experimental cancer studies, tetradecanoyl phorbol acetate (TPA), induces three types of effects in cell culture systems: mimicry of the transformed phenotype, modulation (either inhibition or induction) of differentiation, and membrane effects (Diamond, 1987; Weinstein et al., 1987). Nonionizing EM fields (see following) interact with TPA at cell membranes to modulate its actions on both inward and outward signal streams (Adey, 1986, 1987a,b, 1988a,b).

Available evidence indicates that nonionizing EM fields do not function as classical initiators in the etiology of cancer by causing DNA damage and gene mutation. On the other hand, epidemiological evidence increasingly implicates industrial and domestic exposures to environmental ELF fields and to RF fields amplitude modulated at ELF in increased cancer risk (Wertheimer and Leeper, 1979; Milham, 1985, 1988; Thomas et al., 1987; Savitz et al., 1988). Cell culture studies of the enzyme ornithine decarboxylase (ODC), essential for growth in all eukaryotic cells, have shown its sensitivity to the phorbol ester cancer promoter TPA, and that its response to TPA is enhanced by imposed EM fields (Byus et al., 1987, 1988).

There is also evidence of impaired immune surveillance (see later section), implicating ELF and ELF-modulated RF fields in impairment of immune surveillance through actions on lymphocyte cultures. These include reduced allogeneic cytotoxicity toward tumor cells (Lyle et al., 1983, 1988) and reduced protein kinase activity (Byus et al., 1984).

Membrane Effects of Cancer-Promoting Phorbol Esters

With the discovery that the cancer-promoting phorbol ester TPA activates the membrane-related enzyme phosphatidylserine protein kinase (kinase C, PKC) (Castagna et al., 1982) and subsequent studies indicating that PKC is the major cellular receptor for TPA (Nishizuka, 1983, 1984), a series of bridges now unite research on tumor promotion, growth factors, signal transduction, and the action of specific oncogenes.

There is direct evidence that PKC plays a critical role in normal cellular growth and that it mediates several of the effects of phorbol ester tumor promoters (Weinstein, 1987). PKC is irreversibly activated by phorbol esters. In invertebrate neurons, its activation or injection of the pure enzyme enhances inward Ca^{2+} currents (DeRiemer et al., 1984). Exhaustion of protein kinase C occurs with continued stimulation with phorbol esters and is associated with loss of cell division. This response is restored by intracellular injection of protein kinase C (Pasti et al., 1986), thus linking it to paths that signal from cell membranes to nuclear mechanisms mediating cell division. On the other hand, cells that genetically overexpress PKC exhibit dramatic changes in morphology when exposed to TPA. Unlike control cells which become refractory to TPA because of downregulation of PKC, the overproducer cell lines continue to respond to TPA treatment, and, when maintained at postconfluence, develop small dense foci. Overproduction of a single form of PKC is sufficient to cause anchorage-independent growth and other abnormalities in rat fibroblasts (Housey et al., 1988). PKC belongs to a group of cAMP-independent protein kinases identified as sensitive to weak RF fields amplitude modulated at ELF frequencies (Byus et al., 1984).

Role of Protein Kinase C in Cell Responses to Tumor Promoters

Protein kinase C is a Ca^{2+}- and phospholipid-dependent enzyme that is activated by diacylglycerol. Diacylglycerol is normally almost absent from membranes, but is transiently produced from inositol phospholipids in response to extracellular signals. A wide variety of extracellular signals, which activate cellular functions and cell proliferation through interaction with membrane receptors, stimulate breakdown of inositol phospholipids (Fig. 4). The phorbol ester TPA has a diacylglycerol-like structure and is able to substitute for diacylglycerol at extremely low concentrations. Like diacylglycerol, TPA dramatically increases affinity of the enzyme for Ca^{2+}

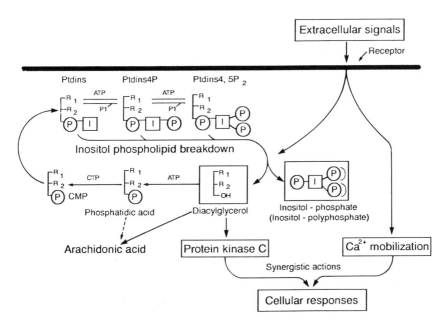

Fig. 4. Inositol phospholipid turnover and role of protein kinase C in signal transduction and cell responses to tumor promoters. Abbreviations: PtdIns, phosphatidylinositol; PtdIns4P, phosphatidylinositol-4-phosphate; PtdIns-4,5P$_2$, phosphatidyl-4-5-*bis*-phosphate; R$_1$ and R$_2$, fatty acyl groups; I, inositol; P, phosphoryl group. (From Nishizuka, 1984.)

to the 10^{+7} M range, resulting in its full activation without detectable mobilization of Ca^{2+} (Nishizuka, 1984).

Phosphatidylserine is indispensable in activation of PKC, and other phospholipids show positive cooperativity with phosphatidylserine. For example, when phosphatidylethanolamine is present as an additional lipid component, PKC is fully activated by diacylglycerol at the 10^{+7} M range of Ca^{2+}, whereas phosphatidylcholine and sphingomyelin are inhibitory. PKC is transferred into cell membranes from the cytosol as an essential step in its activation by TPA. This translocation involves synergism between Ca^{2+} and the phorbol ester (Wolf et al., 1985). As noted, activation of PKC by phorbol esters is irreversible.

The studies of Kikkawa et al. (1983) also "raise the possibility that one molecule of diacylglycerol which is produced *in situ* from inositol phospholipid in membranes may activate every molecule of protein kinase C. Although it is difficult at present to prove the mechanism in which protein kinase C senses such a tiny change in membranes, the results seem to provide an entirely new concept of receptor functions as well as a new approach to understand better the mechanism of cell growth and differentiation."

Isolation of Phorbin, a Gene Induced by PKC Activation

Weinstein (1988) has examined the role of PKC activation in the expression of a number of cellular genes and the underlying molecular mechanisms. These studies have identified phorbin as the gene corresponding to TPA and have shown that activation of PKC is an essential step in its formation. Weinstein concluded that although studies of cancer promotion with TPA are concerned with mouse skin cancers, ". . . they have relevance to other types of tumor promoters and to tumor promotion in other tissues and species, including the process of multistage carcinogenesis in man. Although it is unlikely that all classes of tumor promoters function by directly activating PKC, it is possible that they activate PKC indirectly by enhancing diacylglycerol production or by distorting related pathways of signal transduction."

Experimental Evidence Relating Environmental Electromagnetic Fields to Cell Mechanisms in Immune Surveillance, Growth, and Tumor Promotion

The foregoing discussion has outlined the multistage initiation-promotion model of carcinogenesis and described the known sequence of events in a widely accepted model of chemical promotion arising in the action of phorbol esters. What is the evidence for involvement of environmental EM fields in the promotion phase of carcinogenesis, since there is no evidence to support their action in the initiation phase of damage to nuclear DNA?

Studies in animal models have been extremely limited. Szmiegielski et al. (1982) have reported enhanced tumor rates in mouse skin initiated with benzpyrene and subsequently exposed to microwave fields [2.45 GHz at 5 and 15 mW/cm²; specific absorption ratio (SAR) at 2–3 or 6–8 W/kg; 2 hr/day, 6 days/week for as long as 10 months]. The same exposures accelerated the appearance of spontaneous mammary cancers. Chronic exposure of rats for periods in excess of 2 years to a 2.45-GHz, 1.0-mW/cm² field pulse modulated at 8 Hz resulted in four times the incidence of tumors found in control animals of the same strain (Guy et al., 1985). However, for ELF fields, no comprehensive animal studies on possible tumor formation have so far been reported.

Consequently, current knowledge of possible modification of cell growth regulation by EM fields has rested heavily on cell culture studies. These studies have reported: (1) altered immune responses that may relate to immune responses in the intact organism; and (2) direct effects on growth regulation. Observations in the latter category include modulation of intracellular enzyme activity; altered patterns of protein synthesis attributable to effects on mRNA transcription; interference with intercellular communication either directly or by synergic actions with chemical cancer promoters; and through modulation of cell responses attributable to expression of oncogenes.

EM Field Effects on Cell-Mediated Immune Responses

In cell cultures, the ability of T lymphocytes to destroy tumor cells against which they are targeted (allogeneic cytotoxicity) has been tested in the presence of 60-Hz fields and also in RF fields amplitude modulated at ELF frequencies. Destruction of the target cell depends on cell–cell contact between the lymphocyte and the tumor cell. It results in rupture of the tumor cell membrane (lysis).

Studies with mouse cytotoxic T lymphocytes and human lymphoma cells showed a reduction in lymphocyte-killing capacity of more than 20% in a 450-MHz field (incident energy, 1.5 mW/cm^2) as a function of ELF sinusoidal amplitude-modulation frequency (Fig. 5). This windowed sensitivity was maximal at 60 Hz (Lyle et al., 1983). Tumor cells and lymphocytes in this study were either cocultured for 4 hr or the lymphocytes were first exposed to the field and then mixed with the tumor cells. The reduced cytotoxicity in the preexposed lymphocytes was at the same level immediately after field exposure as in the cocultured study, and gradually returned to normal over a 12-hr period, leading to the conclusion that the reduced cytotoxicity related to EM field actions on the lymphocytes, with their cell membranes as the site of action.

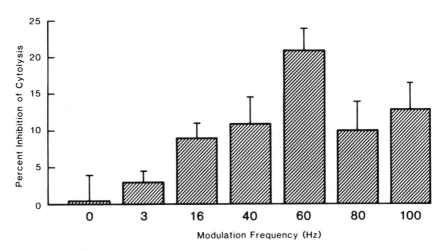

Fig. 5. Inhibition of cytotoxicity of allogeneic T lymphocytes by exposure to 450-MHz field (1.5 mW/cm^2) as sinusoidal amplitude modulation was varied between 0 and 100 Hz shows windowed sensitivity to modulation frequency. (From Lyle et al., 1983.)

Using the same lymphocyte and tumor cell lines, exposure to 60-Hz sinusoidal electric fields led to a 25% inhibition of cytotoxicity in a 4-hr assay, following a 48-hr preexposure of the effector lymphocytes to a 10-mV/cm field ($p < .005$). At 1.0 mV/cm, an inhibition of 19% occurred ($p < .0005$). At 0.1 mV/cm, there was a nonsignificant 7% inhibition of cytotoxicity (Lyle et al., 1988). When the cytotoxicity assay was conducted in the presence of the field using previously unexposed effector lymphocytes, cytotoxicity was not significantly reduced. These data suggest a dose response and a threshold (between 0.1 and 1.0 mV/cm) for inhibition of cytotoxicity in clonal T lymphocytes by exposure to a sinusoidal electric field, and also suggest mechanisms by which these fields could alter immune functions.

Canine leukocytes and human peripheral blood lymphocytes were tested in 60-Hz electric, magnetic, and combined electric and magnetic fields (3–300 mA/cm^2 RMS electric fields; 0.1–1.0 G RMS magnetic fields) for synthesis of DNA, RNA and protein, for surface receptor activity and immunoglobulins, and for blastogenic (transformation) responses (Winters, 1986). No significant changes were reported.

Protein kinase enzymes in cultured human tonsil lymphocytes are sensitive to weak RF fields with ELF modulation. These protein kinases are messengers that activate other enzymes by attaching phosphate groups to them (phosphorylation). Some are activated by cyclic adenosine monophosphate (cAMP), formed from adenosine triphosphate (ATP) with release of metabolic energy. Other protein kinases, including protein kinase C (PKC) discussed previously in connection with chemical cancer promoters, phorbol esters, are activated by signals arising in cell membranes that do not involve this cAMP pathway.

In human tonsil lymphocytes exposed to a 450-MHz field (1.0 mW/cm^2), these cAMP-independent protein kinases showed activity that was windowed with respect to exposure duration and to the modulation frequency (Byus et al., 1984). Reduced enzyme activity only occurred at modulation frequencies between 16 and 60 Hz, and only for the first 15–30 min of field exposure (Fig. 6). Unmodulated fields elicited no responses.

Enzymatic Markers of EM Field Interactions with Cancer Promoters at Cell Membranes

The enzyme ornithine decarboxylase (ODC) is required for growth in all eukaryotic cells. ODC removes carboxyl ($-COOH$) groups (decarboxylation) from the amino acid ornithine, initiating synthesis of long-chain molecules of putrescine, spermatidine and spermine. These polyamines are utilized in DNA and protein synthesis. Clinically, ODC activity in cultures of suspected cancer cells (for example, human prostatic cancer) has proved a useful index of malignancy. All agents that stimulate ODC are not cancer promoting, but all cancer promoters stimulate ODC. Its activation pathways

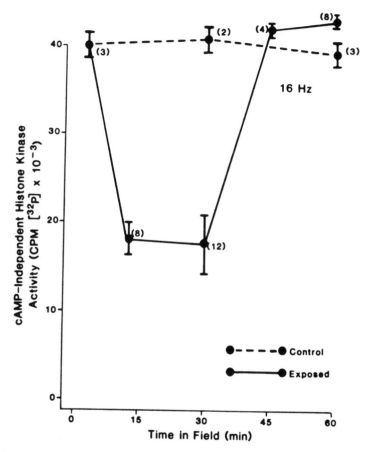

Fig. 6. cAMP-independent protein kinase activity in human lymphocytes is sharply reduced by 450-MHz field (1.0 mW/cm², amplitude-modulated at 16 Hz) 15–30 min after onset of exposure, but returns to control levels at 45–60 min, despite continuing exposure, constituting a window in time. (From Byus et al., 1984.)

are not well understood, but chemical cancer promoters with cell membrane binding sites, including phorbol esters, induce ODC activity. Its activity is also sensitive to ELF fields and to RF fields with ELF amplitude modulation.

A 1-hr exposure to a 60-Hz electric field (10 mV/cm) produced a fivefold increase in ODC activity in human lymphoma CEM cells and a two- to threefold increase in mouse myeloma cells (P3) relative to unexposed cells (Byus et al., 1987). Depending on the cell type, ODC activity increased during the 1-hr exposure period and remained elevated for several hours after exposure. In Reuber H35 hepatoma cells, there was a biphasic response with respect to field intensity. Fields at intensities of 10.0 and 0.1 mV/cm increased ODC activity by 30% immediately after a 1-hr exposure, but fields at intermediate strengths of 1.0 and 5.0 mV/cm had no effect. In these H35 cells, continuous exposure to a 60-Hz, 10-mV/cm field for periods of 2 or 3 hr resulted in either no increase in ODC activity (at 2 hr) or a decrease (at 3 hr) compared to unexposed control cultures.

Byus et al. (1988) reported a similar sensitivity of Reuber H35 hepatoma cells to a 1-hr athermal exposure (< 0.124°C temperature rise) to a 450-MHz field (peak-envelope incident energy, 1.0 mW/cm^2) sinusoidally amplitude modulated at 16 Hz. ODC activity increased by 50%, and this increase lasted for several hours. Fields modulated at 60 and 100 Hz were without effect on ODC activity. Increased ODC activity that followed Reuber H35 hepatoma cell stimulation with the phorbol ester tumor promoter TPA was further enhanced by prior exposure to the same low-energy RF field. This 16-Hz-modulated RF field also increased ODC activity in cultured Chinese hamster ovary (CHO) cells and 294T melanoma cells after a 1-hr exposure; when these CHO cells were exposed to the RF field for 1 hr, they then responded to TPA phorbol ester treatment with a further increase in ODC activity (Fig. 7).

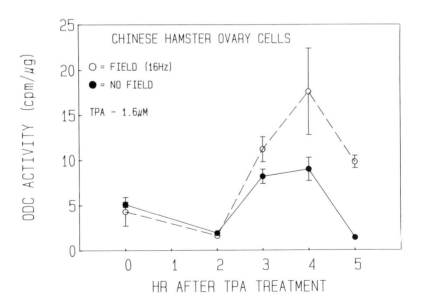

Fig. 7. Ornithine decarboxylase (ODC) activity in Chinese hamster ovary cells stimulated with tumor promoter TPA, with and without concurrent exposure to 450-MHz field (1.0 mW/cm^2) amplitude-modulated at 16 Hz. (From Byus et al., 1988.)

Since these data show that EM field exposure is capable of increasing ODC activity, it is conceivable that exposure to low-energy fields may promote tumor formation in a fashion analogous to the phorbol ester compounds. However, this speculation awaits further evidence from animal or cell culture models of tumor formation. Altered ODC activity involves a key step in growth of individual cells, and it also accompanies the increase in cell numbers characteristic of normal tissue repair and tumor formation. Specifically, increased ODC activity in itself does not necessarily mean that cells have become transformed or are in the process of becoming a tumor.

EM Field Influences on DNA Transcriptional Mechanisms

Exposure of endoreduplicated chromosomes of *Sciara* salivary gland cells to low-frequency EM fields has revealed qualitative and quantitative changes in polypeptide synthesis (Goodman and Henderson, 1988). Transcription autoradiograms showed that previously undetected transcription occurred after short exposures of these cells to ELF EM fields with diverse waveforms, causing the appearance of polypeptides, some of which were signal specific. Synthesis of other polypeptides was also altered from that seen in control cells. Five polypeptides associated with these field exposures were those seen in heat-shock conditions.

In this study, both symmetrical and asymmetrical waveforms were tested, using a Helmholtz coil exposure system at frequencies from 1.5 to 72 Hz, with magnetic-field components from 0.38 to 3.5 mT, and electric fields from 1.5×10^{-5} V/cm to 3×10^{-6} V/cm. Three different types of quasi-rectangular, asymmetric pulsed fields were used (pulse trains at 1.5 and 15 Hz and single pulses at 72 Hz). These were compared with sinusoidal fields at 60 and 72 Hz. All five fields altered polypeptide synthesis. The authors interpret these changes in transcription as a direct measure of a stressed cell state, resulting in activation of a limited number of genes that were either previously silent or not detectable.

Transformation Studies in Cell Cultures Exposed to Environmental EM Fields

Plaque formation and other evidence of unregulated growth in cultured cells exposed to EM fields would constitute important evidence for the ability of these fields to stimulate the promotion phase of carcinogenesis leading to tumor formation, although such data would still fall short of criteria from animal models that might be more directly valid in extrapolation to man. However, data from transformation studies in cultured cells remain scanty.

Two lines of human colon cells (Colo 205 and Colo 320DM) were tested in 60-Hz electric, magnetic and combined electric and magnetic fields (3- to 300-mA/m² electric fields and 0.1- to 1.0-G magnetic fields for transformation responses) (Winters, 1986). Cells were examined for colony counts and growth and for cell-surface transferrin-receptor activity as an index of proliferation. Transferrin receptors were markedly increased in numbers. Natural killer (NK) cell-mediated lysis was markedly decreased. The number of cells that plated from magnetic-field-exposed cultures was significantly higher ($p < .006$ in three of four experiments). However, attempts by Cohen (1987) to replicate the clonogenicity reported in these colon cancer cell studies, using the same agar culture techniques, were unsuccessful.

A possible relationship between activity of transferrin in transporting iron to body tissues is suggested by the finding that serum transferrin saturation is significantly higher in male patients with cancer than in controls, leading to the hypothesis that high body iron stores increase the risk of cancer in men (Stevens et al., 1988). Transferrin is essential for cell proliferation, suggesting that it may trigger a proliferative response following its interaction with receptors and serve as a growth factor.

In studies using an iron chelate (Fe•SIH), which they developed to efficiently supply iron to cells without using the transferrin pathway, Laskey et al. (1988) observed that blocking monoclonal antibodies against transferrin receptors inhibited proliferation of both Ran and murine erythroleukemia cells. This inhibited cell growth was restored by addition of Fe•SIH, which was also shown to deliver iron to Ran cells in the presence of blocking antitransferrin receptor antibodies.

Phillips and coworkers (Phillips, 1986; Phillips et al., 1986) have studied the role of target cell transferrin receptors (TfR) in natural killer (NK) cell-induced cytolysis of human colon cancer cells (Colo 205) in which prior exposure to 60-Hz EM fields produced constitutive expression of maximum numbers of TfR. Cells received a single 24-hr exposure to either a 60-Hz magnetic field (1.0 G) or to a combined 60-Hz electric and magnetic field (current density, 300 mA/m^2, 1.0 G). NK cell cytolysis was decreased in those cells expressing the highest levels of TfR, and cells expressing the lowest levels of TfR were lysed to the greatest extent. In addition to a discordance between TfR expression on target cells and their sensitivity to NK cell lysis, the data provide further evidence of cell membrane changes attributable to exposure to 60-Hz EM fields.

Phorbol ester treatment of embryonic mouse fibroblasts previously irradiated with x rays and microwaves (2.45 GHz, 34 mW/cm^2, 120 pulses/sec, 83-μsec pulse width, 24-hr exposure; SAR, 4.4 W/kg) increased transformation frequencies (mutation with unregulated growth) significantly above cells irradiated only with x rays (Balcer-Kubiczek and Harrison, 1985). These fields also enhanced promotional activity of the phorbol ester tumor promoter TPA (Balcer-Kubiczek and Harrison, 1989). These findings, as with the transferrin system, are consistent with cell membranes as the site of persisting effects of prolonged EM field exposure in interactions with growth factors, chemical tumor promoters, and proteins expressed by oncogenes.

Epigenetic Carcinogenesis: Disruption of Intercellular Communication Through Gap-Junctions by Phorbol Esters and Electromagnetic Fields

Phorbol esters and other chemical cancer promoters disrupt transfer of chemical signals between cells (Yotti et al., 1979; Fletcher et al., 1987b). Trosko has hypothesized that two major types of intercellular communication help to maintain "normal orchestration" of proliferation and differentiation during development and between quiescent stem-progenitor and differentiated cells in the adult (Trosko and Chang, 1986; Trosko, 1987); one involving transfer of molecular signals from cells of one differentiation or tissue type to another over an extracellular space and distance (for example, in the action of hormones, growth factors, and neurotransmitters); and the other mediated by transfer of relatively small molecular weight molecules and ions to neighboring cells via gap-junctions. In Trosko's model, gap-junctional communication would involve at least four steps: recognition by neighboring cells of one another through the action of adhesion molecules; functional gap-junctions; small regulatory or signaling molecules; and transducing protein receptors for these signaling molecules.

Using fluorescent dye transfer techniques, progressive changes in homologous and heterologous gap-junction communication were observed in cell lines derived from selected stages of SENCAR mouse skin carcinogenesis (Klann et al., 1989). Progressive loss of homologous communication occurred as the neoplastic process increased. Tests of heterologous communication of these cells with their normal counterparts were made in cocultures and showed functional communication without selectivity, suggesting that progressive loss of homologous but not heterologous communication accompanies neoplastic development.

Fletcher et al. (1987b) noted that blockage of the entry of natural cytolytic substances, alpha-lymphotoxins (LT) and recombinant tumor necrosis factor (TNF), into Chinese hamster ovary cells depends on their ability to form gap-junctions, a function that varies between different strains of these cells. Fletcher found that the cancer promoter TPA opens gap-junctions to permit entry of LT, leading to cell death (lysis) in a dose-dependent fashion (Fig. 8). Weak RF fields (450 MHz, 1.0–1.5 mW/cm^2 incident energy) with 16-Hz sinusoidal amplitude modulation enhanced this ability of phorbol ester tumor promoters to impair gap-junction communication. Moreover, the enhanced response required the presence of 16-Hz amplitude modulation and did not occur with an unmodulated carrier wave having the same incident energy (Fig. 9).

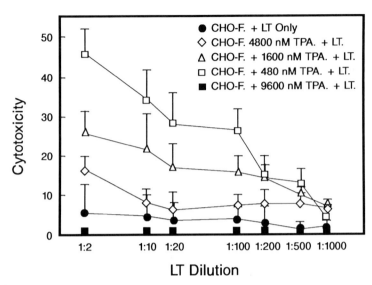

Fig. 8. Effects of cancer-promoting phorbol ester (TPA) on alpha-lymphotoxin- (LT-) mediated cytolysis of Chinese hamster ovary cells (CHO) treated with increasing doses of TPA in presence of graduated concentration of LT. (From Fletcher et al., 1987b.)

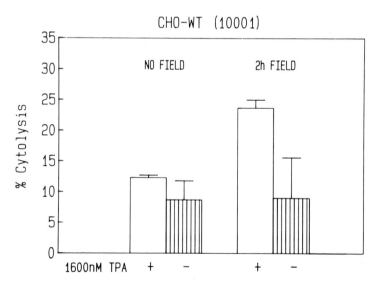

Fig. 9. Separate and combined actions of phorbol ester tumor promoter TPA and 450-MHz RF field to impair gap-junction communication between hamster ovary cells, allowing entry of alpha-lymphotoxin (LT). LT action was enhanced by field with 16-Hz sinusoidal amplitude modulation (2-hr FIELD), but there was no effect of unmodulated carrier wave (NO FIELD) at same incident energy (1.0 mW/cm^2). (From Fletcher et al., 1986.)

SUMMARY

We have reviewed events that couple chemical stimuli from cell-surface receptor sites to the interior. Weak imposed electromagnetic fields with low-frequency components (typically in the spectrum below 100 Hz) have proved to be unique tools in identifying the sequence and energetics in these events. Cell membranes are primary sites of interactions with these fields.

Recent observations have opened doors to new concepts of communication between cells as they whisper together across barriers of cell membranes. Regulation of cell-surface chemical events by these fields indicates a major amplification of initial weak triggers associated with binding of hormones, antibodies and neurotransmitters to their specific binding sites. Calcium ions play a key role in this stimulus amplification. The evidence supports nonlinear, nonequilibrium processes at critical steps in transmembrane coupling.

There is much evidence that the molecular organization in biological systems essential for sensing of weak stimuli, whether thermal, chemical, or electrical, resides in joint functions of molecular assemblies or subsets of these assemblies. On the other hand, it is at the atomic level within these molecular systems that physical rather than chemical events now appear to shape the flow of signals and the transmission of energy.

Communication between cells through gap-junctions is also sensitive to low-frequency electromagnetic fields at athermal intensities. We hypothesize that cancer promotion with tumor formation may involve dysfunctions at cell membranes, disrupting inward and outward signal streams.

In the frame of multistep initiation-promotion models of carcinogenesis, there is no evidence that environmental low-frequency electromagnetic fields, or combinations of electric and magnetic fields, act to damage genetic stores of DNA in cell nuclei. Thus, in the frame of initiation-promotion models of carcinogenesis, there is no evidence that these fields act as initiators.

There is evidence that these fields may participate in the promotion phase of carcinogenesis by at least two mechanisms: effects on immune responses, and direct effects on mechanisms regulating cell growth.

Powerful chemical cancer promoters that act at cell membranes are enhanced by concurrent cell exposures to environmental electromagnetic fields. These chemical cancer promoters direct inward signals from cell membranes to enzymes within cells that regulate growth. They also interfere with outward signals necessary for communication between cells, and interference with these outward signals can lead to unregulated growth in cells that are thereby isolated. Both low-frequency fields and radio-frequency fields with low-frequency modulation can act separately and synergically with chemical cancer promoters at cell membranes.

These actions of low-frequency fields primarily at cell membranes in mechanisms that relate to tumor formation are within the purview of *epigenetic carcinogenesis*, a field of cancer research that considers the role of promoting agents in cells already initiated (transformed, mutated) through prior damage to nuclear DNA stores but not manifesting unregulated growth with tumor formation until the onset of the promotion phase.

ACKNOWLEDGMENTS

We gratefully acknowledge support for studies in our laboratory from the U.S. Department of Energy, the U. S. Environmental Protection Agency, the U.S. Bureau of Devices and Radiological Health (FDA), the U.S. Office of Naval Research, the U.S. Veterans Administration, the Southern California Edison Company, and the General Motors Medical Research Institute.

REFERENCES

Adey, W. R. 1981a. Tissue interactions with nonionizing electromagnetic fields. *Physiol. Rev.* 61:435–514.

Adey, W. R. 1981b. Ionic nonequilibrium phenomena in tissue interactions with nonionizing electromagnetic fields. In: *Biological Effects of Nonionizing Radiation*, Illinger, K. H., ed., pp. 271–298. ACS Symp. Ser. 157. Washington, DC: American Chemical Society.

Adey, W. R. 1984. Nonlinear, nonequilibrium aspects of electromagnetic field interactions at cell membranes. In: *Nonlinear Electrodynamics in Biological Systems*, Adey, W. R., Lawrence, A. F., eds., pp. 2–21. New York: Plenum.

Adey, W. R. 1986. The sequence and energetics of cell membrane transductive coupling to intracellular enzyme systems. *Bioelectrochem. Bioenerg.* 15:447–456.

Adey, W. R. 1987a. Evidence for tissue interactions with microwave and other nonionizing electromagnetic fields in cancer promotion. In: *Biophysical Aspects of Cancer*, Fiala, J., Pokorny, J., eds. Prague: Charles University.

Adey, W. R. 1987b. Cell membranes, electromagnetic fields and intercellular communication. In: *Dynamics of Sensory and Cognitive Processing in the Brain*, Second Symposium, Basar, E., ed. Heidelberg: Springer-Verlag (in press).

Adey, W. R. 1988a. Cell membranes: the electromagnetic environment and cancer promotion. *Neurochem. Res.* 13:671–677.

Adey, W. R. 1988b. Physiological signalling across cell membranes and cooperative influences of extremely low frequency electromagnetic fields. In: *Biological Coherence and Response to External Stimuli*, Frohlich, H., ed., p. 148. Heidelberg: Springer-Verlag.

Adey, W. R. 1988c. Biological effects of radio-frequency radiation. In: *Interaction of Electromagnetic Waves with Biological Systems*, Lin, J. C., ed. New York: Plenum (in press).

Adey, W. R. 1988d. Effects of microwaves on cells and molecules. *Nature (London)* 333:401.

Adey, W. R., Lawrence, A. F., eds. 1984. *Nonlinear Electrodynamics in Biological Systems.* New York: Plenum.

Adey, W. R., Bawin, S. M., Lawrence, A. F. 1982. Effects of weak amplitude-modulated microwave fields on calcium efflux from awake cat cerebral cortex. *Bioelectromagnetics* 3:295–302.

Atkinson M. M., Anderson, S. K., Sheridan, J. D. 1986. Modification of gap junctions in transformed cells by a temperature-sensitive mutant of Rous sarcoma virus. *J. Membr. Biol.* 91:53.

Atkinson, M. M., Menko, A. S., Johnson, R. G., Sheppard, J. R., Sheridan, J. D. 1981. Rapid and reversible reduction of junctional permeability in cells infected with a temperature-sensitive mutant of Rous sarcoma virus. *J. Cell. Biol.* 91:573–578.

Balcer-Kubiczek, E. K., Harrison, G. H. 1985. Evidence for microwave carcinogenesis, in vitro. *Carcinogenesis* 6:859–864.

Balcer-Kubiczek, E. K., Harrison, G. H. 1989. Induction of neoplastic transformation in C3H10T1/2 cells by 2.45-GHz microwaves and phorbol ester. *Radiat. Res.*

Bassett, C. A. L. 1987. Low-energy pulsing electromagnetic fields modify biomedical processes. *BioEssays* 6:36.

Bawin, S. M., Adey, W. R. 1976. Sensitivity of calcium binding in cerebral tissue to weak electric fields oscillating at low frequency. *Proc. Natl. Acad. Sci. USA* 73:1999–2003.

Bawin, S. M., Adey, W. R., Sabbot, I. M. 1978a. Ionic factors in release of $^{45}Ca^{2+}$ from chick cerebral tissue by electromagnetic fields. *Proc. Natl. Acad. Sci. USA* 75:6314–6318.

Bawin, S. M., Sheppard, A. R., Adey, W. R. 1978b. Possible mechanisms of weak electromagnetic field coupling in brain tissue. *Bioelectrochem. Bioenerg.* 45:67–76.

Bawin, S. M., Kaczmarek, L. K., Adey, W. R. 1975. Effects of modulated VHF fields on the central nervous system. *Ann. N.Y. Acad. Sci.* 247:74–80.

Berenblum, I. 1982. Sequential aspects of chemical carcinogenesis: skin. In: *Cancer: A Comprehensive Treatise*, Vol. 1, Ed. 2, Becker, F. F., ed., p. 237. New York: Plenum.

Berridge, M. J. 1987. Inositol triphosphate and diacylglycerol: two interacting second messengers. *Annu. Rev. Biochem.* 56:159–193.

Blackman, C. F., Benane, S. G., Kinney, L. S., Joines, W. T., House, D. E. 1982. Effects of ELF fields on calcium-ion efflux from brain tissue, in vitro. *Radiat. Res.* 92:510–520.

Blackman, C. F., Benane, S. G., House, D. E., Joines, W. T. 1985a. Effects of ELF (1–120 HZ) and modulated (50 Hz) RF fields on the efflux of calcium ions from brain tissue, in vitro. *Bioelectromagnetics* 6:1–12.

Blackman, C. F., Benane, S. G., Rabinowitz, J. R., House, D. E., Joines, W. T. 1985b. A role for the magnetic field in the radiation-induced efflux of calcium ions from brain tissue in vitro. *Bioelectromagnetics* 6:327–338.

Blackman, C. F., Elder, J. A., Weil, C. M., Benane, S. G., Eichinger, D. C., House, D. E. 1979. Induction of calcium ion efflux from brain tissue by radio frequency radiation. *Radio Sci.* 14:93–98.

Blank, M. 1972. Cooperative effects in membrane reactions. *J. Colloid Interface Sci.* 41:97.

Blank, M. 1976. Hemoglobin reactions as interfacial phenomena. *J. Electrochem. Soc.* 123:1653.

Byus, C. V., Pieper, S. E., Adey, W. R. 1987. The effects of low-energy 60-Hz environmental electromagnetic fields upon the growth-related enzyme ornithine decarboxylase. *Carcinogenesis* 8:1385–1389.

Byus, C. V., Kartun, K., Pieper, S., Adey, W. R. 1988. Increased ornithine decarboxylase activity in cultured cells exposed to low energy modulated microwave fields and phorbol ester tumor promoters. *Cancer Res.* 48:4222–4226.

Byus, C. V., Lundak, R. L., Fletcher, R. M., Adey, W. R. 1984. Alterations in protein kinase activity following exposure of cultured lymphocytes to modulated microwave fields. *Bioelectromagnetics* 5:341–352.

Cain, C. D., Luben, R. A. 1987. Pulsed electromagnetic field effects on PTH-stimulated cAMP accumulation and bone resorption in mouse calvaria. In: *Interactions of Biological Systems with Static and ELF Electric and Magnetic Fields*, Anderson, L. E., Kelman, B. J., Weigel, R. J., eds., pp. 269–278, 23rd Annual Hanford Life Sciences Symposium, Richland, Washington. CONF 841041. Springfield, Virginia: National Technical Information Service.

Cain, C. D., Adey, W. R., Luben, R. A. 1987. Evidence that pulsed electromagnetic fields inhibit coupling of adenylate cyclase by parathyroid hormone in bone cells. *J. Bone Miner. Res.* 2:437.

Castagna, M. 1987. Phorbol esters as signal transducers and tumor promoters. *Biol. Cell* 59:3.

Castagna, M., Takai, Y., Kaibuchi, K., Sano, K., Kikkawa, U., Nishizuka, Y. 1982. Direct activation of calcium-activated phospholipid-dependent protein kinase by tumor-promoting phorbol esters. *J. Biol. Chem.* 257:7847–7851.

Cohen, M. M. 1987. *The Effects of Low-Level Electromagnetic Fields on Cloning of Two Human Cancer Cell Lines (colo 205 and colo 320).* Final Report, New York State Power Lines Project. Albany, New York: Wadsworth Center for Laboratories and Research.

Cole, K. S. 1940. Permeability and impermeability of cell membranes for ions. *Cold Spring Harbor Symp. Quant. Biol.* 4:110.

DeRiemer, S. A., Strong, J. A., Albert, K. A., Greengard, P., Kaczmarek, L. K. 1984. Enhancement of calcium current in *Aplysia* neurons by phorbol ester and kinase C. *Nature (London)* 313:313–316.

Diamond, L. 1987. Tumor promoters and cell transformation. In: *Mechanisms of Cellular Transformation by Carcinogenic Agents*, Grunberger, G., Goff, S., eds., pp. 73–134. New York: Pergamon Press.

Dixey, R., Rein, G. 1982. ^{3}H-Noradrenaline release potentiated in a clonal nerve cell line by low-intensity pulsed magnetic fields. *Nature (London)* 296:253–256.

Dutta, S. K., Subramoniam, A., Ghosh, B., Parshad, R. 1984. Microwave radiation-induced calcium efflux from brain tissue, in vitro. *Bioelectromagnetics* 5:71.

Engel, J., Schwarz, G. 1970. Cooperative conformational transitions of linear biopolymers. *Angew. Chem. Int. Ed. Engl.* 9:389.

Fitzsimmons, R. J., Farley, J., Adey, W. R., Baylink, D. J. 1986. Embryonic bone matrix formation is increased after exposure to a low-amplitude capacitively coupled electric field, in vitro. *Biochim. Biophys. Acta* 882:51.

Fletcher, W. H., Shiu, W. W., Haviland, D. A., Ware, C. F., Adey, W. R. 1986. A modulated-microwave field and tumor promoters similarly enhance the action of alpha-lymphotoxin (aLT). In: *Proceedings of the 8th Annual Meeting of the Bioelectromagnetics Society*, Madison, Wisconsin, p. 12. Frederick, Maryland: Bioelectromagnetics Society.

Fletcher, W. H., Byus, C. V., Walsh, D. A. 1987a. Receptor-mediated action without receptor occupancy: a function for cell-cell communication in ovarian follicles. *Adv. Exp. Med. Biol.* 219:299.

Fletcher, W. H., Shiu, W. W., Ishida, T. A., Haviland, D. L., Ware, C. F. 1987b. Resistance to the cytolytic action of lymphotoxin and tumor necrosis factor coincides with the presence of gap junctions uniting target cells. *J. Immunol.* 139:956-962.

Foster, K. R., Guy, A. W. 1986. The microwave problem. *Sci. Am.* 255(3):32-39.

Foulds, L. 1954. The experimental study of tumor progression. *Cancer Res.* 14:327.

Goodman, R., Henderson, A. S. 1988. Exposure of salivary gland cells to low-frequency electromagnetic fields alters polypeptide synthesis. *Proc. Natl. Acad. Sci. USA* 85:3928-3932.

Furshpan, E. J., Furukawa, T. 1962. Intracellular and extracellular responses of several regions of the Mauthner cell of the goldfish. *J. Physiol. (London)* 145: 289-325.

Guy, A. W., Chou, C.-K., Kunz, L. L., Crowley, J., Krupp, J. 1985. *Effects of Long-Term Low-Level Radiofrequency Radiation Exposure on Rats*. Final Report, Vols. 8 and 9, U.S. Air Force Tech. Rep. TR-85-11 and TR-85-64. Brooks Air Force Base, Texas: U.S. Air Force School of Aerospace Medicine.

Hanley, M. R., Jackson, T. 1987. Return of the magnificent seven. *Nature (London)* 329:766.

Henderson, R., Unwin, P. N. T. 1975. Three-dimensional model of purple membrane obtained by electron microscopy. *Nature (London)* 257:28-32.

Herschmann, H. R., Brankow, D. W. 1986. Ultraviolet irradiation transforms C3H10T1/2 cells to a unique suppressible phenotype. *Science* 234:1385-1388.

Herschmann, H. R., Brankow, D. W. 1987. Cell size, cell density and nature of the tumor promoter are critical variables in expression of a transformed phenotype (focus formation) in co-cultures of UV-TDTx and C3H10T1/2 cells. *Carcinogenesis* 8:993-998.

Hiraki, Y., Endo, N., Takigawa, M., Asada, A., Takahashi, H., Suzuki, F. 1987. Enhanced responsiveness to parathyroid hormone and induction of functional differentiation of cultured rabbit costal chondrocytes by a pulsed electromagnetic field. *Biochim. Biophys. Acta* 931:94.

Hodgkin, A. L., Huxley, A. F. 1952. A quantitative description of membrane current and its application to conduction and excitation in nerve. *J. Physiol. (London)* 117:500-544.

Housey, G. M., Johnson, M. D., Hsiao, W.-L., O'Brian, C. A., Murphy, J. P., Kirschmeier, P., Weinstein, I. B. 1988. Overproduction of protein kinase C causes disordered growth control in rat fibroblasts. *Cell* 52:343-354.

Johnson, R. G., Sheridan, J. D. 1971. Junctions between cancer cells in culture: ultrastructure and permeability. *Science* 174:717-719.

Kaczmarek, L. K., Adey, W. R. 1973. The efflux of $^{45}Ca^{2+}$ and 3H-gamma- aminobutyric acid from cat cerebral cortex. *Brain Res.* 63:331-342.

Kaczmarek, L. K., Adey, W. R. 1974. Weak electric gradients change ionic and transmitter fluxes in cortex. *Brain Res.* 66:537–540.

Katchalsky, A. 1974. Concepts of dynamic patterns. Early history and philosophy. *Neurosci. Res. Program Bull.* 12:30.

Katchalsky, A., Curran, P. F. 1965. *Nonequilibrium Thermodynamics in Biophysics.* Cambridge: Harvard University Press.

Katchalsky, A., Rowland, V., Blumenthal, R., eds. 1974. Dynamic patterns of brain cell assemblies. *Neurosci. Res. Program Bull.* 12:1–195.

Kikkawa, U., Takai, Y., Tanaka, Y., Miyake, R., Nishizuka, Y. 1983. Protein kinase C as a possible receptor protein of tumor-promoting phorbol esters. *J. Biol. Chem.* 258:11442.

Klann, R. C., Fitzgerald, D. J., Piccoli, C., Slaga, T. J., Yamasaki, H. 1989. Gap-junctional intercellular communication in epidermal cell lines from selected stages of SENCAR mouse skin carcinogenesis. *Cancer Res.* 49:699.

Kobilka, B. K., Kobilka, T. S., Daniel, K., Regan, J. W., Caron, M. G., Lefkowitz, R. J. 1988. Chimeric alpha$_2$-, beta$_2$-adrenergic receptors: delineation of domains involved in effector coupling and ligand binding specificity. *Science* 240:1310–1316.

Koshland, D. E. 1975. Transductive coupling in chemotactic processes: chemo-receptor-flagellar coupling in bacteria. In: *Functional Linkage in Biomolecular Systems*, Schmitt, F. O., Schneider, D. M., Crothers, D. M., eds., p. 273. New York: Raven Press.

Laskey, J., Webb, I., Schulman, H. M., Ponka, P. 1988. Evidence that transferrin supports cell proliferation by supplying iron for DNA synthesis. *Exp. Cell Res.* 176:87.

Lawrence, A. F., Adey, W. R. 1982. Nonlinear wave mechanisms in interactions between excitable tissue and electromagnetic fields. *Neurol. Res.* 4:115–154.

Lawrence, A. F., McDaniel, J. C., Chang, D. B., Birge, R. R. 1987. The nature of phonons and solitary waves in alpha-helical proteins. *Biophys. J.* 51:785.

Liboff, A. R. 1985. Cyclotron resonance in membrane transport. In: *Interactions Between Electromagnetic Fields and Cells*, Chiabrera, A., Nicolini, C., Schwan, H. P., eds., pp. 281–296. New York: Plenum.

Lin-Liu, S., Adey, W. R. 1982. Low-frequency amplitude-modulated microwave fields change calcium efflux rates from synaptosomes. *Bioelectromagnetics* 3:309–322.

Loewenstein, W. R. 1977. Permeability of membrane junctions. *Ann. N.Y. Acad. Sci.* 137:441.

Loewenstein, W. R. 1979. Junctional intercellular communication and the control of growth. *Biochim. Biophys. Acta* 605:33–91.

Loewenstein, W. R. 1981. Junctional intercellular communication: the cell-to-cell membrane channel. *Physiol. Rev.* 61:829.

Luben, R. A., Cain, C. D. 1984. Use of bone cell hormone responses to investigate biolectromagnetic effects on membranes, in vitro. In: *Nonlinear Electrodynamics in Biological Systems*, Adey, W. R., Lawrence, A. F., eds., p. 23–24. New York: Plenum.

Luben, R. A., Duong, H. P. 1989. A candidate sequence for the mouse 70 kDa mouse osteoblast PTH receptor (PTHR) is homologous to the rhodopsin family of G-protein linked receptors (GLPR). In: *Proceedings of the Annual Meeting of the American Society of Cell Biology*, San Francisco, California, Abstr. 348.

Luben, R. A., Cain, C. D., Chen, M.-Y., Rosen, D. M., Adey, W. R. 1982. Effects of electromagnetic stimuli on bone and bone cells, in vitro: inhibition of responses to parathyroid hormone by low-energy, low-frequency fields. *Proc. Natl. Acad. Sci. USA* 79:4180–4184.

Lyle, D. B., Ayotte, R. D., Sheppard, A. R., Adey, W. R. 1988. Suppression of T-lymphocyte cytotoxicity following exposure to 60-Hz sinusoidal electric fields. *Bioelectromagnetics* 9:303–313.

Lyle, D. B., Schechter, P., Adey, W. R., Lundak, R. L. 1983. Suppression of T-lympho-cyte cytotoxicity following exposure to sinusoidally amplitude-modulated fields. *Bioelectromagnetics* 4:281–292.

Masu, Y., Nakayama, K., Tamaki, H., Harada, Y., Kuno, M., Nakanishi, S. 1987. cDNA cloning of bovine substance-K receptor through oocyte expression system. *Nature (London)* 329:836.

McConnell, H. M. 1975. Coupling between lateral and perpendicular motion in biological membranes. In: *Functional Linkage in Biomolecular Systems*, Schmitt, F. O., Schneider, D. M., Crothers, D. M., eds., p. 123. New York: Raven Press.

Milham, S. 1985. Mortality in workers exposed to electromagnetic fields. *Environ. Health Perspect.* 62:297–300.

Milham, S. 1988. Increased mortality in amateur radio operators due to lymphatic and hemopoietic malignancies. *Am. J. Epidemiol.* 127:50.

Moolenaar, W. H., Aerts, R. J., Tertoolen, L. G. J., DeLast, S. W. 1986. The epider-mal growth factor-induced calcium signal in A431 cells. *J. Biol. Chem.* 261:279–285.

Neumann, E., Katchalsky, A. 1972. Long-lived conformation changes induced by electric pulses in biopolymers. *Proc. Natl. Acad. Sci. USA* 69:993.

Newmark, P. 1987. Oncogenes and cell growth. *Nature (London)* 327:101.

Nishizuka, Y. 1983. Calcium, phospholipid and transmembrane signalling. *Philos. Trans. R. Soc. London B Biol. Sci.* 302:101–112.

Nishizuka, Y. 1984. The role of protein kinase C in cell surface signal transduc-tion and tumour promotion. *Nature (London)* 308:693–698.

Othmer, H. G., Scriven, L. E. 1971. Instability and dynamic pattern in cellular net-works. *J. Theor. Biol.* 32:507.

Pasti, G., Lacal, J.-C., Warren, B. S., Aarenson, S. A., Blumberg, P. M. 1986. Loss of mouse fibroblast resonse to phorbol esters restored by microinjected pro-tein kinase C. *Nature (London)* 324:375.

Phillips, J. L. 1986. Transferrin receptors and natural killer cell lysis. A study using Colo 205 cells exposed to 60-Hz electromagnetic fields. *Immunol. Lett.* 13:295.

Phillips, J. L., Rutledge, L., Winters, W. D. 1986. Transferrin binding to two human colon carcinoma cell lines: characterization and effect of 60-Hz electromagnetic fields. *Cancer Res.* 46:239–244.

Pitts, J. D., Finbow, M. E. 1986. The gap junction. *J. Cell. Sci. (Suppl.)* 4:239–266.

Polk, C. 1984. Time-varying magnetic fields and DNA synthesis: magnitude of forces due to magnetic fields on surface-bound counterions. In: *Proceedings of the 6th Annual Meeting of the Bioelectromagnetics Society*, p. 77 (Abstr.).

Radeke, M. J., Misko, T. P., Hsu, C., Herzenberg, L. A., Shooter, M. 1987. Gene transfer and molecular cloning of the rat nerve growth factor receptor. *Nature (London)* 325:393.

Revel, J.-P., Karnovsky, M. J. 1967. Hexagonal array of subunits in intercellular junctions of the mouse heart and liver. J. Cell Biol. 33:C7–C12.

Riedel, H., Sclessinger, J., Ullrich, A. 1986. A chimeric, ligand binding v-erbB/EGF receptor retains transforming potential. Science 236:197–200.

Robertson, J. D. 1963. The occurrence of a subunit pattern in the unit membranes of club endings in Mauthner cell synapses in goldfish brains. J. Cell Biol. 19:201–221.

Rodbell, M., Lin, M. C., Salomon, Y. 1974. Evidence for interdependent action of glucagon and nucleotides on the hepatic adenylate cyclase system. J. Biol. Chem. 249:59–65.

Savitz, D. A., Wachtel, H., Barnes, F. A., John, E. M., Tvrdik, J. G. 1988. Case-control study of childhood cancer and exposure to magnetic fields. Am. J. Epidemiol. 128:21–38.

Schmitt, F. O., Schneider, D. M., Crothers, D. M., eds. 1975. Functional Linkage in Biomolecular Systems. New York: Raven Press.

Schwarz, G. 1975. Sharpness and kinetics of cooperative transitions. In: Functional Linkage in Biomolecular Systems, Schmitt, F. O., Schneider, D. M., Crothers, D. M., eds., p. 32. New York: Raven Press.

Semm, P. 1983. Neurobiological investigations on the magnetic sensitivity of the pineal gland in rodents and pigeons. Comp. Biol. Physiol. A 76:683–689.

Singer, S. J., Nicolson, G. L. 1972. The fluid mosaic model of the structure of cell membranes. Science 175:720-731.

Slaga, T. J., Butterworth, B. E., eds. 1987. Nongenotoxic Mechanisms in Carcinogenesis. 25th Banbury Report. Cold Spring Harbor, New York: Cold Spring Harbor Laboratory.

Slaga, T. J., Sivak, A., Boutwell, R. K., eds. 1978. Mechanisms of Tumor Promotion and Cocarcinogenesis, Vol. 2. New York: Raven Press.

Stevens, R. G., Jones, D. Y., Micozzi, M. S., Taylor, P. R. 1988. Body iron stores and the risk of cancer. N. Engl. J. Med. 319:1047–1052.

Szmigielski, S., Szudzinski, A., Pietraszek, A., Bialec, M., Janiak, M., Wrembel, J. K. 1982. Accelerated development of spontaneous and benzpyrene- induced skin cancer in mice exposed to 2450-MHz microwave radiation. Bioelectromagnetics 3:179–191.

Takahashi, K., Keneko, I., Dale, M., Fukada, E. 1988. Effect of pulsing electromagnetic fields on DNA synthesis in mammalian cells in culture. Experentia 42:185–186.

Thomas, T. L., Stolley, P. D., Stemhagen, A., Fontham, E. T. H., Bleecker, M. L., Stewart, P. A., Hoover, R. N. 1987. Brain tumor mortality risk among men with electrical and electronics jobs: a case-controlled study. J. Natl. Cancer Inst. 79:233–238.

Trosko, J. E. 1987. Mechanisms of tumor promotion: possible role of inhibited intercellular communication. Eur. J. Cancer Clin. Oncol. 23:599–601.

Trosko, J. E., Chang, C. C. 1986. Oncogene and chemical inhibition of gap-junctional intercellular communication: implications for teratogenesis and carcinogenesis. In: Genetic Toxicology of Environmental Chemicals Part B: Genetic Effects and Applied Mutagenesis, p. 21. New York: Alan R. Liss.

Ullrich, A., Coussens, L., Hayflick, J. S., Dull, T. J., Gray, A., Tam, A. W., Lee, J., Yarden, Y., Libermann, T. A., Schlessinger, J., Downard, J., Mayes, E. L. V.,

Whittle, N., Waterfield, M. D., Seeburg, P. H. 1985. Human epidermal growth factor receptor cDNA sequence and aberrant expression of the amplified gene in A431 epidermoid carcinoma cells. *Nature (London)* 309:425–428.

Weinstein, I. B. 1987. Growth factors, oncogenes and multistage carcinogenesis. *J. Cell. Biochem.* 33:213–224.

Weinstein, I. B. 1988. The origins of human cancer: molecular mechanisms of carcinogenesis and their implications for cancer prevention and treatment. *Cancer Res.* 48:4135–4143.

Weinstein, I. B., Gattoni-Celli, S., Kirschmeier, P., Lambert, M., Hsiao, W.-L., Backer, J., Jeffrey, A. 1984. Multistage carcinogenesis involves multiple genes and multiple mechanisms. In: *Cancer Cells. I. The Transformed Phenotype*, Levine, A., Van de Woude, G., Watson, J. D., Topp, W. C., eds., pp. 229–237. Cold Spring Harbor, New York: Cold Spring Harbor Laboratory Press.

Weinstein, I. B., Lambert, M. E., Garrels, J. E., Ronai, Z., Hsiao, W.-L., Dragani, T. A., Hsieh, L. L., Begemann, M. 1987. The role of asynchronous DNA replication and endogenous retrotransposons in multistage carcinogenesis. In: *Accomplishments in Oncology*, Vol. 2, No. 1, Zur Hausen, H., Schlehofer, J. R., eds., pp. 69–83. Philadelphia: Lippincott.

Welker, H. A., Semm, P., Willig, P., Willschko, W., Vollrath, L. 1983. Effects of an artificial magnetic field on serotonin-*N*-acetyltransferase activity and melatonin content of the rat pineal gland. *Exp. Brain Res.* 50:426–432.

Wertheimer, N., Leeper, E. 1979. Electrical wiring configurations and childhood cancer. *Am. J. Epidemiol.* 109:273–284.

Winters, W. D. 1986. *Biological Functions of Immunologically Reactive Human and Canine Cells Influenced by In Vitro Exposure to 60 Hz Electric and Magnetic Fields.* New York State Power Lines Project, Final Report. Albany, New York: Wadsworth Center for Laboratories and Research.

Wolf, M., LeVine, H., May, W. S., Cuatrecasas, P., Sahyoun, N. 1985. A model for intracellular translocation of protein kinase C involving synergism between Ca^{2+} and phorbol esters. *Nature (London)* 317:546–549.

Wyman, J. 1948. Heme proteins. *Adv. Protein Chem.* 4:407–531.

Wyman, J., Allen, D. W. 1951. The problems of the heme interactions in hemoglobin and the basis of the Bohr effect. *J. Polymer Sci.* 7:499–518.

Yahara, I., Edelman, G. M. 1972. Restriction of the mobility of lymphocyte immunoglobulin receptors by concanavalin A. *Proc. Natl. Acad. Sci. USA* 69: 608–612.

Yamasaki, H. 1987. The role of cell-to-cell communication in tumor promotion. In: *Nongenotoxic Mechanisms in Carcinogenesis*, Butterworth, T. E., Slaga, T. J., eds., pp. 297–307. 25th Banbury Report. Cold Spring Harbor, New York: Cold Spring Harbor Laboratory.

Yotti, L. P., Chang, C. C., Trosko, J. E. 1979. Elimination of metabolic cooperation in Chinese hamster ovary cells by tumor promoter. *Science* 206:1089–1091.

Young, J. Z. 1951. *Doubt and Certainty in Science.* Reith Lectures, British Broadcasting Corporation. Oxford: University Press.

11 Ion Cyclotron Resonance Effects of ELF Fields in Biological Systems

ABRAHAM R. LIBOFF
Department of Physics
Oakland University
Rochester, Michigan

BRUCE R. McLEOD
Department of Electrical Engineering
Montana State University
Bozeman, Montana

STEPHEN D. SMITH
Department of Anatomy
University of Kentucky
Lexington, Kentucky

CONTENTS

THE CRITICAL NEED FOR A PHYSICAL MECHANISM

The public has become increasingly aware of the potential hazard associated with 50- or 60-Hz power transmission, particularly as it involves the magnetic fields derived from the current in these lines and the cancer deaths that may result from associated magnetic exposures in nearby residences. Even though epidemiological data to date have only implicated power-distribution lines (Ahlbom et al., 1987), there is every reason to believe that concern will also be mounting about similar effects from 50- to 60-Hz fields in the workplace, high-voltage lines and residential wiring configurations, and, especially, commonplace home appliances. The scientific community and the electrical power industry have recognized the need to determine the underlying mechanisms, not only to corroborate the epidemiological reports but also to alleviate the widespread concern that always accompanies ignorance in the face of danger.

Unfortunately, however, there is a wide difference of opinion as to what is meant by the term "underlying mechanism." To the oncologist, the critical part of the mechanism may be the altered expression that allows the proto-oncogene to become a transforming gene; to the biochemist, a sharp increase in polyamine biosynthesis; and to the physiologist, alterations in intracellular events that are mediated by receptor activity at the cell surface. To the physical scientist the "underlying mechanism" that might explain the relationship of electromagnetic fields to biological hazard has a totally different meaning. The physicist finds a deeply troublesome question at the heart of the entire problem: What conceivable physical process allows extremely low frequency (ELF) magnetic fields at intensities of 1 to 100 milligauss (mG) to interact with cells to cause the physiological, biochemical, and oncological events that are subsequently observed? Those who simply relegate this physical part of the problem to a vague "transduction mechanism" have greatly underestimated its critical importance. Indeed, it may be scientifically ineffective to pursue the latter events while lacking key information about the primary process. A conglomeration of biological data on ELF field effects obtained for different fields, frequencies, exposure times, model systems, field gradients, and waveshapes by using trial-and-error approaches has been reported. It seems clear that there is a great need to rationalize such studies, to enable the life scientists to use their skills more gracefully. There is also a very practical concern: Once the physical mechanism governing the ELF interaction is properly stated, power companies and regulatory groups can better understand the extent of the problem and plan more effectively.

It is also important to understand the basic physical mechanism underlying field-to-cell coupling because it is frustrating that the larger scientific community in both physics and biology, despite considerable contrary experimental evidence, still finds it difficult to believe that weak ELF electromagnetic fields can interact with living cells. This reluctance to accept this apparent new paradigm has had some disturbing consequences. Not only has basic research been hampered by timid editors and disinterested funding agencies, but discoveries potentially significant to human health have proceeded at a snail's pace pending confirmation and explanation at a basic level. The few attempts to develop clinical regimens involving electromagnetic fields, mainly in bone repair, have been only crudely effective and may actually have retarded more widespread utilization.

There is no doubt that scientific acceptance of the ELF interaction has been agonizingly slow because there is no conceivable mechanism to explain such effects. Without this physical mechanism, the biological evidence has been examined more stringently than "run-of-the-mill" biological research. Although critics insist on replication, more animal studies, more cell studies, etc., these demands would be muted if observed experimental effects agreed better with existing theoretical concepts of electromagnetic interactions. Instead, the biological evidence has been subjected to extraordinary critical analysis.

The critical philosophical difference between the physical and biological aspects of this problem is also significant. Biological effects and properties are based almost completely on careful and reproducible laboratory observations. In the physical sciences, severe constraints are placed upon any hypothetical interaction mechanism, that is, limits based on consideration of physical laws. It is possible to do approximate calculations indicating the improbably low energy available from a 1-G field to volumes of cellular size, the lack of strong magnetic dipole moments in the cell, and the vanishingly small current densities resulting from Faraday induction. These calculations reinforce the disbelief of the skeptics, providing an extra hurdle for life scientists. It is thus of no little importance to identify and develop the physical mechanism or mechanisms that are the basis of these phenomena.

We generalize here the many different observed ELF phenomena, referring to all in the singular: the ELF interaction. We develop the theme that this ELF interaction more often than not obeys simple ion cyclotron resonance (ICR) theory. This is true not only for the body of experimental work usually termed calcium-efflux studies, but also for nearly all the many time-varying magnetic-field experiments. Above all, the 60-Hz biological interactions fit precisely into this ICR framework.

ARE ALL ELF PHENOMENA PHYSICALLY LINKED?

The various ELF interactions reported in tissue remain unexplained or are equally anomalous. It would be remarkable if each phenomenon was the result of separate types of physical interactions. We have yet to unravel a physical mechanism that explains even one of these biological effects, and it is unreasonable to contemplate separate, independent, as yet undiscovered physical mechanisms. The identification of a unifying mechanism may, therefore, be a critical litmus test for any proposed physical process.

To illustrate this difficulty let us compare two widely different experiments: first, cells in culture exposed to time-varying magnetic fields, and second, chick brain specimens exposed to amplitude-modulated (AM) radio-frequency electric fields (Table I). Human fibroblasts and other cells in culture exposed to sinusoidally varying ELF magnetic fields for 24 hr show a robust enhancement in [³H]thymidine incorporation over nonexposed controls (Liboff et al., 1984). If the magnitude of this effect is plotted versus the product of frequency and field, fB, there is no change (Fig. 1). It can be reasoned that if the observed effect resulted from Faraday induction then the response should increase with the level of induced emf, which is in turn directly proportional to dB/dt. For sinusoidal signals, this derivative is directly proportional to the product fB. Although lending support to the concept that low-frequency, weak magnetic fields somehow interact with cells, these observations at the same time argue convincingly that the mode of interaction cannot be explained by eddy currents arising from Faraday induction.

The so-called calcium-efflux experiments provide a totally different framework for developing a basic interaction mechanism. The original experiment at Adey's laboratory (Bawin et al., 1975) was pivotal; others have repeated it and attempted to explain the results. The exposure of radioactively labeled chick brain to weak periodic electric signals revealed that ⁴⁵Ca efflux from these specimens is a function of frequency, with a well-defined extremum at 14 Hz. Although the results (Fig. 2) seem straightforward, they represent a rather complex experiment. For example, the plotted frequency is actually a modulation frequency, with the carrier frequency fixed at 147 MHz. Also, the relative response is not very

TABLE I.
Comparison of Two Very Dissimilar ELF Experiments

	Bawin et al. (1975)	Liboff et al. (1984)
Model	Chick brain samples	Fibroblasts in culture
Exposure	Electric field	Magnetic field
Assay	Calcium efflux	Thymidine incorporation
Conclusion	Response dependent on frequency	Response independent of induced EMF

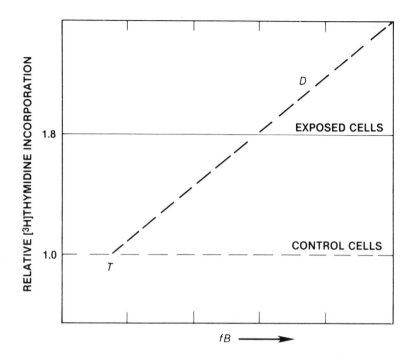

Fig. 1.   ELF interaction does not result from Faraday induction. Liboff et al. (1984) demonstrated that human fibroblastic cells exposed to weak [0.1–1.0 G] low-frequency (15 Hz–4 kHz) magnetic fields for 24 hr exhibit ~80% increase in [³H]thymidine incorporation relative to unexposed cells, over more than four orders of magnitude of product of frequency and field, *fB*. This product is a measure of strength of Faraday induction when sine wave magnetic fields are used, as in this experiment. If cells were responding to interaction involving Faraday induction, a dose–response curve D would have been observed, perhaps with threshold T. Lack of any such dose response is taken to indicate that coupling mechanism is not Faraday induction.

(Reprinted with permission from Liboff et al. 1987a. Experimental evidence for ion cyclotron resonance mediation of membrane transport. In: *Mechanistic Approaches to Interaction of Electric and Electromagnetic Fields with Living Systems*, Blank, M., Findl, E., eds., pp. 109–132. New York: Plenum.)

large, at most 18%–19% over controls (compare to response in Fig. 1). It is still unclear what role the high-frequency component plays in this experiment; without the 147-MHz carrier the extremum occurs in the opposite direction. Thus, at 16 Hz, maximum calcium efflux occurs with the carrier and maximum inhibition occurs without it. In addition, the calcium pathway, and the exact origin of this efflux, are uncertain. The remarkably narrow frequency width (Fig. 2) is a fascinating aspect of these results. Resonance phenomena are often described in terms of the full-width half

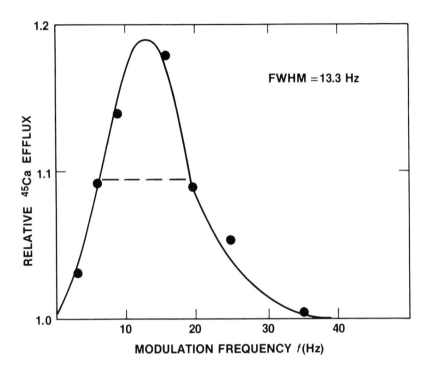

Fig. 2. Representation of results by Bawin et al. (1975). Efflux of ^{45}Ca from chick brain exhibits resonance when plotted as function of modulation frequency of applied electric field. Data indicate peak response at ~14 Hz with half-width of 13.3 Hz. (Reprinted with permission from Liboff et al. 1987a. Experimental evidence for ion cyclotron resonance mediation of membrane transport. In: *Mechanistic Approaches to Interaction of Electric and Electromagnetic Fields with Living Systems*, Blank, M., Findl, E., eds., pp. 109–132. New York: Plenum.)

maximum (FWHM) (also called the half-width), a figure of merit related to dissipative losses in the process. The inverse of the FWHM is an approximate measure of the relaxation time associated with the phenomenon. Results in Fig. 2 correspond to a FWHM of about 13.3 Hz, resulting in an approximate characteristic time of 0.075 sec.

These two experiments (or, more correctly, variations on them) have been independently studied by others: alterations in insect mRNA were found after exposure to either ELF pulsed or sinusoidal magnetic fields by Goodman et al. (1983, 1987); Takahashi et al. (1986) reported that [^{3}H]thymidine uptake in mammalian cells is altered after exposure to ELF pulsed magnetic fields; Parola et al. (1988) observed that chick embryo fibroblasts exposed to sinusoidal magnetic fields appear to undergo a malignant transformation. On the other side of the coin, Blackman's group (1982, 1985a) has repeated and extended the calcium-efflux work on chick brain, establishing the existence of additional extrema at higher modulation frequencies (harmonics) and also reporting intensity "windows" for this effect; Dutta et al. (1984) observed the calcium-efflux effect in neuroblastoma.

Despite the across-the-board differences between the two experiments outlined in Table I, both probably represent different aspects of the same phenomenon. The common thread linking both, as well as the experiments just discussed, is the local DC magnetic field, consistently ignored in all these cases, but without doubt the linchpin that may eventually serve to identify the basic mechanism. This was discovered by Blackman et al. (1985b), who reported at the 1984 BEMS annual meeting that chick brain calcium-efflux results are sharply dependent on the local DC magnetic field. Blackman's results suggested the existence of a series of allowed frequencies, all proportional to the DC field B_0:

$$f_N = (2N + 1)kB_0 \tag{1}$$

where k is a proportionality constant and $N = 0, 1, 2, \ldots$.

Subsequent experiments (Thomas et al., 1986; Smith et al., 1987; Liboff et al., 1987b) have confirmed the critical importance of the DC magnetic field in the ELF interaction; in widely different experiments totally different from the calcium-efflux experiments of Bawin et al. (1975) and Blackman et al. (1982), the local magnetostatic field has been the determining factor. Given the experimental evidence, is there one conceivable physical mechanism that might be the basis for the ELF interaction? The ubiquitous importance of the magnetostatic field in the various ELF experiments makes it likely that some sort of resonance phenomenon is the cause.

INVERSE ION CYCLOTRON RESONANCE

The various magnetic resonance processes require the simultaneous application of both DC magnetic fields and oscillatory electric or magnetic fields. It is interesting to note that when Blackman first presented evidence for the importance of the DC magnetic field at the 1984 annual BEMS meeting (Blackman et al., 1984), the discussion did not explore beyond effects connected to either nuclear magnetic resonance (NMR) or electron spin resonance (ESR). There are probably good reasons why the cyclotron resonance concept was ignored: Somewhat earlier Jafary-Asl et al. (1982) reported a narrow absorption peak in the electrophoretic yield of living cells, occurring in the vicinity of the proton NMR frequency of 2 kHz at the local geomagnetic field, 0.5 G. This work had sparked intense discussion, in part concerning the possibility of an NMR "effect" in living systems.

At the same 1984 BEMS meeting, Polk (1984) had examined, among other things, the possibility of ion cyclotron resonance currents circulating around the cell membrane, concluding that coherent motion of this sort is not possible except in very intense magnetic fields. Also, life scientists are reasonably conversant with the fundamentals of NMR and ESR techniques, in view of their widespread use in biophysics and biochemistry, but may be somewhat less familiar with a third distinct type of resonance interaction in which the orbital motion of charged particles is the key mode, as compared to spin states.

Shortly after the 1984 BEMS meeting, Liboff communicated to Blackman (A. R. Liboff, personal communication) that certain of the combinations of frequency and DC field his group had shown to be successful in enhancing calcium efflux corresponded very closely to the cyclotron resonance condition:

$$f_0 = \frac{q}{2\pi m} B_0 \qquad\qquad [2]$$

when the charge-to-mass ratio q/m in this expression is set equal to that for ionized potassium, K^+. To illustrate this, recall that one DC magnetic field value, B_0, used by Blackman's group (Blackman et al., 1985b) was 0.38 G. If we use this value of B_0 along with the value of q/m for K^+, 2.46×10^6 coulombs (C)/kg, direct substitution into Eq. 2 yields the expected frequency

$$f_0 = \frac{1}{2\pi} \times 2.46 \times 10^6 \times 0.38 \times 10^{-4} = 14.9 \text{ Hz}$$

as compared to the experimental value $f = 15$ Hz reported by Blackman et al. (1985b). We note from Eq. 2 that for a given ionic charge-to-mass ratio q/m, the ratio f_0/B_0 remains a constant. This, too, is consistent with the results of Blackman et al. (1985b): When the frequency was doubled to 30 Hz, the magnetic field for maximum efflux became 0.76 G. The three field combinations found to be successful by Blackman et al. (1985b) are shown in Table II. The first two combinations, 0.38 G at 15 Hz, and 0.76 G at 30 Hz, when substituted in Eq. 2 as shown yield a value for the charge-to-mass ratio close to that of K^+ (Table III). The third combination, 0.255 G at 30 Hz, appears to yield an inconsistent result for q/m, 7.39×10^6 C/kg, that does not correspond to any ion in Table III. However, if one rewrites Eq. 2 to include the possibility of additional harmonics, such that

$$f_n = nf_0 = \frac{nq}{2\pi m} B_0 \qquad\qquad [3]$$

we find that this last field combination does indeed yield the potassium charge-to-mass ratio, 2.46×10^6 C/kg, provided that the value $n = 3$ is used in Eq. 3. As is seen, such *odd* harmonics are consistent with later experimental evidence.

TABLE II.
Combinations of Frequency and DC Magnetic Field Used by Blackman et al. (1985) to Enhance Calcium Efflux and Ionic Charge-to-Mass Ratios Calculated from These Combinations

| | | q/m (calculated in C/kg) | |
f (Hz)	B (G)	$n = 1$	$n = 3$
15	0.38	2.48×10^6	0.83×10^6
30	±0.76	2.48×10^6	0.83×10^6
30	±0.255	7.39×10^6	2.46×10^6

TABLE III.
Charge-to-Mass Ratios for Various Ions and Magnetic Field Required for Cyclotron Resonance

Ion	Atomic weight (u)	q/m Charge-to-mass ratio (C/kg)	B_0 (G)		
			15 Hz	30 Hz	60 Hz
H^+	1.008	9.56×10^7	0.010	0.020	0.039
Li^+	6.94	1.39×10^7	0.068	0.136	0.271
Na^+	22.99	4.19×10^6	0.225	0.450	0.899
Mg^{2+}	24.305	7.93×10^6	0.119	0.238	0.475
Cl^-	35.45	2.72×10^6	0.347	0.693	1.387
K^+	39.10	2.465×10^6	0.382	0.765	1.529
$^{40}Ca^{2+}$	40.08	4.81×10^6	0.196	0.392	0.784
$^{45}Ca^{2+}$	44.95	4.29×10^6	0.220	0.440	0.879

Thus, the introduction of the ion cyclotron resonance (ICR) formalism radically changes the focus of the calcium-efflux experiments. The question of whether the oscillating AC field is electric or magnetic is relegated to lesser importance because the DC magnetic field is the most essential parameter for cyclotron resonance. Also, the carrier frequencies used in the calcium-efflux experiments very likely have little to do with the underlying interaction mechanism, except insofar as they make for a more efficient delivery system. According to the ICR concept, the ELF part of the exposure signal is the interactive component. In addition, there are important implications for 60-Hz studies. From the standpoint of ICR there is no special biological quality attached to frequencies of 14 to 16 Hz. Whatever effect is observed at 15 Hz should also be found at 60 Hz, if the DC field is scaled up by the appropriate factor (see Table III). Further, the interaction is no longer tied to one specific ion (in this case, Ca^{2+}), but involves the possibility of electromagnetic fields affecting a variety of biological ions. Although K^+ was implicated in this first analysis, there is no reason to believe that suitable resonance combinations could not be found for at least some of the other ions listed in Table III.

To test this hypothesis, an experiment involving the cyclotron resonance (CR) condition for Li^+ was initiated in the summer of 1984 in John Thomas' laboratory at the Naval Medical Research Institute (NMRI) in Bethesda. As part of the New York State study on high-voltage 60-Hz power lines, this laboratory developed a protocol in which rats conditioned to perform on multiple fixed-ratio (FR) differential reinforcement of low rate (DRL) schedules were exposed to 60-Hz parallel plate electric fields and/or perpendicularly oriented 60-Hz linearly polarized magnetic fields. A year-long set of exposure data (Thomas et al., 1985) failed to indicate any significant change in schedule performance after 1-hr exposures compared with sham or control baselines. However, after the exposure system was modified to control the DC magnetic field (Fig. 3), striking changes were observed in performance on the DRL schedule when the exposures were "tuned" to the charge-to-mass ratio corresponding to Li^+ (Thomas et al., 1986). Table IV indicates that exposing rats to both DC and AC magnetic fields simultaneously has a profound effect on timing discrimination, but if the exposure is compartmentalized such that *either* the DC or AC field is separately applied, the rat suffers no measurable change in schedule performance; *both* fields are required (Fig. 4). This experiment (1) not only verified the Blackman et al. (1985b) observation that the DC magnetic field plays a role in the ELF interaction, but also showed that (2) the ELF interaction can occur at 60 Hz, (3) it can be extended to other ionic species, (4) entire mammals can be affected, (5) an AC electric field is not needed, and (6) the interaction energies are well below any conceivable thermal process (both DC and AC magnetic-field intensities were less than 1 G).

While this experiment on 60-Hz-mediated rat behavior was still in progress, the first public presentation outlining the ICR hypothesis was delivered at a NATO workshop (Liboff, 1985). In addition to analyzing the results of Blackman et al. (1984), the following points were made:

1. The ICR frequencies for the common biological ions (K^+, Na^+, Cl^-, Mg^{2+}, Ca^{2+}), as they would circulate in the earth's magnetic field, are all in the ELF range (Fig. 5).

2. One excellent candidate for the interaction site of an ICR effect in the cell is the membrane ion channel, mainly because of the intrinsic structural helicity that may result from the α helices making

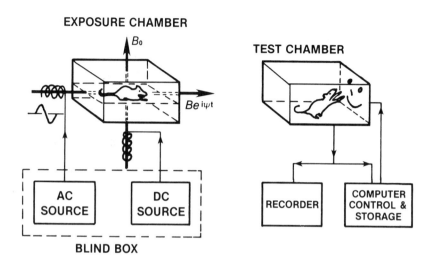

Fig. 3. Schematic of experimental setup of Thomas et al. (1986). Cyclotron resonance exposure (CR) consisting of 60-Hz horizontal magnetic field ($Be^{i\omega t}$) and vertical DC magnetic field (B_0) is applied to previously conditioned rats for 30 min. Existing vertical component of earth's field is thereby reduced, acting also to reduce total DC magnetic intensity from original value, 0.404–0.271 G. Latter is DC value in CR at 60 Hz for lithium ions. Note that this resonance is determined by reduced component of AC field parallel to 0.271-G field. Immediately after exposure, rat is transferred to Skinner box to perform multiple schedule on cue, pushing lever to obtain food pellets. In one category of data analysis, rate of lever pushes is measured and compared to original baseline rate. In all rats studied, this Li^+-resonance exposure acted to significantly disturb schedule performance connected to timing discrimination. (Reprinted with permission from Liboff et al. 1987a. Experimental evidence for ion cyclotron resonance mediation of membrane transport. In: *Mechanistic Approaches to Interaction of Electric and Electromagnetic Fields with Living Systems*, Blank, M., Findl, E., eds., pp. 109–132. New York: Plenum.)

up the channel protein, but also because of the likelihood of reduced scattering in a high-conductance, ion-selective channel.

At the same time the physical difficulties with such a model were also listed by Liboff (1985):

1. The use of Eq. 2 to explain the results of Blackman et al. (1985b) necessarily means that the value that the mass m takes is that of the "naked" ionic mass, that is, the ion stripped of any water molecules. Ions such as K^+ or Ca^{2+} in solution carry hydration complexes that are energetically different to remove. It might therefore

TABLE IV.
Cyclotron Resonance Study of Rat Behavior: Outline of Exposure Conditions[a]

Type of run	Exposure Conditions		FR test response	DRL test response
	Static B (G)	Frequency (Hz)		
Control	By-passed Exposure Chamber		Baseline	Baseline
Sham	0.404	—	Baseline	Baseline
Oscillating magnetic field	0.404	60	Baseline	Baseline
Earth's field changed	0.261	—	Baseline	Baseline
Cyclotron resonance	0.261	60	Baseline	Change

[a] From Thomas et al. (1986).

seem reasonable that an ion entering a channel still has water attached, and therefore if one calculates (q/m) in Eq. 2 then one should use the effective mass as opposed to the smaller naked mass. Of all the various objections to the possibility of magnetic effects within ion channels, this one probably holds the least water. Thus, it is quite possible that ions exchange water clusters in solution for less mobile water clusters inside channels, thereby making the influx or efflux of ions in channels possible at minimal expenditure of hydration energy (Hille, 1984; Kim et al., 1985).

2. A more serious concern involves the rather long characteristic times associated with these cyclotron resonance orbits; for example, at 15 Hz, this time is ~1/15 sec, roughly 70 msec. To sustain the CR condition, the time it takes for an ion to complete one circuital turn must be less than the scattering time, that is, the time between successive collisions. If the ion suffers many collisions before one loop of its orbit is completed, it will be difficult to maintain the resonance conditions. One can estimate the mean time between successive collisions for an ion at a given temperature; at 37°C the collision times are orders of magnitude shorter than milliseconds. Interestingly, this disparity between characteristic time and collision time is not limited to the CR hypothesis; any mechanism proposing to explain the ELF interaction shown (in Fig. 2) must also account for the remarkably long coherence time (75 msec, estimated from the linewidth), orders of magnitude greater than the collision time.

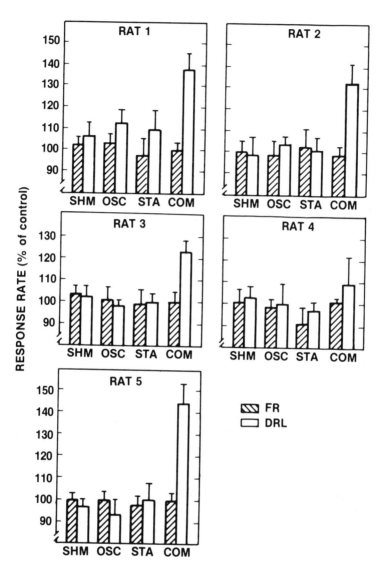

Fig. 4. Change in timing discrimination in five rats after 30-min exposures to cyclotron resonance (CR) fields tuned to Li+. *Shaded bars* represent fixed-ratio (FR) counting schedule in which rat obtains food pellet only after 30 successive lever pushes. This response was unaffected. In low-rate differential timing schedule (DRL) (*unshaded bars*), rat must wait minimum of 18 sec before pushing lever but not more than 24 sec. Performance on this schedule was profoundly affected by CR exposure at 60 Hz. Abbreviations: SHM, sham exposure; OSC, 60-Hz magnetic field alone; STA, 0.271-G magnetostatic field alone; COM, 60-Hz and 0.271-G fields combined.

(Reprinted with permission from Thomas et al. 1986. Low-intensity magnetic fields alter operant behavior in rats. *Bioelectromagnetics* 7:349–357.)

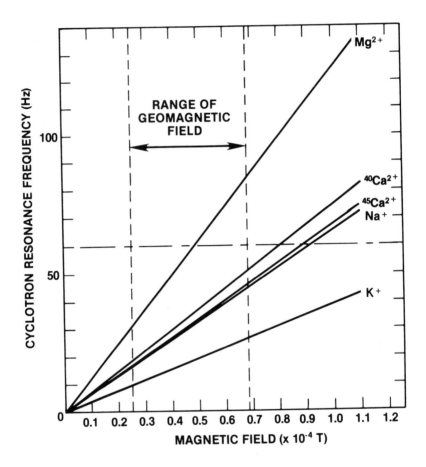

Fig. 5. Cyclotron resonance (CR) frequency as function of magnetic field for various ions. Range of earth's magnetic field (total intensity) over the earth's surface is superposed.
(Reprinted with permission from Liboff et al. 1987a. Experimental evidence for ion cyclotron resonance mediation of membrane transport. In: *Mechanistic Approaches to Interaction of Electric and Electromagnetic Fields with Living Systems*, Blank, M., Findl, E., eds., pp. 109–132. New York: Plenum.)

Lawrence and Adey (1982) attacked this problem in a direct fashion, proposing that the ELF interaction involves low-loss, long-lived waves in the cell membrane that convey energy at acoustic frequencies. These nonlinear waves, also called solitons, could either be coupled directly to the exposure fields or indirectly excited by them. Yet another approach has been taken by Westerhoff et al. (1986),

who suggested that transmembrane enzymes, such as Na⁺-,
K⁺-ATPase, are capable of absorbing free energy from AC electric
fields if the enzyme-driven cyclic charge translocations across the
membrane are in phase with the time-varying field. The most interesting part of this concept is the intrinsic coherence that would tend
to overcome thermal collisions if the process runs long enough. It is
difficult to imagine, on the other hand, exactly how very weak ELF
time-varying *magnetic* fields could play a role in such a mechanism.
3. The energy associated with any CR interaction in the cell membrane
of necessity is extremely small, far below kT, because the energy
of a particle in cyclotron resonance varies as the square of the radius
of the orbit. Therefore, one can compare the kinetic energy of an
ion experiencing no friction constrained to move within a transmembrane protein channel having a diameter, say, of 7 nm, and the same
ion circulating round the periphery of a spherical cell of diameter
35 μm. For the same frequency and field, the kinetic energy of this
ion inside the channel will be reduced by a factor of about 10^{-10}
of what it would be if it were orbiting around the perimeter of the
cell.

It is important to note that later versions of this hypothesis (Mcleod
and Liboff, 1986, 1987; Liboff and McLeod, 1987, 1988) have stressed the
helical constraints on the ion moving in the channel. It is argued that any
regular helical motion of a charged particle, as derived from structural
potential variations, will exhibit CR eigenstates when subjected to a DC
magnetic field. The magnetic field in this case does not "cause" the helical
motion, but serves to select or tune for frequencies appropriate to the
charge-to-mass ratio of the particle, as derived from Eq. 2. To distinguish
between the more common case in CR, where one deals with an unconstrained particle whose radius is determined by its energy, the present mechanism has been termed inverse ion cyclotron resonance (Liboff
and McLeod, 1987). There is excellent reason to believe that, at least in
some channels, ions may be structurally constrained to move only in helical
paths (Kim et al., 1985; Skerra and Brickmann, 1987).

The original calcium-efflux data reported by Bawin et al. (1975) (see
Fig. 2) bear a remarkably close resemblance to the sort of functional
behavior exhibited by a damped harmonic oscillator driven by a signal that
is close to resonance (Liboff et al., 1987a). This observation led McLeod
and Liboff (1986; Liboff and McLeod, 1988) to reexamine the earlier
calcium-efflux data of both Bawin et al. (1975) on chick brain and Dutta
et al. (1984) on neuroblastoma. Assuming that a channel ion under
transport can be modeled as moving through a viscous medium under the
simultaneous application of a DC magnetic field B and a sinusoidal electric field E, one obtains the simple force equation for an ion having a
charge-to-mass ratio q/m and a velocity u,

$$m \frac{du}{dt} = qE + q(u \times B) - \frac{mu}{\tau} \qquad [4]$$

where the last term is the retarding force, determined by the mean time between collisions τ. From this, it can be shown (Liboff and McLeod, 1988) that the ion conductivity σ assumes a resonant form dependent on the angular frequency ω of the applied electric field:

$$\sigma = \sigma' \frac{1 + (\omega_0 + \omega)^2 \tau^2}{1 + [(\omega_0^2 - \omega^2)\tau^2]^2 + 4\omega^2\tau^2} \qquad [5]$$

where σ' is a constant and ω_0 is the angular frequency corresponding to the CR frequency f_0 in Eq. 2, i.e., $\omega_0 = 2\pi f_0$. Note that, in this expression, the value of σ is maximized when the applied frequency ω equals ω_0. The results of the various calcium-efflux experiments, exhibiting a variation in ionic transport as a function of frequency, can be directly compared to Eq. 5 if one assumes that the variation in charge transport is equivalent to a variation in conductivity. By assuming a value for the steady-state tissue conductivity σ', it is possible to reduce the comparison between theory and data to only three parameters: the collision time τ, the charge-to-mass ratio q/m, and the local magnetic field B. The collision time is obtained directly from the linewidth (see Fig. 2), and trials of q/m corresponding to different ionic species again indicate, as with the Blackman et al. (1985b) data, that the K^+ ion is implicated.

With this approach one obtains values for B to give a best fit that provides estimates for the local magnetic field at the time the experiment was originally done. Although the value of the DC magnetic field was not reported in either the Bawin et al. (1975) or the Dutta et al. (1984) results, the estimates resulting from curve-fitting both sets of data are reasonable. Indeed, Fig. 2 shows a remarkably good fit between the calcium-efflux data on the one hand and Eq. 5 on the other. Note that this analysis also implies that the observed ^{45}Ca assay is secondary to a primary magnetic stimulation of K^+ transport. It is fair to say, all in all, that CR modeling has proven useful in providing a framework with which to explain the functional nature of previous calcium-efflux studies.

Unlike the calcium-efflux experiments for which attempts have been made to use CR to retrospectively explain the results, two totally independent experiments have been directly designed to test for CR. If one measures the motility of the benthic diatom *Amphora coffeaeformis* as a function of local calcium concentration, a sharp threshold is observed (Cooksey and Cooksey, 1980); the diatoms are essentially unable to move at concentrations less than about 0.25 mM. Smith et al. (1987) took advantage of this calcium threshold to determine the effect of CR tuning for Ca^{2+} on motility when diatoms are maintained below this calcium threshold. The results of this experiment have made the strongest case that the ELF interaction is caused by ICR in the cell membrane.

In this model, diatoms cultured on agar are exposed to carefully deployed combinations of AC and DC magnetic fields (Fig. 6) for 1 hr, and the plate is then examined under a microscope to measure the fraction of motile diatoms in the population. This experiment has yielded some remarkable results, not the least of which is shown in Fig. 7. The motility is plotted against the AC magnetic-field frequency for one value of the dc magnetic field (0.209 G). It is immediately seen (Fig. 7) that the motility peaks at precisely that frequency (16 Hz) which is determined by Eq. 2, when the DC magnetic field (0.209 G) is substituted into this equation as well as the charge-to-mass ratio for Ca^{2+} (4.81 × 10^6 C/kg). This can be checked from either Eq. 2 or Table III; in the latter case, note that the ratio of frequency to field for the case of Ca^{2+} is 60/0.784 Hz per gauss, which is the same ratio as the values used in Smith et al. (1987).

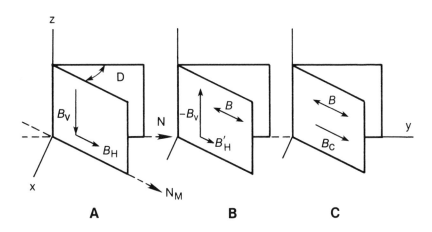

A   **B**   **C**

Fig. 6. Configuration of magnetic-field components. (**A**) Ambient field. Geomagnetic-field component vectors are contained in plane defined by magnetic north N_M and vertical z. This plane is at magnetic declination angle D to geographic north N. Vertical component of earth's field is B_V and horizontal component is B_H. (**B**) Applied fields. Three sets of coils are oriented to produce magnetic-field vectors solely in geomagnetic plane. Three vectors are applied: (1) vertical field $-B_V$ equal and opposite to the earth's field B_V shown in **A**; (2) horizontal static field B'_H; and (3) horizontal oscillatory field B. (**C**) Net field. Combining all vertical and horizontal components shown in **A** and **B**, only two components remain, both horizontal: (1) oscillatory field B, characterized by frequency f; and (2) horizontal static field B_C. Latter is sum of separate horizontal components in **A** and **B**, given by $B_C = B_H + B'_H$. To achieve cyclotron resonance (CR) condition for a given ionic species having the charge-to-mass ratio q/m, the net magnetic field B_C and the frequency f are adjusted to fit the relationship $2\pi f = (q/m)B_C$.

(Reprinted with permission from Rozek et al., 1987. Nifedipine is an antagonist to cyclotron resonance enhancement of ^{45}Ca incorporation in human lymphocytes. *Cell Calcium* 8:413–427.)

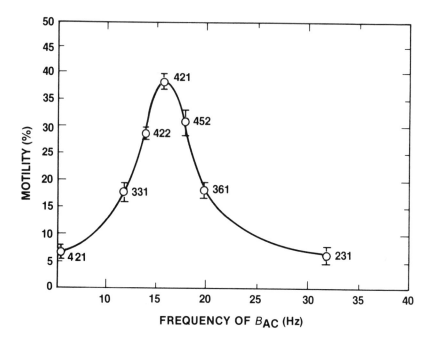

Fig. 7. Variation of diatom motility with AC magnetic field. Magnetic fields corresponding to arrangement in Fig. 6C are B = 0.148 G rms and B_c = 0.209 G. Frequency of
B, the AC field, is set at various levels from 6 through 32 Hz, and motility is measured
after 1-hr exposure. Attached to each point is number of diatoms measured at each
frequency. Two distinct results are observed: first, motility variation exhibits resonance
shape; second, peak in motility occurs exactly at frequency (16.0 Hz) corresponding
to Ca^{2+} cyclotron resonance (CR) condition for B_c = 0.209 G, in full agreement with
Eq. 2. Background calcium concentration is 0.25 mM. When the concentration is reduced very much further, as close to zero as possible, resonance variation disappears,
and all diatoms remain immobile. When concentration is increased to 5 mM, diatoms
exhibit maximum motility regardless of frequency. We concluded that calcium ions
are stimulated by CR tuning to move across cell membrane into cell.
(Reprinted with permission from Smith et al. 1987. Calcium cyclotron resonance and diatom mobility. *Bioelectromagnetics* 8:215–227.)

At frequencies lower and higher than 16 Hz, the motility drops off,
following the theoretical prediction using Eq. 5. This is a startling result.
A simple organism not only will move at low calcium concentrations for
a specific, *predictable*, magnetic frequency, but also follows in its movement a classical resonance tuning curve, behaving precisely in the same
mathematical way as would an oscillating spring or a tuned electrical cir-
cuit. Apart from the fascinating questions raised by the fact that these
results are mediated by DC fields of the same magnitude as the geomagnetic field, there is an obvious comparison to the calcium-efflux results.

It takes little imagination to realize that the results reported by Bawin et al. (1975), Dutta et al. (1984), Blackman et al. (1985b), and Smith et al. (1987) are very likely different aspects of the same phenomenon. In the first three experiments, we deduced after the facts the magnetic field required to produce peak efflux; in the fourth experiment, Smith et al. (1987) predicted that the peak occurs at a certain magnetic field. By tuning to Ca^{2+} there is more efficient utilization of local calcium external to the membrane. Because the motility does not exhibit this resonant variation when the external concentration of calcium is completely minimized (Smith et al., 1987), one must infer that the resonant magnetic field transfers external calcium to the interior of the cell, where it takes part in the utilization of energy stores required for movement. By contrast, in the calcium-efflux experiments, the ELF field couples to the K^+ ion and drives it across the membrane, resulting in a net release of calcium.

This K^+ argument is bolstered by additional experiments using the diatom model (McLeod et al., 1987). Figure 8 again indicates the change in motility as a function of magnetic-field frequency, but in this case the external calcium concentration in the agar plates was adjusted to 5 mM, so that the diatoms enjoyed maximum motility. The DC magnetic field was also different from that in Fig. 7, changed from 0.209 to 0.41 G. Under these new conditions, the motility can be reduced to zero simply by tuning to 16 Hz. Instead of the clear peak shown in Fig. 7, we now find a sharp minimum in the motility in Fig. 8. Despite the fact that the frequency was 16 Hz in both cases, the two different DC magnetic fields correspond to two different ionic resonances, the first for Ca^{2+} and the second for K^+. One can use either Eq. 2 to calculate the charge-to-mass ratio at 16 Hz and 0.41 G, or alternatively note from Table III that the ratio of frequency to field used in this experiment, 16/0.41, is extremely close to the predicted ratio for K^+, namely, 60/1.53 Hz per gauss. It seems clear that tuning for the Ca^+ ion drives calcium into the cell, enhancing motility; if one instead tunes for the K^+ ion, the motility is reduced, in turn implying that calcium is being removed from the cell. Thus, the original arguments, suggesting first that K^+ CR explains the Blackman et al. (1985b) results (Liboff, 1985), and then implicating the K^+ ion (McLeod and Liboff, 1986) to explain the calcium-efflux results of Bawin et al. (1975) and Dutta et al. (1984), now appear to have a much firmer basis. This analysis of the interplay of K^+ and Ca^{2+} ions revealed in the ELF interaction with diatoms should make it quite clear that without the insights that can be quickly obtained with access to the underlying physical mechanism, the road to understanding "mechanisms" from a nonphysical point of view will be unnecessarily difficult.

Additional independent evidence tends to support the hypothesis that ICR excitation is the basis for the ELF interaction. On exposing freshly obtained mixed human lymphocytes for 1 hr to a DC magnetic field of 0.209 G, Liboff et al. (1987b) reported that the peak incorporation of [45]Ca

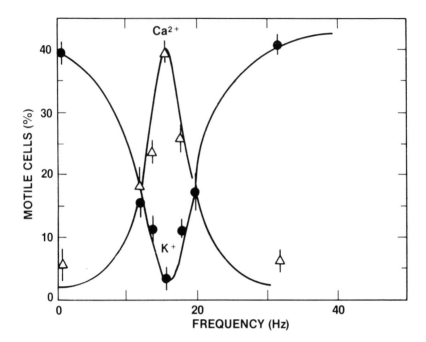

Fig. 8. Comparison of motility changes: Ca^{2+} (*triangle symbols*) and K^+ (*solid circles*) resonances. Ca^{2+} results obtained as in Fig. 7: DC magnetic field, 0.209 G; external Ca^{2+} concentration, 0.25 mM. K^+ results are obtained for DC magnetic field of 0.41 G and external Ca^{2+} concentration of 5 mM. Latter concentration means that diatoms ordinarily enjoy maximum motility. As frequency is changed, one now finds a resonance inhibition of motility, again at 16.0 Hz. This experiment demonstrates conclusively that a physiological event (in this case motility) can be enhanced or inhibited by exposing the system to different DC magnetic fields, even when the AC frequency is unchanged. This may be significant relative to question of efflux reversals observed in different calcium-efflux experiments.

[Reprinted with permission from McLeod et al. 1987. Calcium and potassium cyclotron resonance curves and harmonics in diatoms (*A. coffeaeformis*). *J. Bioelectr.* 6:153–168.]

into these lymphocytes occurs at a frequency of 14.3 Hz, as compared to 16 Hz, that frequency required to obtain CR in chemical calcium, ^{40}Ca, as described in the diatom experiments. This observed downshift in frequency is precisely what is expected on the basis of the difference in charge-to-mass ratios for ^{40}Ca and ^{45}Ca. This confirms that the critical cellular parameter in the ELF interaction is the ionic charge-to-mass ratio. One sees that the CR condition (Eq. 2) governs an interaction so specific that different isotopes of the same element require different tuning frequencies.

The key role of the charge-to-mass ratio in the ELF interaction was further examined in yet another experiment involving diatom motility

(McLeod et al., 1987). Motility tuning curves similar to Fig. 7 were obtained in eight separate DC magnetic fields. Diatoms in each of these different fields exhibited motilities that peaked at eight different frequencies ranging from 8 to 64 Hz. In each case, the ratio of frequency to DC field was found to correspond very closely to the charge-to-mass ratio for Ca^{2+}, namely, 4.81 × 10⁶ C/kg (Fig. 9). Finally, the specific mode of ion transport was indicated by use of the calcium blocker nifedipine, which completely negates the CR incorporation of ⁴⁵Ca in human lymphocytes (Rozek et al., 1987). These results (Fig. 10) suggest that the resonance effect occurs within the same channels as are affected by the blocking action, namely the slow inward calcium channels.

It is clear that accumulated ICR studies go far toward confirming the hypothesis we suggested at the outset of this chapter—that all the ELF phenomena, no matter how different in external appearance or in biological endpoint, are actually caused by a single physical mechanism of interaction. In making this connection we carefully avoid the question of the complete underlying explanation for an ICR mechanism. Obviously, far more

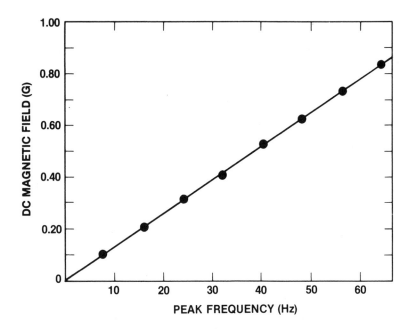

Fig. 9. Linearity of cyclotron resonance (CR) dependence. Data in Fig. 7 indicate peak in motility when DC field is 0.209 G and AC frequency is 16.0 Hz. When experiment is repeated at other DC fields, peak motility shifts to other frequencies. Straightline dependence of this peak, expected on the basis of Eq. 2, yields slope corresponding to charge-to-mass ratio of Ca^{2+}. This provides strong evidence in favor of CR mechanism but also shows 16-Hz frequency is not biologically unique.

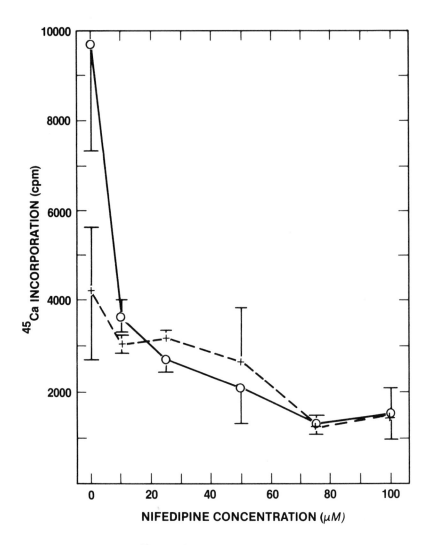

Fig. 10.   Incorporation of ^{45}Ca as β activity (cpm) as function of nifedipine concentration for unexposed cells (*dashed line*) and exposed cells (*solid line*). Exposed cells were subjected for 1 hr to cyclotron resonance (CR) field tuned to isotopic mass of ^{45}Ca, corresponding to 14.3 Hz and 0.209 G. Note large increase in ^{45}Ca incorporation in exposed cells compared to controls is nullified by addition of calcium blocker; for concentration range 10–100 μM nifedipine, mean ratio of exposed to unexposed cells is 0.99, compared to ratio of 2.3 when nifedipine is not present.

(Reprinted with permission from Rozek et al. 1987. Nifedipine is an antagonist to cyclotron resonance enhancement of ^{45}Ca incorporation in human lymphocytes. *Cell Calcium* 8:413–427.)

theoretical work, as well as experiments at a basic biophysical level such as in reconstituted membranes, are necessary. Nonetheless, the experimental evidence indicates that there is, indeed, only one ELF interaction. We point out the remarkable agreement with Eq. 2 found in a diverse group of ELF experiments, and emphasize that this simple ICR condition is the *only* analytic relation ever offered for any one of these experiments, to say nothing of the group as a whole.

WINDOWS

Any attempt to provide a physical basis for the ELF interaction must include two distinct types of functional behavior. Experimental evidence indicates that the interaction is restricted to relatively narrow regions of AC intensity on the one hand and to certain frequency ranges on the other. The intensity "windows" are complicated by the fact that under certain conditions one observes an efflux of Ca^{2+} and under other conditions an influx. This reversal, reported by Bawin and Adey (1976), Bawin et al. (1978), and Blackman et al. (1982), is difficult to explain using the framework we developed for a possible ICR mechanism. Although one can argue that tuning to a different ionic charge-to-mass ratio may result in a reversal of efflux, as described for diatom motility (McLeod et al., 1987), the work in Adey's and in Blackman's laboratories tends to indicate that these efflux reversals are tied to intensity and not to frequency.

There is also a hint of this in the CR studies on human lymphocytes (Liboff et al., 1987b). Figure 11 shows the relative incorporation of ^{45}Ca in lymphocytes following a 1-hr exposure to two different AC magnetic intensities over the same frequency range. Although there is a peak in incorporation at 0.21 G rms, one finds a minimum in ^{45}Ca incorporation at 1.5 G rms at the CR frequency (14.3 Hz). Note that this really is not equivalent to a reversal, because even at the minimum (Fig. 11) the incorporation in the exposed cells does not fall below that of the controls. The data do suggest, nevertheless, that when one is tuned to the ^{45}Ca point (in this case 14.3 Hz) the incorporation of ^{45}Ca is sharply dependent on the AC intensity.

Quite apart from this reversal problem, there is evidence that the ELF interaction occurs within certain ranges of AC electric- or magnetic-field intensity. Bawin and Adey (1976) found that electric fields between 3.5 and 21 V/m rms (fields external to specimens) caused significant changes in calcium efflux. A subsequent study by Blackman et al. (1982) revealed *two* distinct windows, one centered at approximately 2 V/m rms and the other at 14 V/m rms. At first glance, it might seem that there is rough agreement between the two laboratories: Both results were obtained at a frequency of 16 Hz, and there seems to be evidence for one or more electric-field

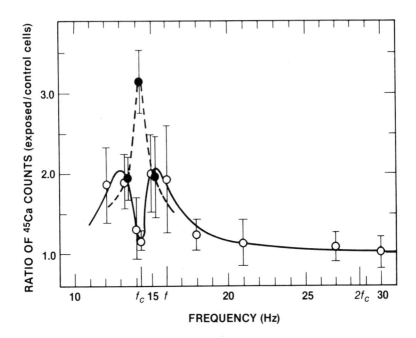

Fig. 11. Relative ^{45}Ca incorporation in magnetically exposed compared to nonexposed lymphocytes, after 1-hr exposure. Ratio of β counts in exposed to control cells is plotted against frequency of applied AC field B. Error flags are shown as standard deviations using pooled ratios from six runs. Each point in given run was determined in triplicate. Frequency f_c (14.27 Hz) is determined from cyclotron resonance (CR) expression $2\pi f_c = (q/m)B_c$, where (q/m) is charge-to-mass ratio for $^{45}Ca^{2+}$, using isotopic mass of 44.95 u and $B_c = 0.209$ G. Frequency f_c (16.00 Hz) is calculated using normal calcium mass, 40.08 u, and $B_c = 0.209$ G. *Open circles* represent data obtained for AC magnetic signal of 1.5 G rms; *solid circles* correspond to 0.21 G rms.
(Reprinted with permission from Liboff et al. 1987b. $Ca^{2+}-45$ cyclotron resonance in human lymphocytes. *J. Bioelectr.* 6:13–22.)

ELF interactions in the range of 2 to 20 V/m rms. However, this agreement may have been fortuitous. The Bawin and Adey (1976) data were obtained for "pure" electric-field exposures while the Blackman et al. (1982) results were obtained for an electromagnetic signal consisting of combined electric and magnetic fields. When this latter experiment was repeated (Blackman et al., 1985a) in a manner that drastically reduced the AC magnetic field so as to use only the electric field, the ELF interaction vanished. On the other hand, using an equivalent setup with the magnetic-field exposures that were present in the first Blackman et al. (1982) experiment (i.e., 0.089 and 0.595 mG rms), Blackman's group obtained a verification of the earlier results.

As such, there is some confusion concerning the interpretation of the results on intensity windows. The experiment by Blackman et al. (1985b) seems to clearly indicate that these windows are tied to the intensity of the AC magnetic-field component and are independent of the time-varying electric field. However, this is very much at odds with Bawin and Adey (1976); because they used an exposure setup that precluded any oscillatory magnetic field, they have claimed that the ELF interaction window is a function of the electric field only. The physics embodied in the CR hypothesis provides one way to resolve this puzzle. Presumably a finite DC magnetic field was present in both laboratories – in Blackman et al. (1985b) careful reference is made to the measurement and control of this parameter – even though the AC magnetic field was not present in Bawin and Adey (1976). Although it has not yet been documented in a carefully designed experiment, there is every reason to believe that an AC electric field can be oriented and adjusted to be in cyclotron resonance with the local DC magnetic field, following Eq. 2 in the same way as the AC magnetic field, namely through the frequency ω.

The main difference, in principle, should be the relative orientation of DC and AC field vectors. For the AC magnetic field, the DC and AC fields should be parallel for optimal interaction, whereas in the case of the AC electric field, the DC and AC vectors should be normal to one another. It is then pertinent to ask why Blackman et al. (1985b) saw no interaction for the case of AC electric-field exposures. As a pure speculation, this might be explained in terms of the relative orientations of the respective local DC magnetic fields in the two laboratories. Alternatively, it is conceivable that perhaps the Bawin and Adey (1976) experimental setup unknowingly included an AC magnetic-field source, perhaps resulting from a nearby power supply or the like. Judging from the very small AC magnetic field (0.089 mG rms) that Blackman et al. (1985b) reported for the ELF interaction, this is possible.

It is interesting to note that a distinctly different type of intensity window has been reported by Liboff et al. (1987b) when human lymphocytes are exposed to CR-tuned magnetic fields. The incorporation of ^{45}Ca as a function of magnetic-field intensity (Fig. 12) appears to follow a linear dose response to 0.2 G rms and is then followed by a relatively sharp cutoff at higher intensities. There is a rather large difference in intensities between the Blackman et al. (1985b) and the Liboff et al. (1987b) experiments (0.6 and 200 mG rms, respectively), and there may be problems in attempting to scale the two observations. Nonetheless, it is worth pointing out that the Blackman et al. (1982) data were sufficiently thorough that any dose–response effect at intensities below the lower edge of the window would have been discerned, if such an effect was indeed present.

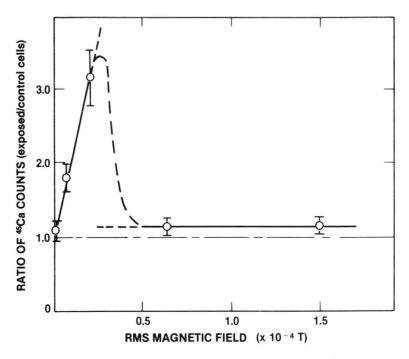

Fig. 12. ^{45}Ca-incorporation ratio of exposed to control cells as function of B_{AC}. Note appearance of two distinct modes of variation at low and at high fields. Dip in resonance in Fig. 11 corresponds to $B_{AC} = 1.5$ G rms, and resonance maximum in Fig. 5 to much lower effective field $B_{AC} = 0.21$ G rms.
(Reprinted with permission from Liboff et al. 1987b. Ca^{2+} −45 cyclotron resonance in human lymphocytes. J. Bioelectr. 6:13–22.)

HARMONICS

Just as the ELF interaction appears to function solely within certain intensity windows, it also exhibits frequency windows, that is, the responses to ELF signals are limited to certain frequencies. One can argue that the possibility of frequency windows may be more reasonable than intensity windows, because resonance effects, in general, are often accompanied by additional frequencies, termed harmonics or overtones, that are in simple ratios to some basic fundamental frequency. For example, there was a strong indication of such harmonics in ELF-induced calcium ion efflux enhancement studies by Blackman et al. (1985a). They found "highly significant" enhancements induced by 15-, 30-, 45-, 60-, 75-, 90-, and 105-Hz fields, implying an $n \times 15$ pattern, where $n = 1, 2, 3, 4, \ldots$. Furthermore, there was limited evidence in the statistical groupings tending to show that this series was actually made up of two interleaved frequency groups: 15, 45,

75, and 105 Hz, as well as 30, 60, and 90 Hz. The 15-Hz frequency series was further studied in subsequent work (Blackman et al., 1985c), at constant intensity with the following reported as effective frequencies: 15, 45, 75, 105, and 135 Hz in one group, and 195, 225, 255, 285, and 315 Hz in a higher group. Except for the hiatus at 165 Hz, these two groups correspond to the series $n \times 15$ Hz where n is restricted to odd integers.

Experiments involving diatom motility (Smith et al., 1987; McLeod et al., 1987) provide strong evidence that such frequency series are probably related to ICR harmonics. The resonance curve shown in Fig. 7 was extended, at constant DC magnetic field and constant AC magnetic intensity, to cover a range of frequencies through the seventeenth harmonic ($n = 17$). The results (Fig. 13) are at once similar to and different from the results of Blackman et al. (1985a). The similarity lies in the fact that only odd harmonics are observed in both sets of data. The difference is in those odd harmonics that are missing in each case. These are grouped, for convenient comparison, in Table V.

The results in Fig. 13 correspond to the specific case in which the AC and DC magnetic fields are arranged, using Eq. 2, to obtain CR tuning

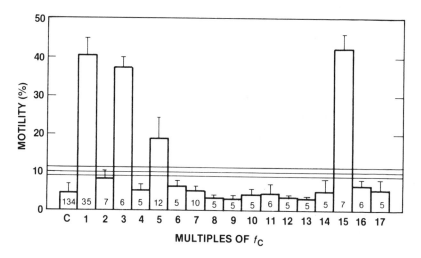

Fig. 13. Calcium harmonics, $n = 1$ through $n = 17$. Response of diatoms (percent motility) shown as function of harmonic number n. All cells in this series were tested at concentration of 0.25 mM Ca^{2+}. Upper, middle, and lower horizontal lines represent, respectively, 0.01, 0.02, and 0.05 confidence limits relative to controls. Number inside each data bar is cell count divided by 100. Harmonic number of n listed beneath each data bar is multiple of f_c, the fundamental cyclotron resonance (CR) frequency (in this case, 8 Hz). Value of B_c for each of 17 harmonics was 0.1045 G, corresponding to charge-to-mass ratio for Ca^{2+}. Note that only four harmonics ($n = 1, 3, 5, 15$) act to enhance calcium transport.

[Reprinted with permission from McLeod et al. 1987. Calcium and potassium cyclotron resonance curves and harmonics in diatoms (*A. coffeaeformis*). *J. Bioelectr.* 6:153–168.]

TABLE V.
Comparison of Harmonic Frequencies Observed in
ELF Calcium Efflux and Diatom Motility Experiments

Experiment	Observed Harmonics
Calcium efflux	1, 3, 5, 7, 9, —, 13, 15, 17, 19, 21
Diatom motility	1, 3, 5, —, —, —, —, 15, —

to the charge-to-mass ratio of Ca^{2+}, or suitable multiples (harmonics) of this tuning frequency as given in Eq. 3:

$$nf_0 = \frac{nq}{2\pi m} B_0 \qquad [3]$$

It is also possible to choose f_0 and B_0 so that one tunes to the K^+ ion. As explained, McLeod et al. (1987) observed that this tuning condition leads to inhibition of motility as compared to the enhancement following Ca^{2+} tuning. Interestingly, one can also obtain (McLeod et al., 1987) a set of harmonics for the K^+-inhibition tuning condition, again using Eq. 3. (Fig. 14). It is a remarkable fact that the sequence of harmonics $n = 1, 3, 5, 15$ applies equally well for enhancement tuning to Ca^{2+} or inhibition tuning to K^+. One possible conclusion from these homologous results is that the ion channels corresponding to these two ionic species may have very similar structures (or are perhaps even identical) and that the selectivity of these channels for the two species may be determined not by changes in protein configuration but rather by the charge-to-mass ratio of the ion itself. In this regard, note that the charge-to-mass ratios of K^+ and Ca^{2+} are almost integrally related: 2.46×10^6 and 4.81×10^6 C/kg, respectively. A periodic structure designed to operate according to Eq. 3 might perhaps also be expected to accommodate to a factor of 2 in the charge-to-mass ratio.

It should be pointed out that rationales have been presented, using CR arguments, for both the appearance of harmonics in the ELF interaction (Liboff and McLeod, 1988) and the restriction of these harmonic frequencies to odd values (McLeod and Liboff, 1987). This prediction, coupled to the experimental verification, again lends strength to an ICR mechanism as the physical basis for the ELF interaction. The fact that harmonics are also observed in the calcium-efflux experiments and further that only odd harmonics are observed is highly suggestive that the calcium-efflux and diatom experiments are closely related, most likely both sharing a common CR mechanism. The most intriguing aspect to the comparison drawn in Table V is the great difference in AC intensities; less than 1 mG rms

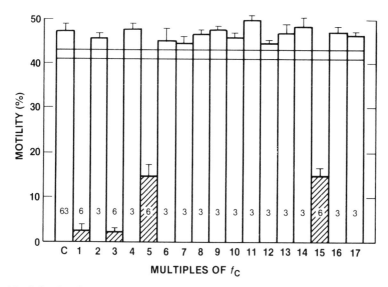

Fig. 14. Potassium harmonics, $n = 1$ through $n = 17$. As in Fig. 13, response of diatoms (percent motility) is shown as function of n. All tests at 5.0 mM Ca^{2+} concentration; control bar (C) at left reflects maximum motility. Upper and lower horizontal lines represent 0.05 and 0.01 confidence limits, respectively, against controls. Fundamental frequency set to cyclotron resonance (CR) tuning condition for K^+ charge-to-mass ratio: $f_c = 8$ Hz, $B_c = 0.2045$ G. Shaded harmonics act to inhibit calcium transport. Thus, identical set of harmonics ($n = 1, 3, 5, 15$) enhance calcium transport at one resonance and inhibit it at another.

[Reprinted with permission from McLeod et al. 1987. Calcium and potassium cyclotron resonance curves and harmonics in diatoms (*A. coffeaeformis*). *J. Bioelectr.* 6:153–168.]

for the calcium-efflux exposures and approximately 200 mG rms for the diatoms.

ICR AND 50- or 60-HZ POWER TRANSMISSION

The relationship that is currently emerging between high-voltage power lines and cancer in humans is only one aspect of a much wider set of biological phenomena, all of which are very likely tied to merely one physical effect, the ELF interaction. Thus, the biochemical sequence that leads to leukemia or brain tumors must surely be derived from the same initial physical insult that can also lead to behavioral changes, hormone imbalances, or problems in fetal development. This is certainly the case in vitro, where cells exposed to ELF fields have altered mitotic cycles (Marron et al., 1975), respond differently to mitogenic stimulation (Conti et al., 1986), suffer changes in motility (Smith et al., 1987), show developmental changes (Cameron et al., 1985), changes in DNA synthesis (Liboff et al., 1984), changes in nRNA production (Goodman et al., 1983), and either incorporate calcium more rapidly (Rozek et al., 1987) or exhibit enhanced calcium efflux (Bawin et al., 1975). The wide range of biological responses that can follow a single type of physical event is already well established for ionizing radiation. Furthermore, as with ionizing radiation, the question of the ELF interaction with living tissue can be a two-edged sword. For example, one can use therapeutic x rays to attack a malignancy that might have been originally caused by x rays. Similarly, it is probable that once the physical mechanism underlying the ELF interaction is better understood, important therapeutic benefits will be derived from the very same physical interaction that currently concerns the power industry.

Interestingly, the realization that these various effects all stem from one specific physical interaction also implies that potential hazards are not limited to high-voltage transmission lines. Power-distribution systems, residential wiring, home appliances, electric industrial devices, electric transportation systems, video display terminals – all are suspect. This wider perspective makes one realize the ELF fields also appear naturally in a geophysical context; the earth's magnetic field has a weak time-varying character, with daily, seasonal, and longer periods. The range of rms magnetic fields that have been biologically implicated extends down to almost unbelievably small intensities, for example, ~0.1 mG (Blackman et al., 1985b) and ~ 1.0 mG (Leal et al., 1986). These amplitudes are within the range of variations on the earth's surface caused by natural changes in the geomagnetic environment.

Judging from the epidemiological data (Ahlbom et al., 1987), AC magnetic-field intensity appears to be the key parameter involved in potential hazards from power-distribution lines. The CR explanation of the power-line magnetic-field interaction with living systems is that the 60-Hz magnetic field from such a line may conform to one of the resonant configurations shown in Table III. This resonance depends on the DC magnetic field (i.e., the local geomagnetic field) that is already present (Fig. 15), but it also depends on the specific ion that may be physiologically active. There is no reason to believe that the ELF interaction is limited to calcium. Experimental data have been obtained to date for Li^+, $^{40}Ca^{2+}$, $^{45}Ca^{2+}$, K^+, and Mg^{2+}. Obviously, although the 60-Hz frequency can in principle excite any of the ions listed in Table III, it is far from clear what biological change might be expected from, say, effecting a Mg^{2+} or K^+ resonance in the lymph system or the brain as compared to a Ca^{2+} resonance. Most of the present research on "mechanisms" may be fruitless if indeed the CR model is even partially correct. There can be a wide range of ionic resonances at 60 Hz depending on the orientation and intensity of the local DC field. The latter will not only vary from house to house but also will take different values within any one residence. The possibility, therefore, arises of having the AC field from a nearby power line combine with different intensities of the local DC field at different locations within a single dwelling to simultaneously give more than one type of ionic resonance. It seems fairly clear that the ICR model provides both the epidemiologist and the

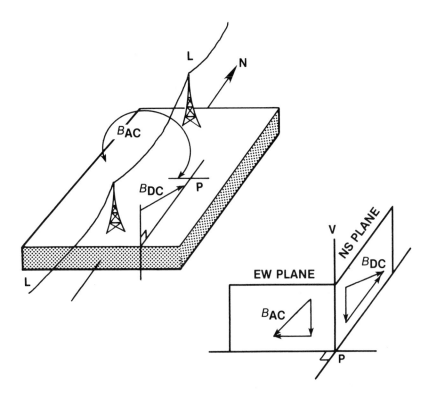

Fig. 15. Cyclotron resonance (CR) interaction of powerline magnetic field with geomagnetic field. Axis of power line L is oriented for simplicity along geomagnetic NS direction. This results in magnetic field circuital B_{AC} contained completely within EW plane, at 90° to the magnetic NS direction. Also shown is earth's magnetic field B_{DC} totally in NS plane. At arbitrary point P there will be magnetic contributions from both B_{AC} and B_{DC}. To right are drawn EW and NS planes that intersect at P and AC and DC vectors. Magnitude of B_{AC} is indicated at maximum amplitude; one half-cycle later it will point in reverse direction with same amplitude. Regardless of phase of B_{AC}, it will always have two components, one that is vertical, pointing along V, and, as indicated, a horizontal one within EW plane. Local static field at P, B_{DC}, also has two components, one vertical and one horizontal, within NS plane. Ion cyclotron resonance (ICR) interaction will be finite only for AC and DC components. AC and DC components at 90° will not interact. For this orientation, therefore, the only possible CR interaction is between vertical component of vector B_{AC} and vertical component of vector B_{DC}. CR interaction possibilities become more complicated for different orientation of line axis L: When L is EW, both vertical AC and DC components and horizontal AC and DC components can interact independently. For orientations between 0° and 90°, additional CR interactions will be found. An interesting observation is that for orientation pictured, where line axis is parallel to magnetic compass heading, there will be no CR interaction at any point P located directly *beneath* line, because B_{AC} directly beneath line has only horizontal and no vertical component. Interaction when L is parallel to NS is limited to AC and DC vertical components. Thus one should expect no effect directly under line.

experimentalist with a highly testable hypothesis. Without some such rational framework in which the life scientist can operate, neither the public nor the power companies will be served.

The relative effectiveness of the AC magnetic-field intensity is critical to the powerline question. Are there indeed intensity windows? Is there a dose–response change with increasing field intensity? Is there a lower threshold below which the ELF interaction disappears? Although data at present are insufficient to fully answer these questions, it is informative to compare the functional dependence on AC intensity for those experiments that examine the effects of exposures to time-varying magnetic fields. Three such experiments are grouped together in Fig. 16, with the ordinates appropriately scaled for ease of comparison. The results of Smith et al. (1987) on diatom motility and Liboff et al. (1987b) on ^{45}Ca incorporation in human lymphocytes are shown as a function of rms magnetic-field intensity; both curves were obtained under CR conditions, tuned for calcium ions. Although the biological systems under study are radically

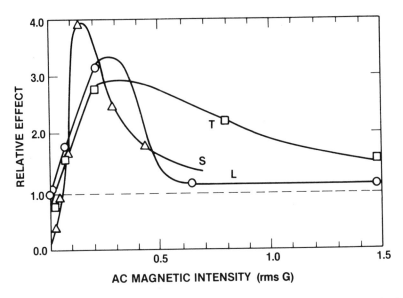

Fig. 16. ELF interaction as function of AC magnetic intensity. Three independent experiments are shown: T (*squares*) (Takahashi et al., 1986), S (*triangles*) (Smith et al., 1987), and L (*circles*) (Liboff et al., 1987b). Functional dependence on B_{AC} was studied for [^{3}H]thymidine incorporation in fibroblasts (T), diatom motility (S), and ^{45}Ca incorporation in lymphocytes (L). "Relative effect" is arbitrarily scaled for each set of results to accommodate all results on one graph. An interesting functional similarity in these results shows a peak response for 150–300 mG rms, followed by fall-off at higher intensities. Note different exposure conditions in each case: T, 100-Hz magnetic pulses; S, 16-Hz magnetic sine waves; L, 14.3-Hz magnetic sine waves. S and L used cyclotron resonance (CR) condition; T did not.

different, there is a certain similarity in the type of variation, with peaks all occurring in the vicinity of 150 to 200 mG rms. On the same graph we plot the results of Takahashi et al. (1986) for [³H]thymidine uptake in fibroblasts; in this experiment magnetic-field pulses at a repetition rate of 100 Hz were used, as compared to the 14- and 16-Hz sine waves used in the lymphocyte and diatom work. It is important to note that Takahashi et al. (1986) made no attempt to achieve any sort of CR condition during their exposures. A remarkably similar picture emerges from these three independent experiments on three totally different biological systems: The response in each case peaks in the region of 150 to 300 mG rms but falls off sharply at lower intensities, and falls also at higher intensities, in the region of 1 G rms and higher. The CR study on rat behavior by Thomas et al. (1986) was extended (Liboff et al., 1989) to include the effect of higher and lower AC magnetic intensities (Fig. 17). Note that a 30-min exposure to a field combination tuned to the Li⁺ ion causes aberrant behavioral response (as measured by the rate of lever pushes), but only above a certain

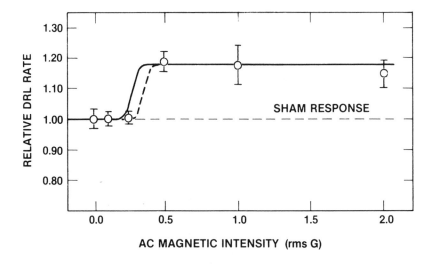

Fig. 17. 60-Hz magnetic intensity threshold for behavioral change. Rate of DRL lever pushes shown in Fig. 4 was measured at various exposure intensities without disturbing Li⁺ cyclotron resonance (CR) condition. Intensity scale (rms G) represents 60-Hz magnetic exposure intensity measured in horizontal direction (see Fig. 3). However, true or effective AC intensity is that component pointing in direction of DC magnetic field. When this correction is made (Liboff et al., 1989), results are shifted left such that threshold region moves from *dotted curve* to *solid curve*. Midpoint of resulting threshold intensity is 270 mG rms. Although this result is functionally different from results in Fig. 16, it is interesting to note that this threshold level also falls within range of maximum response for experiments in Fig. 16, 150–300 mG rms.

threshold AC intensity. This threshold value is centered at 270 mG rms, within the intensity region of 150 to 300 mG rms found for the experiments shown in Fig. 16. Can we directly compare the results in Figs. 16 and 17? A peak in cellular response occurs in the former case; the rat behavior experiment is by contrast an intensity threshold. Regardless of the functional nature of the bioresponse, one can argue that intensities between 150 and 300 mG rms appear to be effective in enhancing the ELF interaction.

It is, therefore, surprising to find other reports in the literature indicating ELF interactions at substantially reduced intensities, as much as 200 times weaker. It was pointed out that Blackman et al. (1985a) found two intensity ranges, one centered at 0.09 mG rms and the other at 0.6 mG rms. These values are consistent with the lower levels associated with high-voltage power lines at right-of-way distances (~ 0.1–20 mG rms). We list the following possible reasons for these interesting discrepancies:

1. There are multiple intensity windows, as suggested by the data of Blackman et al. (1985a). The data shown in Fig. 16 may constitute still another intensity window, in this case centered at about 200 mG rms as compared to the lower windows at 0.09 and 0.6 mG rms.

2. This approach may be inappropriate at powerline intensities in terms of studying the powerline/cancer correlation. All the results we have discussed [Blackman et al. (1985a), as well as those in Figs. 16 and 17], are for relatively short exposures, 0.5 hr (Thomas et al., 1986) to 24 hr (Takahashi et al., 1986). However, residential exposures from powerline fields are much longer in duration, extending for years or even decades. One interesting hypothesis, not yet tested, is that the biological effectiveness of the ELF interaction is a function of the exposure time, such that higher AC intensities over short exposure times are equivalent to much smaller intensities over very long exposure times.

3. It is conceivable that the discrepancy in effective intensity is not strictly a physical phenomenon, but rather a biological one. Thus, the intensity required to make the ELF interaction physiologically effective may vary with downstream factors such as enzymatic rate constants and nonequilibrium charge transfer.

4. Finally, one may again invoke the CR concept to suggest that different types of resonance may, for reasons as yet unknown, require different intensities. If one examines Table III, there is no a priori reason to indicate that the same AC intensity applies for each ion listed. In the same vein, there is also the possibility that the various harmonic frequencies derived from Eq. 3 may exhibit differing strengths at the same intensity.

This wide assortment of ELF intensity windows makes the problem of sorting out powerline hazards that much more difficult. We have seen that the connection of the ELF interaction to ICR seems readily apparent when considering the parameters of DC magnetic field, AC frequency, ionic charge-to-mass ratio, and frequency harmonics. Unfortunately, the picture is far from clear when one looks at the available data on AC intensity levels.

SUMMARY

Public concern over cancer promotion in residences exposed to powerline magnetic fields is increasing. To allay fears among consumers and provide direction for power companies and regulators, it is essential that the mechanism underlying this effect be better understood. The basic physical process involved has been all but ignored in the attempt to pinpoint biological effects. It is abundantly clear, however, that without a firm understanding of the initial interaction there can be very little real progress for the life scientists studying this problem.

The central theme of the present approach is that, even though a wide variety of biological phenomena result from weak, low-frequency electromagnetic exposures, all are ultimately connected to the same initial physical process. We call this process *the ELF Interaction*.

Very good evidence now indicates that the ELF interaction obeys simple ICR theory, notwithstanding the readily apparent difficulties involved in describing this model at the molecular level. The ICR approach has been particularly effective in explaining why the ELF interaction requires the presence of a DC magnetic field in addition to a low-frequency magnetic or electric field. It has been strikingly successful in explaining previous ELF-modulated calcium-efflux studies in chick brain, as well as the more recent ELF/DC magnetic work on the same model. It has been used to successfully predict the outcomes in three specially designed experiments: on operant-conditioned rats, on benthic diatoms, and on human lymphocytes. The ICR approach has revealed a dependence on a parameter that was heretofore unsuspected, the ionic charge-to-mass ratio, which has important significance for the physiologist. It is now apparent that the ELF interaction can directly affect Ca^{2+}, K^+, Li^+, and Mg^{2+} ions as opposed to merely the Ca^{2+} ion, as had been suggested. Indeed, at least some of the work on calcium efflux may directly involve the K^+ ion. Experiments on diatoms reveal a remarkable pattern of odd harmonics of the CR fundamental frequency. To some degree this bears a similarity to earlier "frequency-window" reports in calcium efflux. The restriction of these harmonics to $n = 1, 3, 5, 15, \ldots$ for both the Ca^{2+} and K^+ resonances is taken to indicate an intrinsic structural periodicity in the particular protein that serves as the focus of the ELF interaction.

The ELF interaction has been found to vary with AC magnetic intensity in each of the three ICR experiments, with an overall effective intensity region of approximately 150 to 300 mG rms. This intensity region is centered at a value much larger than that reported for the calcium-efflux studies (in one case, <0.1 mG rms). The range of interesting intensities associated with powerline exposures (~ 0.1–20 mG rms) in right-of-way residences may be compatible with these higher effective ICR intensities, if one assumes a linear dose response and a biological effectiveness that is enhanced by longer exposure times. The oscillatory magnetic fields that lend themselves to the ELF interaction are by no means limited to high-voltage power lines. One should also expect biological responses to result from exposure to power-distribution systems, residential electric wiring, home appliances, electric industrial devices, electric transportation systems, and video display terminals. There is no intrinsic reason why only devices operating at 60 Hz should result in the ELF interaction. There are many lower and higher frequencies that can satisfy ICR conditions at any given DC magnetic field. Because the geomagnetic field has a weak time-varying character, at times encompassing amplitudes that may reach 0.1-mG levels, one can safely assume that similar effects will result from variations in the earth's magnetic field if indeed powerline magnetic fields constitute a hazard.

Further, it is important to place the cancer question into proper perspective. Just as there are many devices and sources that can yield the fields required for the ELF interaction, it is difficult to believe that only one pathology will result from exposure to this interaction. The biochemical chain of events that leads to leukemia or glioblastoma stems from the same initial physical interaction that may lead also to behavioral changes, mood swings, hormonal or electrolyte imbalance, jet lag, problems in fetal development, genetic transformations, or immunosuppressive breakdowns. An argument can be made that the net human morbidity related to the ELF interaction, considering all possibilities, may far outweigh that from cancer alone. On the other hand, there also is good reason to believe that the physiologist and clinician will be enabled, once the physical nature of the ELF interaction is understood, to sort out these varied effects and devise exposure regimens that can be used therapeutically. In the past, pharmaceutical procedures were used to this end. For the ELF interaction, the physical process is at the heart of it all.

REFERENCES

Ahlbom, A., Albert, E. N., Fraser-Smith, A. C., Grodzinsky, A. J., Marron, M. T., Martin, A. O., Persinger, M. A., Shelanski, M. L., Wolpow, E. R. 1987. *Biological Effects of Power Line Fields*. Final Report. Albany, New York: New York State Department of Health.

Bawin, S. M., Adey, W. R. 1976. Sensitivity of calcium binding in cerebral tissue to weak environmental electric fields oscillating at low frequencies. *Proc. Natl. Acad. Sci. USA* 73:1999–2003.

Bawin, S. M., Kazmarek, K. L., Adey, W. R. 1975. Effects of modulated VHF fields on the central nervous system. *Ann. N.Y. Acad. Sci.* 247:74–81.

Bawin, S. M., Sheppard, A., Adey, W. R. 1978. Possible mechanisms of weak electromagnetic field coupling in brain tissue. *Bioelectrochem. Bioenerg.* 5:67–76.

Blackman, C. F., Benane, S. G., Kinney, L. S., Joines, W. T., House, D. E. 1982. Effects of ELF fields on calcium-ion efflux from brain tissue in vitro. *Radiat. Res.* 92:510-520.

Blackman, C. F., Benane, S. G., House, D. E., Rabinowitz, J. R., Joines, W. T. 1984. A role for the magnetic component in the field-induced efflux of calcium ions from brain tissue. In: *Abstracts, Proceedings of the Sixth Annual Meeting BEMS*, Atlanta, Georgia, Abstr. SA-40. Frederick, Maryland: Bioelectromagnetics Society.

Blackman, C. F., Benane, S. G., House, D. E., Joines, W. T. 1985a. Effects of ELF (1–120 Hz) and modulated (50 Hz) RF fields on the efflux of calcium ions from brain tissue in vivo. *Bioelectromagnetics* 6:1–11.

Blackman, C. F., Benane, S. G., Rabinowitz, J. R., House, D. E., Joines, W. T. 1985b. A role for the magnetic field in the radiation-induced efflux of calcium ions from brain tissue in vivo. *Bioelectromagnetics* 6:327–337.

Blackman, C. F., Benane, S. G., House, D. E. 1985c. A pattern of frequency-dependent responses in brain tissue to ELF electromagnetic fields. In: *Abstracts, Proceedings of the Seventh Annual Meeting of the Bioelectromagnetics Society*, San Francisco, California, Abstr. C-90. Frederick, Maryland: Bioelectromagnetics Society.

Cameron, I. L., Hunter, K. E., Winters, W. D. 1985. Retardation of embryogenesis by extremely low frequency 60-Hz electromagnetic fields. *Physiol. Chem. Phys. Med. NMR* 17:135–138.

Conti, P., Gigante, G. E., Cifone, M. G., Alesse, E., Fieschi, C., Bologna, M., Angeletti, P. U. 1986. Mitogen dose-dependent effect of weak pulsed electromagnetic field on lymphocyte blastogenesis. *FEBS Lett.* 199:130-134.

Cooksey, B., Cooksey, K. E. 1980. Calcium is necessary for mobility in the diatom *Amphora coffeaeformis*. *Plant Physiol.* 65:129–131.

Dutta, S. K., Subramoniam, A., Ghosh, B., Parshad, R. 1984. Microwave radiation-induced calcium ion efflux from human neuroblastoma cells in culture. *Bioelectromagnetics* 5:71–78.

Goodman, R., Abbott, J., Henderson, A. 1987. Transcriptional patterns in the X chromosome of *Sciara coprophila* following exposure to magnetic fields. *Bioelectromagnetics* 8:1–7.

Goodman, R., Bassett, C. A., Henderson, A. 1983. Pulsing electromagnetic fields induce cellular transcription. *Science* 220:1283–1285.

Hille, B. 1984. *Ionic Channels of Excitable Membranes*, pp. 188–300. Sunderland, Massachusetts: Sinauer.

Jafary-Asl, A. H., Solanki, S. N., Aarholt, E., Smith, C. W. 1982. Dielectric measurements on live biological materials under magnetic resonance conditions. *J. Biol. Phys.* 11:15–22.

Kim, K. S., Nguyen, H. L., Swaminathan, P. K., Clementi, E. 1985. Na^+ and K^+ ion transport through a solvated gramicidin A transmembrane channel: molecular dynamics studies using parallel processors. *J. Phys. Chem.* 89:2870–2876.

Lawrence, A. F., Adey, W. R. 1982. Nonlinear wave mechanisms in interactions between excitable tissue and electromagnetic fields. *Neurol. Res.* 4:115–153.

Leal, J., Trillo, M. A., Ubeda, A., Abraira, V., Shamsaifar, K., Chacon, L. 1986. Magnetic environment and embryonic development: a role of the earth's field. *IRCS Med. Sci.* 14:1145–1146.

Liboff, A. R. 1985. Cyclotron resonance in membrane transport. In: *Interactions Between Electromagnetic Fields and Cells*, Chiabrera, A., Nicolini, C., Schwan, H. P., eds., pp. 281–296. London: Plenum.

Liboff, A. R., McLeod, B. R. 1987. Cyclotron resonance eigenfrequencies and ion channel periodicity. In: *Ninth Annual Conference of IEEE Engineering in Medicine and Biological Science*, Boston, Massachusetts. Washington, D.C.: Institute of Electrical and Electronic Engineers.

Liboff, A. R., McLeod, B. R. 1988. Kinetics of channelized membrane ions in magnetic fields. *Bioelectromagnetics* 9:39–51.

Liboff, A. R., Thomas, J. R., Schrot, J. 1989. Intensity threshold for 60-Hz magnetically induced behavioral changes in rats. *Bioelectromagnetics* 10:111–113.

Liboff, A. R., Smith, S. D., McLeod, B. R. 1987a. Experimental evidence for ion cyclotron resonance mediation of membrane transport. In: *Mechanistic Approaches to Interactions of Electric and Electromagnetic Fields with Living Systems*, Blank, M., Findl, E., eds., pp. 109–132. New York: Plenum.

Liboff, A. R., Rozek, R. J., Sherman, M. L., McLeod, B. R., Smith, S. D. 1987b. Ca^{2+}-45 cyclotron resonance in human lymphocytes. *J. Bioelectr.* 6:13–22.

Liboff, A. R., Williams, T., Jr., Strong, D. M., Wister, R., Jr. 1984. Time-varying magnetic fields: effects on DNA synthesis. *Science* 223:818–820.

Marron, M. T., Goodman, E. M., Greenebaum, B. 1975. Mitotic delay in the slime mold *Physarum polycephalum* induced by low intensity 60- and 75-Hz electromagnetic fields. *Nature (London)* 254:66–67.

McLeod, B. R., Liboff, A. R. 1986. Dynamic characteristics of membrane ions in multi-field configurations of low-frequency electromagnetic radiation. *Bioelectromagnetics* 7:177–189.

McLeod, B. R., Liboff, A. R. 1987. Cyclotron resonance in cell membranes: the theory of the mechanism. In: *Mechanistic Approaches to Interactions of Electric and Electromagnetic Fields with Living Systems*, Blank, M., Findl, E., eds., pp. 97–108. New York: Plenum.

McLeod, B. R., Smith, S. D., Liboff, A. R. 1987. Calcium and potassium cyclotron resonance curves and harmonics in diatoms (*A. coffeaeformis*). *J. Bioelectr.* 6:153–168.

Parola, A. H., Porat, N., Kiesow, L. A. 1988. Time-varying magnetic field causes cell transformation. *Biophys. J.* 53:488a.

Polk, C. 1984. Time-varying magnetic fields and DNA synthesis: magnitude of forces due to magnetic fields on surface bound counter-ions. In: *Abstracts, Proceedings of the Sixth Annual Meeting of the Bioelectromagnetics Society*, Atlanta, Georgia, Abstr. P-36. Frederick, Maryland: Bioelectromagnetics Society.

Rozek, R. J., Sherman, M. L., Liboff, A. R., McLeod, B. R., Smith, S. D. 1987. Nifedipine is an antagonist to cyclotron resonance enhancement of [45]Ca incorporation in human lymphocytes. *Cell Calcium* 8:413–427.

Skerra, A., Brickmann, J. 1987. Structure and dynamics of one-dimensional ionic solutions in biological transmembrane channels. *Biophys. J.* 51:969–976.

Smith, S. D., McLeod, B. R., Liboff, A. R., Cooksey, K. 1987. Calcium cyclotron resonance and diatom mobility. *Bioelectromagnetics* 8:215–227.

Takahashi, K., Keneko, I., Date, M., Fukada, E. 1986. Effect of pulsing electromagnetic fields on DNA synthesis in mammalian cells in culture. *Experentia* 42:185–186.

Thomas, J. R., Schrot, J., Liboff, A. R. 1985. *Investigation of Potential Behavioral Effects of Exposure to 60-Hz Electromagnetic Fields.* NMRI 85-63. Bethesda, Maryland: Naval Medical Research and Development Command.

Thomas, J. R., Schrot, J., Liboff, R. 1986. Low-intensity magnetic fields alter operant behavior in rats. *Bioelectromagnetics* 7:349–357.

Westerhoff, H. V., Tsong, T. Y., Chock, P. B., Chen, Y.-D., Astumanian, R. D. 1986. How enzymes can capture and transmit free energy from an oscillating electric field. *Proc. Natl. Acad. Sci. USA* 1986:4734–4737.

12 Biological Interactions and Human Health Effects of Extremely Low Frequency Magnetic Fields

THOMAS S. TENFORDE
Life Sciences Center
Pacific Northwest Laboratory
Richland, Washington

CONTENTS

Magnetic fields in the extremely low frequency (ELF) range below 300 Hz are present throughout the environment and originate from both natural and man-made sources (Polk, 1974; Grandolfo and Vecchia, 1985). The naturally occurring, time-varying fields of the atmosphere have several origins, including diurnally varying fields of the order of 30 nanotesla

[1 nT: 1 tesla (T) = 10,000 gauss (G)] associated with solar and lunar influences on ionospheric currents. The largest time-varying, atmospheric magnetic fields arise intermittently from intense solar activity and thunderstorms, and reach intensities of the order of 0.5 mT during a large magnetic storm. Superimposed on the magnetic fields associated with irregular atmospheric events is a weak ELF field resulting from the Schumann resonance phenomenon. These fields are generated by lightning discharges and propagate in the resonant atmospheric cavity formed by the surface of the earth and the lower boundary of the ionosphere.

ELF magnetic fields originating from man-made sources generally have much higher intensities than the naturally occurring atmospheric fields, and in some occupational settings reach levels that approach 0.1 T. Two sources of ELF fields that have been the topics of considerable public interest are high-voltage transmission lines and land-based naval communication systems. The field at ground level beneath a 765-kV, 60-Hz power line carrying 1 kA per phase is 15 μT (Scott-Walton et al., 1979). The maximum field at ground level associated with the ELF antennae used in submarine communications is 14 μT (American Institute of Biological Sciences, 1985). Household appliances operated from a 60-Hz line voltage source produce local fields in their immediate vicinity with intensities as high as 2.5 mT (Grandolfo and Vecchia, 1985). However, the magnetic-field strength decreases rapidly as a function of distance from the surfaces of household devices (Gauger, 1984), and the ambient field levels at most locations within a household environment are generally less than 0.3 μT (Caola et al., 1983; Male et al., 1987). The video display terminals present in most offices generate local ELF magnetic fields with intensities up to 5 μT (Stuchly et al., 1983). A number of industrial processes that involve induction heating produce ELF magnetic fields of high intensity within the occupational environment. For example, based on a survey of electrosteel and welding industries in Sweden, it was reported that the local fields near 50-Hz ladle furnaces reached intensities of 8 mT, and intensities as high as 0.07 T were measured near induction heating devices operating in the 50-Hz to 10-kHz range (Lövsund et al., 1982).

In this chapter, a summary and critical evaluation are given of studies that describe the biological interactions and human health effects of ELF magnetic fields. Five aspects of this subject are discussed: (1) mechanisms of interaction of ELF magnetic fields at the tissue, cellular, and molecular levels; (2) magnetophosphenes and other visual phenomena; (3) laboratory studies on ELF magnetic-field interactions; (4) biological effects of combined static and ELF magnetic fields; and (5) epidemiological studies on human health effects of exposure to ELF fields.

INTERACTION MECHANISMS

The primary physical interaction of time-varying magnetic fields with living systems is the induction of electric fields and currents in tissue. For the specific case of a circular loop of tissue with radius R intersected by a spatially uniform, time-varying magnetic field orthogonal to the loop, Faraday's law gives for the magnitude of the average electric field tangent to the loop surface:

$$E = (R/2)\frac{dB}{dt} \qquad [1]$$

If the magnetic field is sinusoidal with an amplitude B_0 and a frequency f, then $B = B_0 \sin(2\pi ft)$ and from Eq. 1:

$$E = \pi fRB_0 \sin(2\pi ft) \qquad [2]$$

From Ohm's law, the current density J induced in tissue with an average conductivity σ is given by

$$J = \sigma E \qquad [3]$$

Various physical characteristics of time-varying magnetic fields are important in assessing their biological effects, including the fundamental field frequency, the maximum and average flux densities, the presence of harmonic frequencies, and the waveform and polarity of the signal. Several types of waveforms have been used in biological research with ELF magnetic fields, including sinusoidal, square-wave, and pulsed waveforms. Two characteristics that are of key importance in analyzing the effects of square-wave and pulsed fields are the rise and decay times of the magnetic-field waveform. These parameters determine the maximum time rate of change of the magnetic field, and hence the maximum instantaneous electric field and resulting current density induced in tissue (see Eqs. 1–3).

Although the initial physical interaction of ELF fields with living systems is the induction of electric currents in tissue, a number of secondary events may occur that involve biochemical and structural alterations at the cellular and subcellular levels. At the present time, there is convincing evidence that ELF fields do not produce cytogenetic alterations and are not directly mutagenic (Cohen et al., 1986; Livingston et al., 1986). In contrast, a substantial amount of evidence indicates that the pericellular currents established by ELF fields produce alterations in components of the cell membrane surface (Adey, 1981; Tenforde and Kaune, 1987). For example, ELF fields have been reported to alter Ca^{2+} binding to cell surfaces (Bawin and Adey, 1976; Blackman et al., 1985a), to suppress T-lymphocyte cytotoxicity, which is dependent on cell-surface antigen binding (Lyle et al., 1983), to inhibit human lymphocyte activation by mitogenic compounds that bind to the cell surface (Grattarola et al., 1985), to alter the response of adenylate cyclase to the binding of parathyroid hormone molecules at receptor sites on bone cell surfaces (Luben et al., 1982), to alter the distribution of cell-surface receptors and the lifetimes of ligand–receptor complexes (Chiabrera et al., 1984), to influence the release of insulin molecules from pancreatic cells (Jolley et al., 1985), and to alter cellular partitioning in an aqueous two-phase system that is sensitive to changes in membrane composition (Marron et al., 1983).

The various mechanisms through which ELF fields could influence membrane properties have recently been reviewed (Tenforde and Kaune, 1987). Two general classes of phenomena have been proposed: (1) long-range cooperative interactions at the cell surface, for example, coherent electric dipole oscillations, and (2) effects on specific membrane structures or localized membrane functions, for example, alterations in ligand–receptor binding or ion transport channels. These changes in membrane properties are envisioned to set up transmembrane signaling events, possibly mediated by Ca^{2+} or cyclic AMP, that trigger abnormal biochemical and cell growth states. The existence of intracellular responses to ELF fields has been clearly demonstrated by several lines of experimental evidence. For example, it has been demonstrated that exposure to pulsed and sinusoidal magnetic fields leads to altered RNA transcription patterns in dipteran salivary gland cells (Goodman and Henderson, 1986a,b, 1988; Goodman et al., 1983, 1987). This effect was accompanied by a significant change in the spectrum of cellular proteins synthesized by the exposed cells relative to control cells (Goodman and Henderson, 1986b, 1988). A total of 248 polypeptides in the control cells were resolved by two-dimensional gel electrophoresis, while 326 were observed in cells exposed to a 72-Hz pulsed magnetic field. The polypeptides synthesized in the dipteran salivary gland cells were specific to the characteristics of the ELF field to which these cells were exposed, with various polypeptides being either enhanced in quantity or suppressed relative to those observed for unexposed cells. The pattern of polypeptide synthesis was also found to differ from that in cells exposed to heat shock; only five of the new polypeptides synthesized by cells exposed to ELF fields overlapped the polypeptides induced by heat shock. Further, the suppression of total protein synthesis characteristic of heat shock was not observed in the cells exposed to electromagnetic fields. A new electrochemical model that describes the stimulation of protein biosynthesis by ELF fields has been developed by Blank and Goodman (1988).

Another finding of relevance to ELF field effects on macromolecular synthesis patterns is the demonstration that ornithine decarboxylase levels are increased in several lines of cultured cells after exposure to a 60-Hz electric field (Byus et al., 1987). This enzyme is essential for polyamine synthesis and cell growth, and experimental findings suggest that ELF field interactions may alter cellular biochemistry in a manner similar to the effects of tumor promoters such as phorbol esters. These compounds have been shown to produce large elevations in ornithine decarboxylase activity via their binding to membrane-associated phosphokinase C receptors and the subsequent production of new mRNA specific for ornithine decarboxylase (Verma et al., 1986).

A hypothetical sequence of interactions of ELF fields with living tissues that involves transmembrane coupling events and cytoplasmic responses is shown in Fig. 1. At the present time, the types of ELF signal transduction and amplification processes that occur in cell membranes have not been defined experimentally. However, it is clear that membrane interactions play a central role in mediating cellular and tissue responses to ELF fields. Several innovative new theoretical models have been proposed to explain the type of signal transduction processes that might occur at membrane surfaces. In one of these models, electric fields are postulated to

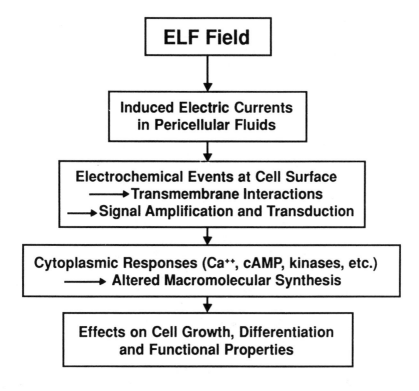

Fig. 1. Schematic representation of the cascade of physical and biochemical cellular events that have been hypothesized to be triggered by exposure to an ELF field.

change the intrinsic conformational equilibrium of membrane-bound proteins such as ATPases (Tsong and Astumian, 1988) with resultant effects on enzyme kinetics. Another model of electric-field interactions focuses on electrical double-layer processes at the cell membrane, and describes the ionic fluxes that occur across excitable membranes in terms of electrodiffusion equations (Blank, 1987). When the effects of a time-varying electric field are incorporated into these equations, it is predicted that ion concentration changes occur in the electrical double layers that surround charged groups at the outer and inner membrane surfaces. These concentration changes are predicted to influence ion transport through membrane channels, with the effects being maximal for imposed fields in the ELF range.

MAGNETOPHOSPHENES AND OTHER VISUAL PHENOMENA

An effect of time-varying magnetic fields on humans that was first described by d'Arsonval (1896) is the induction of a flickering illumination within the visual field known as magnetophosphenes. This phenomenon occurs as an immediate response to stimulation by either pulsed or sinusoidal magnetic fields with frequencies less than 100 Hz, and the effect is completely reversible with no apparent influence on visual acuity. The maximum visual sensitivity to sinusoidal magnetic fields has been found at a frequency of 20 Hz in human subjects with normal vision. At this frequency, the threshold magnetic-field flux density found by Lövsund et al. (1980a) to elicit phosphenes is approximately 10 mT. The corresponding time rate of change of the sinusoidal field is 1.26 T/sec. In recent studies Silny (1986) has observed thresholds for magnetophosphene perception in human volunteers as low as 5 mT with 18-Hz sinusoidal fields. In studies with pulsed fields having a rise time of 2 msec and a repetition rate of 15 Hz, the threshold values of dB/dt for eliciting phosphenes ranged from 1.3 to 1.9 T/sec in five adult subjects (Budinger et al., 1984). A trend in the data suggested that the threshold was lower among younger subjects. In related studies it was also observed that the stimulus duration is an important parameter, because pulses of 0.9-msec duration with $dB/dt = 12$ T/sec did not evoke phosphenes.

Several types of experimental evidence indicate that the magnetic-field interaction leading to magnetophosphenes occurs in the retina: (1) magnetophosphenes are produced by time-varying magnetic fields applied in the region of the eye, and not by fields directed toward the visual cortex in the occipital region of the brain (Barlow et al., 1947); (2) pressure on the eyeball abolishes sensitivity to magnetophosphenes (Barlow et al., 1947); (3) the threshold magnetic-field flux density required to elicit magnetophosphenes in human subjects with defects in color vision was found to have a different dependence on the field frequency than that observed for subjects with normal color vision (Lövsund et al., 1980a); and (4) in a patient in whom both eyes had been removed as the result of severe glaucoma, phosphenes could not be induced by time-varying magnetic fields, thereby precluding the possibility that magnetophosphenes can be initiated directly in the visual pathways of the brain (Lövsund et al., 1980a).

Although experimental evidence has clearly implicated the retina as the site of magnetic-field action leading to phosphenes, it is not as yet resolved whether the photoreceptors or the neuronal elements of the retina are the sensitive substrates that respond to the field. In a series of experiments on in vitro frog retinal preparations, extracellular electrical recordings were made from the ganglion cell layer of the retina immediately following termination of exposure to a 20-Hz, 60-mT field (Lövsund et al., 1981). It was found that the average latency time for response of the ganglion cells to a photic stimulus increased by 5 msec ($p < 0.05$) in the presence of the magnetic field. In addition, the ganglion cells that exhibited electrical activity during photic stimulation ("on" cells) ceased their activity during magnetic-field stimulation (i.e., they became "off" cells). The converse behavior of ganglion cells was also observed. These observations indicate that stimulation of the retina by light and by a time-varying magnetic field elicits responses in similar postsynaptic neural pathways.

The direct involvement of electric currents induced in the retina by ELF magnetic fields in the stimulation of postsynaptic neurons is also suggested by the close similarity of electrophosphenes and magnetophosphenes (Lövsund et al., 1980b). As shown in Fig. 2, the frequency response of the human eye to electrically generated phosphenes is similar to that observed for magnetically induced phosphenes. In a recent study with human subjects, Carstensen et al. (1985) found that electrophosphenes were generated by a 25-Hz electric current of 4 mA passing along a path from the left eye to the right hand. At 60 Hz the current required to generate electrophosphenes increased by a factor of 5. These investigators estimated that the threshold 60-Hz electric field that must be induced in the retina to produce phosphenes is approximately 1 V/m. Based on a finite-element analysis of the field distribution in the head, the threshold 60-Hz current density required to induce electrophosphenes was estimated to be of the order of 10 mA/m^2.

Other phenomena related to the sensitivity of the visuosensory system to time-varying magnetic fields were also studied by Silny (1986). In experiments with human subjects it was found that distinct flickering could be elicited in the visual field by sinusoidal magnetic fields in the frequency range of 5 to 60 Hz. The threshold field intensity varied with the field frequency and background light level, but was as low as 5 mT under optimal conditions. Alterations in visually evoked potentials (VEP) were also reported to occur in sinusoidal ELF magnetic fields at intensity levels that

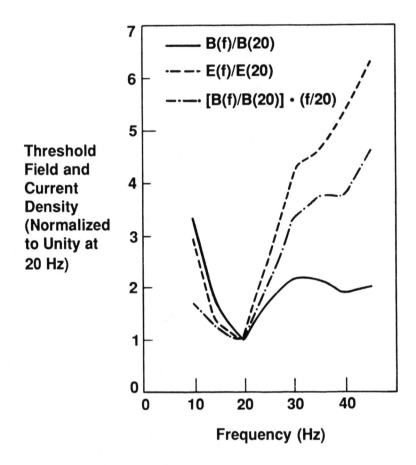

Fig. 2. Threshold magnetic flux densities and electric-field intensities that produce magneto- and electrophosphosphenes are plotted as function of field frequency. Also shown is product of the threshold magnetic flux density and frequency, which is proportional to the electric field and current density induced in the retina during exposure. Curve for product $B \cdot f$ closely parallels the curve for electric fields produced by electrodes placed on the temple near the eyes. For both magnetically induced currents and directly applied electric currents, maximum visual sensitivity was observed at 20 Hz. (Adapted from Fig. 4 of Lovsund et al., 1980b.)

are 5 to 10 times greater than those which produce magnetophosphenes (Silny, 1986). The change in VEP was characterized by a reversal of polarity and a decreased amplitude of the three major evoked potentials. These effects were observed within 3 min after onset of the magnetic-field exposure, and the VEP returned to normal only after a recovery period of approximately 30–70 min following termination of the exposure. The relationship of these changes in the VEP to the mechanism of magnetophosphene induction is not clear from the evidence that is presently available.

GENERAL SUMMARY OF LABORATORY STUDIES ON ELF MAGNETIC-FIELD EFFECTS

Time-varying magnetic fields that induce current densities above 1 A/m^2 in tissue lead to neural excitation and are capable of producing irreversible biological effects such as cardiac fibrillation. Several investigators have achieved direct neural stimulation using pulsed or sinusoidal magnetic fields that induced tissue current densities in the range of 1 to 10 A/m^2. In one study involving electromyographic recordings from the human arm (Polson et al., 1982), it was found that a pulsed field with dB/dt greater than 10^4 T/sec was required to stimulate the median nerve trunk. The duration of the magnetic stimulus has also been found to be an important parameter in the excitation of nerve and nerve-muscle specimens. Using a 20-kHz sinusoidal field applied in bursts of 0.5- and 50-msec duration, Öberg (1973) found that a progressive increase in the magnetic flux density was required to stimulate the frog gastrocnemius neuromuscular preparation when the burst duration was reduced to less than 2–5 msec. A similar rise in threshold stimulus strength has been observed for frog neuromuscular stimulation using pulsed magnetic fields with pulse durations less than approximately 1 msec (Ueno et al., 1978, 1984).

Time-varying magnetic fields that induce tissue current densities less than approximately 1–10 mA/m^2 have been found to produce few, if any, biological effects. This general observation is not surprising, since the endogenous current densities present in many organs and tissues lie in the range of 0.1 to 10 mA/m^2 (Bernhardt, 1979). In contrast, ELF magnetic fields that induce peak current densities greater than approximately 1–10 mA/m^2 have been reported to produce various alterations in the biochemistry and physiology of cells and organized tissues. One example is the effect of the bidirectional pulsed fields used to facilitate bone fracture reunion in humans (Bassett et al., 1982). A large number of laboratory investigations have also led to reports of a broad spectrum of alterations in cellular, tissue, and animal systems in which current densities exceeding 1–10 mA/m^2 were induced by ELF magnetic fields (Tenforde, 1986a,b, 1988, 1989; Tenforde and Budinger, 1986). These effects include (1) altered cell growth rate, (2) decreased rate of cellular respiration, (3) altered metabolism of carbohydrates, proteins, and nucleic acids, (4) effects on gene expression and genetic regulation of cell functions, (5) teratological and developmental effects, (6) morphological and other nonspecific tissue changes in animals, frequently reversible with time following exposure, (7) endocrine alterations, (8) altered hormonal responses of cells and tissues, including effects on cell-surface receptors, and (9) altered immune response to antigenic stimulation.

In assessing these reported effects of ELF magnetic fields, it is important to recognize that very few of the observations have been independently replicated in a second laboratory. In many cases where attempts at replication were carried out, the results were contradictory. One notable example of this variability in results is the attempt by several groups of investigators to determine whether teratological effects result from the exposure of chicken embryos to pulsed magnetic fields of low intensity, as originally reported by Delgado et al. (1981, 1982). The results of experiments described in several publications on this subject from 1981 to 1988 are summarized in Table I. Maffeo et al. (1984) reported an unsuccessful attempt to replicate the findings of Delgado et al. (1982), but the pulsed fields used in these two different studies did not have identical rise times. Rooze and Hinsenkamp (1985) and Siskin et al. (1986) were also unable to detect any teratological effects on chick embryos caused by the pulsed magnetic fields that are used clinically to facilitate bone fracture repair. Using a 20-kHz magnetic field with a sawtooth waveform typical of the fields that emanate from video display terminals, Sandström et al. (1987) were unable to find any influence of exposure on the early development of chicken embryos. In another study Juutilainen and associates (Juutilainen, 1986; Juutilainen and Saali, 1986; Juutilainen et al., 1986) observed significant teratological effects of bidirectional magnetic fields with sinusoidal, square-wave, and pulsed waveforms. Leal et al. (1986), working in Delgado's laboratory, were able to successfully replicate their earlier results. Martin (1988) recently reported that pulsed ELF fields produce abnormalities in chicken embryos when the exposure is carried out during the first 24 hr of incubation. However, exposure to a pulsed field during the interval from 24 to 48 hr after the start of incubation was found to have no influence on embryo development.

In an effort to resolve the question of whether the original experimental results of Delgado et al. (1981, 1982) are replicable under controlled laboratory conditions, the Office of Naval Research and the U.S. Environmental Protection Agency recently sponsored an international cooperative effort involving six independent laboratories (Berman et al., 1988). Laboratories located in Spain, Sweden, Canada, and the United States were equipped with identical pulsed magnetic-field exposure systems that had been constructed and tested by the same engineering team. The pulse parameters chosen for this study were 100-Hz repetition frequency, 500-μsec pulse duration, 2-μsec pulse rise and fall times, and peak magnetic flux density of 1.0 μT. Each of the six laboratories conducted 10 separate experiments with 20 chick eggs, 10 of which were exposed to the pulsed field for the first 48 hr of incubation while the remaining 10 eggs were sham-exposed for the same time interval. Two of the six laboratories observed a statistically

TABLE I.
Pulsed Magnetic Fields (PMF) and
Developmental Abnormalities in Chicken Embryos

Delgado et al. (1981, 1982): Significant increase in abnormal embryos (AE) after 48-hr exposure to square-wave PMF relative to controls (10-, 100-, and 1000-Hz pulse repetition frequencies; 500-μsec pulse duration; 2-μsec pulse rise time; and 0.12-, 1.2-, and 12-μT flux densities).

Ubeda et al. (1983): Strongest teratogenic effect observed after 48-hr exposure to a square-wave PMF with a 100-Hz repetition frequency, 500-μsec pulse duration, 1.7-μsec pulse rise time, and 1.0-μT flux density.

Maffeo et al. (1984): Could not replicate the earlier studies using a 48-hr exposure to a square-wave PMF (100- and 1000-Hz repetition frequencies; 500-μsec pulse duration; 10-μsec pulse rise time; and 1.2- and 12-μT flux densities).

Rooze and Hinsenkamp (1985): No effect of clinical PMF with 5-msec bidirectional pulses at 15-Hz repetition rate, applied for 100-150 hr.

Siskin et al. (1986): No effect of two types of clinical PMF applied for 1 or 7 days (50-msec bidirectional pulses repeated at 2 Hz, and 5-msec bidirectional pulses repeated at 15 Hz).

Juutilainen et al. (1986): Consistent increase in AE observed with 48-hr exposure to sinusoidal, square-wave, and pulsed magnetic fields. Only bidirectional waveforms were effective. Frequencies of 16.7 Hz to 100 kHz and flux densities of 1.25–125 μT were most effective in producing AE.

Leal et al. (1986): Verified earlier findings that PMF increase the percent of AE relative to controls (100-Hz repetition frequency; 500-μsec pulse duration; 2-μsec pulse rise time; and 0.4–1.0 μT flux density).

Sandström et al. (1987): No effect of bidirectional magnetic fields with sawtooth waveforms (20-kHz pulse repetition frequency; 45-μsec pulse rise time and 5-μsec fall time; and 0.1-, 1.5-, and 16-μT flux densities).

Martin (1988): Observed AE when PMF applied during first day of incubation, but not at later times (100-Hz repetition frequency; 500-μsec pulse duration; 2-μsec pulse rise and fall times; and 1.0-μT flux density).

---

significant increase in the proportion of abnormal embryos ($p < 0.001$ and $p = 0.03$), while the other four laboratories did not observe a significant difference between the exposed and sham-exposed embryos. The overall data from the six different laboratories, however, did show a statistically significant increase in the proportion of abnormal embryos in the exposed groups of eggs. The interlaboratory variations observed in this series of experiments are indicative of the difficulty encountered in replicating the results of biological studies on ELF field effects, even when exceptional efforts are made to control the relevant experimental variables.

ION CYCLOTRON RESONANCE PHENOMENA IN COMBINED STATIC AND ELF MAGNETIC FIELDS

Several experimental studies have provided experimental evidence that the combination of a weak static magnetic field, comparable in strength to the geomagnetic field, and a time-varying magnetic field in the ELF frequency range can produce resonance interactions that influence ion movements through membrane channels and other biological phenomena. The physical mechanism underlying this effect has been suggested to be ion cyclotron resonance (Liboff, 1985; McLeod and Liboff, 1986; Liboff and McLeod, 1988; Durney, 1988). In this process a resonant transfer of energy from a time-varying magnetic field occurs when its frequency matches the cyclotron resonance frequency of an ion moving within a static magnetic field (Fig. 3). The resonance condition is formally expressed by the equation:

$$f_c = qB/2\pi m \qquad [4]$$

where f_c is the ion cyclotron resonance frequency, q = ion charge, m = ion mass, and B = flux density of the static magnetic field. For the typical range of the geomagnetic field over the surface of the earth (30–70 µT), the resonant frequencies of many biologically important ions such as Na^+, K^+, and Ca^{2+} fall within the ELF range.

Several lines of experimental evidence suggest that ion cyclotron resonance interactions can influence biological processes. Four recent publications on this subject have reported that certain combinations of static magnetic-field flux density and time-varying magnetic-field frequency can alter (1) the rate of calcium ion release from brain tissue (Blackman et al., 1985b), (2) the operant behavior of rats in a timing discrimination task (Thomas et al., 1986), (3) calcium-dependent diatom mobility (Smith et al., 1987), and (4) calcium ion uptake by human lymphocytes (Liboff et al., 1987). The frequencies of the time-varying fields used in these various experiments matched the ion cyclotron resonance frequencies for potassium, lithium, and calcium ions.

Although these experimental results suggest a resonance mechanism through which weak static and ELF fields could produce measurable biological effects, the interpretation of this work presents several theoretical difficulties. There are four major problems with the ion cyclotron resonance theory: (1) the collision frequency of ions undergoing cyclotron resonance motion in membrane channels is required to be orders of mag-

Ion Cyclotron Resonance in Combined Static and AC Magnetic Fields

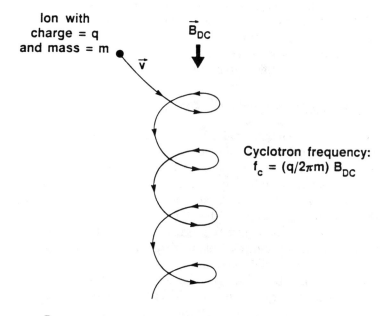

Ion with
charge = q
and mass = m

$\vec{B}_{DC}$

$\vec{v}$

Cyclotron frequency:
$f_c = (q/2\pi m)\, B_{DC}$

Resonant energy transfer occurs when a
time-varying magnetic field, $\vec{B}_{AC}$, is
also present and has a frequency matching f_c.

Fig. 3. Representation of ion cyclotron resonance conditions for an ion moving in a helical trajectory.
(Adapted from Fig. 1 of Liboff, 1985.)

nitude less than the typical collision frequency in an aqueous solution at physiological temperatures; (2) the interaction energy of the weak static magnetic field with biological ions is several orders of magnitude less than the Boltzmann thermal energy, kT ($= 4.28 \times 10^{-21} J$ at 310 K); (3) the thermally generated electrical noise (Nyquist noise) present in ion transport channels that traverse biological membranes is approximately two orders of magnitude greater than the electric field established in these channels by the resonant time-varying magnetic field (Tenforde and Kaune, 1987); and (4) for ion motion that is constrained to lie along a prescribed path, such as the helical path envisioned by Liboff (1985) for ion transport through membrane channels, it follows directly from the equation of motion for the particle that a static magnetic field cannot influence the ion movement and establish a resonance condition (Halle, 1988). The ion cyclotron resonance interaction is thus limited to unconstrained ion movements through membrane channels. All these factors would interfere with the establishment of ion cyclotron resonance conditions in combined static and time-varying magnetic fields. Obviously there is a need to refine the theoretical description of this phenomenon before it can form a plausible basis for weak field interactions with biological systems.

HUMAN HEALTH STUDIES OF ELF FIELD EFFECTS

One of the most controversial issues related to the interaction of ELF fields with humans is the reported link between residential and occupational exposure to power-frequency fields and cancer risk. The first report on this subject was published by Wertheimer and Leeper (1979), who found that cancer deaths (primarily leukemia and nervous system tumors) in children less than 19 years of age in the Denver, Colorado, area were correlated with the presence of high-current primary and secondary wiring configurations near their residences. This retrospective epidemiological study was based on 344 fatal childhood cancer cases during the period 1950 to 1973 and an equal number of age-matched controls chosen from birth records. The electrical power lines near the birth and death residences of the cancer cases and the residences of the controls were inspected and classified as being either high-current configurations (HCC) or low-current configurations (LCC), which were assumed to reflect the local intensity of the 60-Hz magnetic fields within the homes of the subjects. The percentage of the cancer cases whose birth and death residences were near HCC was found to be significantly greater than for the control subjects, from which Wertheimer and Leeper concluded that an association may exist between the strength of magnetic fields from the residential power-distribution lines and the frequency of childhood cancer. In a subsequent publication, these authors reported that a similar association exists for the incidence of adult cancer (Wertheimer and Leeper, 1982). This later study was based on 1179 cancer cases (78% fatal cancers) in Denver, Boulder, and Longmont, Colorado, during the period 1967 to 1977.

Following the initial report of Wertheimer and Leeper on childhood cancer, four other epidemiological studies have been made to determine whether a relationship exists between residential magnetic fields from powerline sources and the incidence of leukemia in children. In the first of these studies, Fulton et al. (1980) used methodology that was matched as closely as possible to that of Wertheimer and Leeper, including the designation of HCC and LCC power lines. This study involved 119 leukemia patients with ages of onset from 0 to 20 years, whose address histories were obtained from medical records at Rhode Island Hospital, and 240 control subjects chosen from Rhode Island birth certificates. In their study, Fulton et al. (1980) concluded that no statistically significant correlation existed between the incidence of leukemia and the residential powerline configurations. Wertheimer and Leeper (1980) were critical of the study by Fulton et al. (1980) on the basis that the case and control groups had not been matched for interstate migration, for years of occupancy at residences, or for the ages of the children at the time their residential addresses were determined from birth records and hospital medical records. In a reevaluation of the data obtained by Fulton et al. (1980), Wertheimer and Leeper (1980) excluded cases and controls aged 8 and older, which allowed them to define a complete residential history for the remaining subjects (53 cases and 71 controls). In this subset of the total population studied by Fulton and associates, Wertheimer and Leeper found a weakly significant correlation ($p = 0.05$) between the incidence of leukemia and residential HCC wiring configurations.

Another study of childhood leukemia incidence was conducted in the county of Stockholm by Tomenius (1986), who analyzed the residential 50-Hz magnetic fields for 716 cases that had a stable address from the time of birth to the time of leukemia diagnosis, and for 716 controls who were matched for age, sex, and birth location. An evaluation was made of the electrical wiring configurations near the residences of the study population, and measurements were also made at the entrance door to each residence of the magnetic-field flux density in the frequency range above 30 Hz. Among the residences where a magnetic-field level exceeding 0.3 μT was recorded, the incidence of leukemia was greater by a statistically significant amount than the expected level. The most frequently observed types of cancer were nervous system tumors and leukemia.

In contrast to the findings of Wertheimer and Leeper (1979) and Tomenius (1986), Myers et al. (1985) found no relationship between the risk of childhood cancer and residential proximity to overhead power lines. This study was conducted in the Yorkshire Health Region in England, and included 376 cancer cases diagnosed in children less than 15 years of age during the period 1970 to 1979. A total of 590 age-matched controls were included in the study. Magnetic fields at the birth addresses were calculated on the basis of data from the electrical load records for the overhead lines. The results of this epidemiological study showed no significant elevation in the cancer risk ratio with increasing field strength, and no dependence of the risk ratio on distance from the overhead lines was observed.

Another recent case-control epidemiological study by Savitz et al. (1988) attempted to verify the initial findings of Wertheimer and Leeper (1979) on childhood cancer in the Denver, Colorado, area. This study involved a total of 357 cancer cases diagnosed between 1976 and 1983, and the cancer incidence data were analyzed on the basis of both the Wertheimer/Leeper wiring code and spot measurements of power-frequency magnetic fields in the homes. A correlation between cancer risk in children less than 14 years of age and the proximity of their residences to high-current wiring configurations was found. However, no statistically significant association was observed between the measured household fields and childhood cancer incidence. The results of this study are therefore ambiguous, and indicate that a more precise estimate of exposure than the powerline configuration or a point-in-time measurement of household fields is needed.

Two other recent epidemiological surveys failed to detect an association between residential exposure to power-frequency fields and cancer risk. An epidemiological study conducted by McDowall (1986) in England found no correlation between cancer mortality and residential exposure to the fields from electrical utility installations (substations and overhead power lines). This study involved a retrospective analysis of mortality from 1971 to 1983 among a population of 7631 persons in East Anglia who were identified as living near electrical installations. The standardized mortality ratios for this large study population were computed for three major causes of death: cancer, cardiovascular disease, and respiratory disease. The overall mortality rates were lower than expected for all disease classes, and the results of this study did not support the earlier claims of an elevated cancer risk associated with residential exposure to power-frequency fields.

A recent case-control study of the incidence of acute nonlymphocytic leukemia (ANL) in three counties in Washington State also failed to find a correlation between residential exposure to power-frequency fields and cancer risk (Stevens et al., 1986; Severson et al., 1988). This epidemiological survey involved 164 cases of ANL and 204 controls from the same geographic area. Residential wiring codes were analyzed by the Wertheimer/ Leeper technique, and this information was supplemented by direct measurements of the residential electric and magnetic fields (Kaune et al., 1987). In addition, confounding variables such as smoking habits and socioeconomic status of the case and control subjects were analyzed. The overall results of this study provided no evidence for a possible association between residential exposure to power-frequency fields and the risk of contracting ANL.

The controversy surrounding the issue of ELF field exposure and cancer risk has been increased by a large number of epidemiological reports published since 1982 in which an apparent association was found between employment in various electrical occupations and cancer risk (primarily leukemia). Many of these studies have been reviewed previously (Savitz, 1986; Tenforde, 1986b, 1989). Savitz and Calle (1987) recently attempted to collate the data from 11 of these published studies to estimate the average relative risk of all leukemias, acute leukemias, and acute myelogenous leukemias among workers in 12 different classes of electrical occupations. The overall relative risk and 95% confidence intervals for leukemia mortality were found to be the following: (1) total leukemias: 1.2 (1.1–1.3); (2) acute leukemias: 1.4 (1.2–1.6); and (3) acute myelogenous leukemias: 1.5 (1.2–1.8). Savitz and Calle (1987) concluded from their analysis that a correlation exists between employment in electrical occupations and leukemia risk. However, they pointed out that none of the epidemiological surveys conducted thus far have established that exposure to power-frequency electromagnetic fields is the causal factor leading to an elevated cancer risk among electrical workers.

In addition to the reported correlation between mortality from leukemia and occupational exposure to ELF fields, four separate studies have reported an increased risk of brain tumors among workers in electrical occupations (Milham, 1985; Lin et al., 1985; Thomas et al., 1987; Speers et al., 1988). These epidemiological studies were based on data obtained from death certificates of workers in electrical occupations in different geographic areas within the United States. Although the findings were reasonably consistent, none of these studies established a true causal relationship between exposure to ELF fields and the risk of brain malignancies. In addition, the potential contribution to cancer risk of exposure to other agents such as organic solvents was not assessed in any of the epidemiological studies on brain tumor incidence among electrical workers.

Overall, the epidemiological studies on the possible correlation between cancer risk and residential exposure to electromagnetic fields do not support the conclusion of a strong association. In the earlier studies on this subject, especially those conducted by Wertheimer and Leeper (1979, 1982) in the Denver, Colorado area, the control groups were chosen in a nonblind manner. In addition, quantitative measurements of the power-frequency fields within the residences of the case and control subjects have been made only in the recent studies by Savitz et al. (1988) and by Severson et al. (1988). Finally, with the exception of these two studies, no attempt was made to analyze the role of confounding variables (for example, smoking habits) in the overall cancer risk of the case and control populations. These considerations, as well as the primarily negative outcomes of the most recent epidemiological studies, suggest that ELF fields in the residential environment pose little, if any, cancer risk. For workers in electrical occupations, the available information indicates that these individuals have a small elevation in the risk of cancer, especially leukemia and brain tumors. However, a causal link between occupational exposure to power-frequency fields and cancer has not been established. There exists a clear need for additional epidemiological surveys on large populations of electrical workers, in which efforts are made to analyze the possible role of confounding variables and to conduct proper dosimetry measurements for exposure assessment.

Several studies have also been made of the general health profiles of individuals who work in electrical occupations or who were exposed to ELF magnetic fields under controlled laboratory conditions. Medical examinations of 379 workers in electrical substations in Italy revealed no adverse clinical symptoms relative to a control group of 133 workers (Baroncelli et al., 1986). Laboratory studies on humans exposed to ELF magnetic fields have also failed to reveal any adverse physiological or psychological symptoms in the exposed subjects. The strongest field used in these experiments was a 5-mT, 50-Hz field to which subjects were exposed for 4 hr by Sander et al. (1982). No field-associated changes were observed in serum chemistry, blood cell counts, blood gases and lactate concentration, electrocardiogram, pulse rate, skin temperature, circulating hormones (cortisol, insulin, gastrin, thyroxin), and various neuronal measurements including visually evoked potentials recorded in the electroencephalogram.

A study by Wertheimer and Leeper in 1986 led to evidence of seasonal changes in fetal growth and in abortion rate among women that used electrically heated beds during the winter months. The authors contended that these adverse effects on fetal development could result from exposure to the power-frequency electromagnetic fields present at the surfaces of electrically heated beds. They pointed out, however, that the potentially harmful effect of excessive heat on fetal growth cannot be excluded on the basis of their data.

Goldhaber et al. (1988) recently reported that the rate of miscarriages increased by approximately 80% among women who worked on video display terminals for more than 20 hr per week as compared to women who did similar work without the use of these devices. No statistically significant risk of miscarriage was found among women who worked at video display terminals for less than 20 hr per week. Overall, the increase in miscarriage rate was about 40% for women who worked at video display terminals for more than 5 hr per week, but this increase was not statistically significant. Although the results of this study suggest that the fields from video display terminals may enhance the risk of miscarriage, the possible role of job stress or other unidentified factors cannot be excluded on the basis of the available information. In distinct contrast to the results of the study by Goldhaber et al. (1988), four other epidemiological surveys have not obtained evidence for a significant elevation in abortion rate or birth defects as the result of prolonged exposure during pregnancy to the electromagnetic fields from video display terminals (Kurppa et al., 1985; Ericson and Kallen, 1986a,b; McDonald et al., 1986).

In general, there is no convincing evidence at the present time that exposure to ELF fields influences fetal development or leads to adverse human health effects when endpoints other than cancer are considered.

SUMMARY AND CONCLUSIONS

A number of different effects of ELF magnetic fields have been reported to occur at the cellular, tissue, and animal levels. Certain effects, such as the induction of magnetophosphenes in the visual system, have been established through replication in several laboratories. A number of other effects, however, have not been independently verified or, in some cases, replication efforts have led to conflicting results. A substantial amount of experimental evidence indicates that the effects of ELF magnetic fields on cellular biochemistry, function, and structure can be related to the induced current density, with a majority of the reported effects occurring at current-density levels in excess of 10 mA/m^2. These effects therefore occur at induced current-density levels that exceed the endogenous currents normally present in living tissues. From this perspective, it is extremely difficult to interpret the results of recent epidemiological studies that have found an apparent correlation between exposure to ELF fields and cancer incidence. The levels of current density induced in tissue by occupational or residential exposure to ELF fields are, in nearly all circumstances, significantly lower than the levels found in laboratory studies to produce measurable perturbations in biological functions. There is a clear need for additional epidemiological research to clarify whether exposure to ELF fields is in fact causally linked to cancer risk. In addition, more studies of both a theoretical and experimental nature are needed to elucidate the molecular and cellular mechanisms through which low-intensity ELF fields can influence living systems. A growing body of evidence indicates that cell membranes play a key role in the amplification of ELF field signals. Elucidation of the physical and biochemical pathways that mediate these transmembrane signaling events will represent a major advance in our understanding of the molecular basis of ELF field effects on biological systems.

ACKNOWLEDGMENT

The skillful assistance of Bonnie J. Williams in the preparation of this manuscript is gratefully acknowledged.

REFERENCES

Adey, W. R. 1981. Tissue interactions with nonionizing electromagnetic fields. *Physiol. Rev.* 61:435–514.

American Institute of Biological Sciences. 1985. *Biological and Human Health Effects of Extremely Low Frequency Electromagnetic Fields.* Arlington, Virginia: American Institute of Biological Sciences.

Barlow, H. B., Kohn, H. I., Walsh, E. G. 1947. Visual sensations aroused by magnetic fields. *Am. J. Physiol.* 148:372–375.

Baroncelli, P., Battisti, S., Checcucci, A., Comba, R., Grandolfo, M., Serio, A., Vecchia, P. 1986. A health examination of railway high-voltage substation workers exposed to ELF electromagnetic fields. *Am. J. Ind. Med.* 10:45–55.

Bassett, C. A. L., Mitchell, S. N., Gaston, S. R. 1982. Pulsing electromagnetic field treatment in ununited fractures and failed arthrodeses. *JAMA* 247:623–628.

Bawin, S. W., Adey, W. R. 1976. Sensitivity of calcium binding in cerebral tissue to weak environmental electric fields oscillating at low frequency. *Proc. Natl. Acad. Sci. USA* 73:1999–2003.

Berman, E., House, D. E., Koch, W. E., Leal, J., Martin, A. H., Martucci, G., Mild, K. H., Monahan, J. C. 1988. "The Henhouse Project": Effect of pulsed magnetic fields on early chick embryos. In: *Abstracts of the Tenth Annual Bioelectromagnetics Society Meeting*, p. 14, 19–23 June 1988, Stamford, Connecticut. Frederick, Maryland: Bioelectromagnetics Society.

Bernhardt, J. 1979. The direct influence of electromagnetic fields on nerve and muscle cells of man within the frequency range of 1 Hz to 30 MHz. *Radiat. Environ. Biophys.* 16:309–323.

Blackman, C. F., Benane, S. G., House, D. E., Jones, W. T. 1985a. Effects of ELF (1–120 Hz) and modulated (50 Hz) fields on the efflux of calcium ions from brain tissue in vitro. *Bioelectromagnetics* 6:1–11.

Blackman, C. F., Benane, S. G., Rabionowitz, J. R., House, D. E., Jones, W. T. 1985b. A role for the magnetic field in the radiation-induced efflux of calcium ions from brain tissue in vitro. *Bioelectromagnetics* 6:327–337.

Blank, M. 1987. The surface compartment model: a theory of ion transport focused on ionic processes in the electrical double layers at membrane protein surfaces. *Biochim. Biophys. Acta* 906:277–294.

Blank, M., Goodman, R. 1988. An electrochemical model for the stimulation of biosynthesis by external electric fields. *Bioelectrochem. Bioenerg.* 19:569–580.

Budinger, T. F., Cullander, C., Bordow, R. 1984. Switched magnetic field thresholds for the induction of magnetophosphenes. In: *Abstracts of the Society of Magnetic Resonance in Medicine 3rd Annual Meeting*, pp. 118–119, 13–17 August 1984, New York, New York.

Byus, C. V., Pieper, S. E., Adey, W. R. 1987. The effects of low-energy 60-Hz environmental fields upon the growth-related enzyme ornithine decarboxylase. *Carcinogenesis* 8:1385–1390.

Caola, R. J., Deno, D. W., Dymek, V. S. W. 1983. Measurements of electric and magnetic fields in and around homes near a 500 kV transmission line. *IEEE Trans. Power Appar. Syst.* PAS-102:3338–3347.

Carstensen, E. L., Buettner, A., Genberg, V. L., Miller, M. W. 1985. Sensitivity of the human eye to power frequency electric fields. *IEEE Trans. Biomed. Eng.* BME-32:561–565.

Chiabrera, A., Grattarola, M., Viviani, R. 1984. Interaction between electromagnetic fields and cells: microelectrophoretic effect on ligands and surface receptors. *Bioelectromagnetics* 5:173–191.

Cohen, M. M., Kunska, A., Astemborski, J. A., McCulloch, D., Paskewitz, D. A. 1986. Effects of low-level, 60-Hz electromagnetic fields on human lymphoid cells: I. Mitotic rate and chromosome breakage in human peripheral lymphocytes. *Bioelectromagnetics* 7:415–423.

d'Arsonval, M. A. 1896. Dispositifs pour la mesure des courants alternatifs à toutes frequences. *C. R. Soc. Biol. (Paris)* 3 (100 Ser.): 450-451.

Delgado, J. M. R., Leal, J., Monteagudo, J. L., Garcia, M. G. 1982. Embryological changes induced by weak, extremely low frequency electromagnetic fields. *J. Anat.* 134:533-551.

Delgado, J. M. R., Monteagudo, J. L., Garcia-Gracia, M., Leal, J. 1981. Teratogenic effects of weak magnetic fields. *IRCS Med. Sci.* 9:392.

Durney, C. H., Rushforth, C. K., Anderson, A. A. 1988. Resonant AC-DC magnetic fields: calculated response. *Bioelectromagnetics* 9:315-336.

Ericson, A., Kallen, B. 1986a. An epidemiological study of work with video display screens and pregnancy outcome: I. A registry study. *Am. J. Ind. Med.* 9:447-457.

Ericson, A., Kallen, B. 1986b. An epidemiological study of work with video display screens and pregnancy outcome: II. A case-control study. *Am. J. Ind. Med.* 9:459-475.

Fulton, J. P., Cobb, S., Preble, L., Leone, L., Forman, E. 1980. Electrical wiring configurations and childhood leukemia in Rhode Island. *Am. J. Epidemiol.* 111:292-296.

Gauger, J. R. 1984. *Household Appliance Magnetic Field Survey.* IIT Research Institute Report No. E06549-3. Washington, DC: U. S. Naval Electronics System Command.

Goldhaber, M. K., Polen, M. R., Hiatt, R. A. 1988. The risk of miscarriage and birth defects among women who use video display terminals during pregnancy. *Am. J. Ind. Med.* 13:695-706.

Goodman, R., Henderson, A. S. 1986a. Sine waves enhance cellular transcription. *Bioelectromagnetics* 7:23-29.

Goodman, R., Henderson, A. S. 1986b. Some biological effects of electromagnetic fields. *Bioelectrochem. Bioenerg.* 15:39-55.

Goodman, R., Henderson, A. S. 1988. Exposure of salivary gland cells to low-frequency fields alters polypeptide synthesis. *Proc. Natl. Acad. Sci. USA* 85:3928-3932.

Goodman, R., Abbott, J., Henderson, A. S. 1987. Transcriptional patterns in the X chromosome of *Sciara coprophila* following exposure to magnetic fields. *Bioelectromagnetics* 8:1-7.

Goodman, R., Bassett, C. A. L., Henderson, A. S. 1983. Pulsing electromagnetic fields induce cellular transcription. *Science* 220:1283-1285.

Grandolfo, M., Vecchia, P. 1985. Natural and man-made environmental exposures to static and ELF electromagnetic fields. In: *Biological Effects and Dosimetry of Static and ELF Electromagnetic Fields,* Grandolfo, M., Michaelson, S. M., Rindi, A., eds., pp. 49-70. New York: Plenum.

Grattarola, M., Chiabrera, A., Bonanno, G., Viviani, R., Raveane, A. 1985. Electromagnetic field effects on PHA-induced lymphocyte activation. In: *Interaction Between Electromagnetic Fields and Cells,* Chiabrera, A., Nicolini, C., Schwan, H., eds., pp. 401-421. New York: Plenum.

Halle, B. 1988. On the cyclotron resonance mechanism for magnetic field effects on transmembrane ion conductivity. *Bioelectromagnetics* 9:381-385.

Jolley, W. B., Hinshaw, D. B., Knierim, K., Hinshaw, D. B. 1985. Magnetic field effects on calcium efflux and insulin secretion in isolated rabbit islets of Langerhans. *Bioelectromagnetics* 4:103-106.

Juutilainen, J. P. 1986. Effects of low frequency magnetic fields on chick embryos: dependence on incubation temperature and storage of the eggs. Z. Naturforsch. 41:1111–1115.

Juutilainen, J. P., Saali, K. 1986. Development of chick embryos in 1 Hz to 100 kHz magnetic fields. Radiat. Environ. Biophys. 25:135–140.

Juutilainen, J. P., Harris, M., Saali, K., Lahtinen, T. 1986. Effects of 100-Hz magnetic fields with various waveforms on the development of chick embryos. Radiat. Environ. Biophys. 25:65–74.

Kaune, W. T., Stevens, R. G., Callahan, N. J., Severson, R. K., Thomas, D. B. 1987. Residential magnetic and electric fields. Bioelectromagnetics 8:315–335.

Kurppa, K., Holmberg, P. C., Rantala, K., Nurminen, T., Saxen, L. 1985. Birth defects and exposure to video display terminals during pregnancy. Scand. J. Work Environ. Health 11:353–356.

Leal, J., Trillo, M. A., Ubeda, A., Abraira, V., Shamsaifar, K., Chacon, L. 1986. Magnetic environment and embryonic development: a role of the earth's field. IRCS Med. Sci. 14:1145–1146.

Liboff, A. R. 1985. Geomagnetic cyclotron resonance in living cells. J. Biol. Phys. 13:99–102.

Liboff, A. R., McLeod, B. R. 1988. Kinetics of channelized membrane ions in magnetic fields. Bioelectromagnetics 9:39–51.

Liboff, A. R., Rozek, R. J., Sherman, M. L., McLeod, B. R., Smith, S. D. 1987. $^{45}Ca^{++}$ cyclotron resonance in human lymphocytes. J. Bioelectr. 6:13–22.

Lin, R. S., Dischinger, P. C., Conde, J., Farrell, K. P. 1985. Occupational exposure to electromagnetic fields and the occurrence of brain tumors: an analysis of possible associations. J. Occup. Med. 27:413–419.

Livingston, G. K., Gandhi, O. P., Chatterjee, I., Witt, K., Roti Roti, J. L. 1986. Reproductive Integrity of Mammalian Cells Exposed to 60 Hz Electromagnetic Fields. Final report for New York State Power Lines Project #218209. Albany, New York: NYSPLP, Wadsworth Center for Laboratories and Research.

Lövsund, P., Nilsson, S. E. G., Öberg, P. Å. 1981. Influence on frog retina of alternating magnetic fields with special reference to ganglion cell activity. Med. Biol. Eng. Comput. 19:679–685.

Lövsund, P., Öberg, P. Å., Nilsson, S. E. G., Reuter, T. 1980a. Magneto-phosphenes: a quantitative analysis of thresholds. Med. Biol. Eng. Comput. 18:326–334.

Lövsund, P., Öberg, P. Å., Nilsson, S. E. G. 1980b. Magneto- and electrophosphenes: a comparative study. Med. Biol. Eng. Comput. 18:758–764.

Lövsund, P., Öberg, P. Å., Nilsson, S. E. G. 1982. ELF magnetic fields in electrosteel and welding industries. Radio Sci. 17(5S):35S–38S.

Luben, R. A., Cain, C. D., Chen, M. C.-Y., Rosen, D. M., Adey, W. R. 1982. Effects of electromagnetic stimuli on bone and bone cells in vitro: inhibition of responses to parathyroid hormone by low-energy low-frequency fields. Proc. Natl. Acad. Sci. USA 79:4180-4184.

Lyle, D. B., Schechter, P., Adey, W. R., Lundak, R. L. 1983. Suppression of T-lymphocyte cytotoxicity following exposure to sinusoidally amplitude-modulated fields. Bioelectromagnetics 4:281–292.

Maffeo, S., Miller, M., Carstensen, E. L. 1984. Lack of effect of weak low frequency electromagnetic fields on chick embryogenesis. J. Anat. 139:613–618.

Male, J. D., Norris, W. T., Watts, M. W. 1987. Exposure of people to power-frequency electric and magnetic fields. In: *Interaction of Biological Systems with Static and ELF Electric and Magnetic Fields*, Anderson, L. E., Kelman, B. J., Weigel, R. J., eds., pp. 407–418. Proceedings of the 23rd Annual Hanford Life Sciences Symposium, Richland, Washington. CONF-841041. Springfield, Virginia: National Technical Information Service.

Marron, M. T., Greenebaum, B., Swanson, J. E., Goodman, E. M. 1983. Cell surface effects of 60 Hz electromagnetic fields. *Radiat. Res.* 94:217–220.

Martin, A. H. 1988. Magnetic fields and time dependent effects on development. *Bioelectromagnetics* 9:393–396.

McDonald, A., Cherry, N., Delorme, C., McDonald, J. D. 1986. Visual display units and pregnancy: evidence from the Montreal survey. *J. Occup. Med.* 28:1126–1131.

McDowall, M. E. 1986. Mortality of persons resident in the vicinity of electricity transmission facilities. *Br. J. Cancer* 53:271–279.

McLeod, B. R., Liboff, A. R. 1986. Dynamic characteristics of membrane ions in multifield configurations of low-frequency electromagnetic radiation. *Bioelectromagnetics* 7:177–189.

Milham, S., Jr. 1985. Mortality in workers exposed to electromagnetic fields. *Environ. Health Perspect.* 62:297–300.

Myers, A., Cartwright, R. A., Bonnell, J. A., Male, J. C., Cartwright, S. C. 1985. Overhead powerlines and childhood cancer. In: *Abstracts of the International Conference on Electric and Magnetic Fields in Medicine and Biology*, 4–5 December 1985, London, England.

Öberg, P. Å. 1973. Magnetic stimulation of nerve tissue. *Med. Biol. Eng.* 11:55–64.

Polk, C. 1974. Sources, propagation, amplitude and temporal variation of extremely low frequency (0–100 Hz) electromagnetic fields. In: *Biological and Clinical Effects of Low Frequency Magnetic and Electric Fields*, Llaurado, J. G., Sauces, A., Jr., eds., pp. 21–48. Springfield, Illinois: Charles C. Thomas.

Polson, M. J. R., Barker, A. T., Freeston, I. L. 1982. Stimulation of nerve trunks with time-varying magnetic fields. *Med. Biol. Eng. Comput.* 20:243–244.

Rooze, M., Hinsenkamp, M. 1985. In vivo modifications induced by electromagnetic stimulation of chicken embryos. *Reconstr. Surg. Traumatol.* 19:87–92.

Sander, R., Brinkmann, J., Kuhne, B. 1982. Laboratory studies on animals and human beings exposed to 50 Hz electric and magnetic fields. In: *Proceedings of the International Conference on Large High Voltage Electric Systems*, Paper 36-01, 1–9 September 1982, Paris, France.

Sandström, M., Hansson-Milk, K., Lovtrup, S. 1987. Effects of weak pulsed magnetic fields on chick embryogenesis. In: *Work with Display Units 86*, Knave, B., Widebäck, P.-G., eds., pp. 135–140. New York: Elsevier.

Savitz, D. A. 1986. Human health effects of extremely low frequency electromagnetic fields: critical review of clinical and epidemiological studies. In: *Biological Effects of Power Frequency Electric and Magnetic Fields*, Feero, W. E., ed., pp. 49–64. IEEE Spec. Publ. No. 86TH0139-6-PWR. Piscataway, New Jersey: Institute of Electrical and Electronics Engineers.

Savitz, D. A., Calle, E. E. 1987. Leukemia and occupational exposure to electromagnetic fields: review of epidemiologic surveys. *J. Occup. Med.* 29:47–51.

Savitz, D. A., Wachtel, H., Barnes, F. A., John, E. M., Tvrdik, J. G. 1988. Case-control study of childhood cancer and exposure to 60-Hz magnetic fields. *Am. J. Epidemiol.* 128:21–38.

Scott-Walton, B., Clark, K. M., Holt, B. R., Jones, D. C., Kaplan, S. D., Krebs, J. S., Polson, P., Shepherd, R. A., Young, J. R. 1979. *Potential Environmental Effects of 765-kV Transmission Lines: Views Before the New York State Public Service Commission, Cases 26529 and 26559, 1976–1978.* DOE/EV-0056, p. II-7. Springfield, Virginia: National Technical Information Service.

Severson, R. K., Stevens, R. G., Kaune, W. T., Thomas, D. B., Henser, L., Davis, S., Sever, L. E. 1988. Acute nonlymphocytic leukemia and residential exposure to power frequency magnetic fields. *Am. J. Epidemiol.* 128:10–20.

Silny, J. 1986. The influence threshold of a time-varying magnetic field in the human organism. In: *Biological Effects of Static and Extremely Low Frequency Magnetic Fields*, Bernhardt, J. H., ed., pp. 105–112. München: MMV Medizin Verlag.

Siskin, B. F., Fowler, I., Mayand, C., Ryaby, J. P., Pilla, A. A. 1986. Pulsed electromagnetic fields and normal chick development. *J. Bioelectr.* 5:25–34.

Smith, S. D., McLeod, B. R., Liboff, A. R., Cooksey, K. 1987. Calcium cyclotron resonance and diatom mobility. *Bioelectromagnetics* 8:215–227.

Speers, M. A., Dobbins, J. G., Miller, V. S. 1988. Occupational exposures and brain cancer mortality: a preliminary study of East Texas residents. *Am. J. Ind. Med.* 13:629–638.

Stevens, R. G., Severson, R. K., Kaune, W. T., Thomas, D. B. 1986. Epidemiological study of residential exposure to ELF electric and magnetic fields and risk of non-lymphocytic leukemia. In: *Abstracts of the 1986 DOE/EPRI/NY State Power Lines Project Contractors Review*, 18–20 November 1986, Denver, Colorado.

Stuchly, M. A., Lecuyer, D. W., Mann, R. D. 1983. Extremely low frequency electromagnetic emissions from video display terminals and other devices. *Health Phys.* 45:713–722.

Tenforde, T. S. 1986a. Interaction of ELF magnetic fields with living matter. In: *Handbook of Biological Effects of Electromagnetic Radiation*, Polk, C., Postow, E., eds., pp. 197–225. Boca Raton, Florida: CRC Press.

Tenforde, T. S. 1986b. Biological effects of extremely-low-frequency magnetic fields. In: *Biological Effects of Power Frequency Electric and Magnetic Fields*, Feero, W. E., ed., pp. 21–40. IEEE Spec. Publ. No. 86TH0139-6-PWR. Piscataway, New Jersey: Institute of Electrical and Electronics Engineers.

Tenforde, T. S. 1988. Interaction mechanisms, biological effects and biomedical applications of static and extremely-low-frequency magnetic fields. In: *Proceedings of the Twenty-Second Annual Meeting of the National Council on Radiation Protection and Measurements: Nonionizing Electromagnetic Radiations and Ultrasound*, pp. 181–217, 2–3 April 1986, Bethesda, Maryland. Washington, D.C.: National Council on Radiation Protection and Measurements.

Tenforde, T. S. 1989. Biological responses to static and time-varying magnetic fields. In: *Interaction of Electromagnetic Waves in Biological Systems*, Lin, J. C., ed., pp. 83–107. New York: Plenum.

Tenforde, T. S., Budinger, T. F. 1986. Biological effects and physical safety aspects of NMR imaging and in vivo spectroscopy. In: *NMR in Medicine: Instrumentation and Clinical Applications*, Thomas, S. R., Dixon, R. L., eds., pp. 493–548. New York: American Association of Physicists in Medicine.

Tenforde, T. S., Kaune, W. T. 1987. Interaction of extremely-low-frequency electric and magnetic fields with humans. *Health Phys.* 53:585–606.

Thomas, J. R., Schrot, J., Liboff, A. R. 1986. Low-intensity magnetic fields alter operant behavior in rats. *Bioelectromagnetics* 7:349–357.

Thomas, T. L., Stolley, P. D., Stemhagen, A., Fontham, E. T. H., Bleecker, M. L., Stewart, P. A., Hoover, R. N. 1987. Brain tumor mortality risk among men with electrical and electronics jobs: a case-control study. *J. Natl. Cancer Inst.* 79:233–238.

Tomenius, L. 1986. 50-Hz electromagnetic environment and the incidence of childhood tumors in Stockholm County. *Bioelectromagnetics* 7:191–207.

Tsong, T. Y., Astumian, R. D. 1988. Electroconformational coupling: how membrane-bound ATPase transduces energy from dynamic electric fields. *Annu. Rev. Physiol.* 50:273–290.

Ubeda, A., Leal, J., Trillo, M. A., Jimenez, M. A., Delgado, J. M. R. 1983. Pulse shape of magnetic fields influence chick embryogenesis. *J. Anat.* 137:513–536. (Data correction: Ubeda et al., 1985. Erratum. *J. Anat.* 140:721.)

Ueno, S., Lövsund, P., Öberg, P. Å. 1978. Capacitative stimulatory effect in magnetic stimulation of nerve tissue. *IEEE Trans. Mag.* MAG–14:958–960.

Ueno, S., Harada, K., Ji, C., Oomura, Y. 1984. Magnetic nerve stimulation without interlinkage between nerve and magnetic flux. *IEEE Trans. Mag.* MAG–20:1660-1662.

Verma, A. K., Pong, R.-C., Erickson, D. 1986. Involvement of protein kinase C activation in ornithine decarboxylase gene expression in primary culture of newborn mouse epidermal cells and in skin tumor promotion by 12-O-tetradecanoylphorbol-13-acetate. *Cancer Res.* 46:6149–6155.

Wertheimer, N., Leeper, E. 1979. Electrical wiring configurations and childhood cancer. *Am. J. Epidemiol.* 109:273–284.

Wertheimer, N., Leeper, E. 1980. RE: Electrical wiring configurations and childhood leukemia in Rhode Island. *Am. J. Epidemiol.* 111:461–462.

Wertheimer, N., Leeper, E. 1982. Adult cancer related to electrical wires near the home. *Int. J. Epidemiol.* 11:345–355.

Wertheimer, N., Leeper, E. 1986. Possible effects of electric blankets and heated waterbeds on fetal development. *Bioelectromagnetics* 7:13–22.

PART V.
Possible
Consequences

13 The Emerging Role of the Pineal Gland and Melatonin in Oncogenesis

DAVID E. BLASK
Department of Anatomy
University of Arizona College of Medicine
Tucson, Arizona

CONTENTS

One hallmark of the pineal gland and melatonin function is the pineal gland's ability to inhibit reproductive organ growth in prepubertal animals as well as induce a marked regression in reproductive organ size in adults. This is typically accomplished either by placing animals in short photoperiods or by injecting long-photoperiod-exposed animals with melatonin in the late afternoon on a daily basis for several weeks (Reiter, 1980). Therefore, it is not surprising that investigators have begun to apply these same principles to examine the question of whether the pineal gland, and more specifically its primary hormone melatonin, influences the growth of malignant neoplasms (Lapin, 1976; Tapp, 1982; Blask, 1984). It is interesting to note that the initial work on pineal effects on oncogenesis (Georgiou, 1929) actually predates the seminal work of Reiter and colleagues in the 1960s and 1970s, which demonstrated that the pineal gland (under short-day conditions) and injections of melatonin inhibit the growth of (or cause the regression of) the reproductive organs in the rat and the Syrian hamster.

The basic and clinical aspects of the pineal gland and cancer have been exhaustively reviewed elsewhere by the author (Blask, 1984; Blask and Hill, 1988); therefore, in this chapter the focus is on more recent work that strongly implicates an inhibitory role for the pineal gland and melatonin in the processes of oncogenesis. Of all neoplasms studied to date, the experimental results obtained on breast cancer and melanoma are perhaps the most compelling with respect to pineal and melatonin influences. Therefore, these two tumor types, both of which are endocrine responsive, are featured as model systems currently holding the most promise for elucidating an important role for the pineal gland and melatonin in oncogenesis.

MELANOMA

In Vivo Studies

Reports by Das Gupta and co-workers (Das Gupta and Terz, 1967; Stanberry et al., 1983) indicated that the removal of the pineal gland from Syrian hamsters, before their inoculation with the transplantable and hormone-responsive melanotic melanoma No. 1 (MM1), resulted in a rather dramatic increase in the growth and metastases of the tumor in a 5-week period of observation during which the animals were maintained on a long photoperiod. Using a somewhat different system, Aubert and colleagues (1970) also found that pinealectomy promoted the growth of melanomas induced in long-photoperiod-exposed Syrian hamsters by the chemical carcinogen 7,12-dimethyl-benzanthracene (DMBA). However, when performed on hamsters (with MM1 melanomas) maintained in short photoperiods, pinealectomy further exacerbates the tumor growth-retarding effects of short days (Stanberry et al., 1983). This seemingly paradoxical finding was explained on the basis that the photoperiod in which hamsters are kept governs the melanoma growth response to pinealectomy.

Das Gupta and associates also reported that daily melatonin injections had no effect on melanoma growth in hamsters unless they were pinealectomized. Under these circumstances, melatonin administration inhibited the pinealectomy-induced promotion of tumor growth and metastases (El-Domeiri and Das Gupta, 1973, 1976; Ghosh et al., 1973). However, a more recent investigation by this same group revealed that daily, morning, or late-afternoon injections of melatonin or subcutaneous silastic melatonin implants stimulated melanoma growth in pineal-intact hamsters maintained in a long photoperiod. Conversely, in short-day-exposed hamsters, melatonin implants had a tumor-inhibiting effect, suggesting that both the type of photoperiodic regimen employed and the method of melatonin administration are important factors affecting the responsiveness of this type of melanoma to melatonin (Stanberry et al., 1983). Whether the tumor-promoting or tumor-inhibiting effects of melatonin result from an influence on the neuroendocrine axis or directly on the tumor tissue itself is difficult to ascertain from these results.

In addition to melatonin injections or implants, oral administration of melatonin was recently found to be highly effective in slowing the growth of B16 melanoma cells transplanted into BALB/c athymic mice (Narita and Kudo, 1985). Narita and co-workers postulated that oral melatonin inhibited the growth of these hormone-responsive melanomas by inhibiting gonadal and/or adrenal function, by decreasing the pituitary secretion of melanocyte-stimulating hormone (MSH), or by blocking cell proliferation directly (Narita, 1988).

In Vitro Studies

A study by Walker et al. (1978) demonstrated that millimolar concentrations of melatonin caused an approximately 25% inhibition of the growth of a cloned melanoma cell line (B7) derived from a tumor that had spontaneously developed in a male golden hamster. However, a biphasic effect of this molecule was suggested by the fact that micromolar concentrations of melatonin actually stimulated cell proliferation by about 37%. In a similar study, Bartsch et al. (1986) found that micromolar concentrations of melatonin caused about 60% inhibition of the growth of early-passaged (slower growing) human melanoma cells, whereas in late-passaged cells (faster growing) these concentrations were ineffective. Millimolar melatonin concentrations led to 20% stimulation of cell proliferation in the early-passaged cells; in late-passaged cells, these extremely high melatonin concentrations induced 90% inhibition of cell growth by the end of a 7-day culture period; lower concentrations in the physiological range were without effect.

In another study using human melanoma cells, Meyskens and Salmon (1981) demonstrated varied responses of these cells to a wide range of concentrations of melatonin; the cells were clonogenic melanoma cells derived from biopsy material taken from metastatic nodules of malignant melanoma. For example, in slightly more than half the biopsy specimens evaluated in the clonogenic assay, total cloning efficiency decreased as the melatonin concentration increased from $10^{-15}\ M$ to $10^{-5}\ M$, with no accompanying change in either colony type or mean colony diameter. However, in one particular patient biopsy, a physiological concentration of melatonin ($10^{-9}\ M$) caused a fourfold increase in cloning efficiency. In a clone from another patient, $10^{-11}\ M$ melatonin induced a dramatic ninefold increase in the number of dark, small-cell colonies over the total colonies present in the control plates, which consisted predominantly of the light, large-cell variant; however, the number of colonies decreased as the concentration of melatonin increased. The remaining 27% of biopsy specimens exhibited no response to melatonin in the clonogenic assay.

BREAST CANCER

In Vivo Studies

Experimental breast cancer, typically induced in female rats with the chemical carcinogen 9,10-dimethylbenz(a)anthracene (DMBA), represents the most extensively studied malignant neoplasm with respect to the effects of pinealectomy, photoperiod, and/or melatonin administration (Blask, 1984). The development and growth of DMBA-induced breast cancers are primarily responsive to the direct mitogenic effects of prolactin (PRL) secreted from the pituitary gland and secondarily to estrogen (E) through its ability to stimulate PRL release (Welsch, 1985).

Effects of Pinealectomy

The initial studies on the effects of pinealectomy on DMBA-induced mammary cancer growth were essentially negative. For example, Lapin (1976) found that neonatal pinealectomy before DMBA administration had no effect on mammary oncogenesis as compared with control animals unless the rats were also treated with reserpine (a catecholamine-depleting drug) for 40 consecutive days after an initial intragastric feeding of DMBA. Three years later, Aubert and colleagues (1980) also reported that pinealectomy in adult female rats before DMBA administration had no effect on subsequent tumor development. The following year, Tamarkin et al. (1981) were the first to report that pinealectomy before intragastric administration of DMBA promoted tumor growth when a low dose of the carcinogen was given; at a high dose, however, no effect of pinealectomy was observed. The mechanism by which pinealectomy promoted tumor development at low doses of DMBA could not be ascertained from this investigation. Subsequently, Kothari et al. (1984) found that pinealectomy before DMBA administration had no effect on mammary oncogenesis in short-photoperiod-exposed rats (10:14 light:dark cycle) when compared with short-photoperiod-exposed, intact animals. A surprising result by Chang et al.

(1985) was that the denervation of the pineal gland by bilateral superior cervical ganglionectomy before DMBA administration, which creates a functional pinealectomy, actually inhibited tumor growth and development. Taking these discrepancies into account, we reexamined the effects of pinealectomy before tumor initiation in the N-methylnitrosourea (NMU) model of breast cancer (Pelletier et al., 1987; Blask et al., 1988). The advantages of using this model system in lieu of the DMBA model is that NMU-induced mammary cancers are more akin to human breast cancers in terms of their growth characteristics. Furthermore, NMU tumors, like human breast cancers, respond to the direct mitogenic effects of E in addition to PRL (Welsch, 1985). In our study, pinealectomy before tumor initiation with NMU led to a modest enhancement of mammary cancer development and growth (Fig. 1). This perhaps occurred via a pinealectomy-induced increase in PRL levels and tumor E receptors (Pelletier et al., 1987; Blask et al., 1988; Blask and Grosso, unpublished results). In most of the studies cited, the hypothesis has been promulgated that pinealectomy promotes breast cancer growth by eliminating the daily, endogenous nocturnal melatonin signal (Fig. 1) (Blask et al., 1988).

Effects of Photoperiodic Manipulations

Another approach by which the effects of the pineal gland on neoplastic growth have been assessed involves the manipulation of the photoperiod in such a way that pineal activity is either diminished or enhanced. A number of investigators, with the notable exception of Aubert and associates (1980), have shown that prolonged exposure of rats with DMBA mammary tumors to constant light stimulates breast cancer development and growth as compared with animals maintained on an alternating, daily light:dark cycle (Hamilton, 1969; Kothari et al., 1982, 1984; Shah et al., 1984). However, the problem in interpreting the data of Kothari's group (Kothari et al., 1982, 1984; Shah et al., 1984) is that oncogenesis in constant light-exposed rats was compared with that in rats kept on short photoperiod (10:14 light:dark); no long-photoperiod control group (14:10 light:dark) was included. Thus, it is difficult to assess whether tumor growth was actually greater under constant light or merely less in a short photoperiod; the inclusion of a long-photoperiod control group could have resolved this issue.

Constant light exposure, because of suppression of melatonin production, has been regarded as being tantamount to a "physiological pinealectomy" (Reiter, 1980). Not only does constant light eliminate the melatonin signal but it also stimulates the secretion of PRL (Vaticon et al., 1980), which may in turn be part of the mechanism by which tumorigenesis is enhanced, particularly since DMBA breast cancers are PRL dependent (Welsch, 1985).

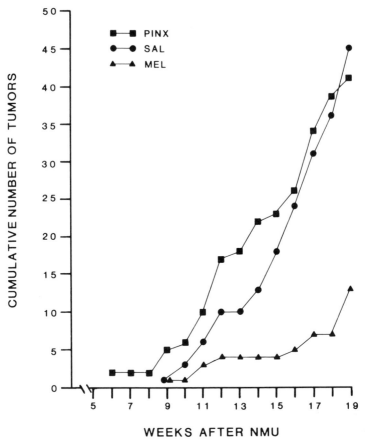

WEEKS AFTER NMU

Fig. 1. Effect of pinealectomy (PINX), daily afternoon melatonin injections (MEL) (500 μg/day), or saline (SAL) injections on N-methylnitrosourea- (NMU-) induced mammary carcinogenesis in female rats. Treatments were initiated before tumor induction with NMU and continued for 19 weeks.

As most strains of albino rat are less photoperiodically sensitive than some other species such as the hamster, light deprivation does not inhibit the neuroendocrine–reproductive axis unless it is combined with a potentiating factor such as olfactory bulbectomy or underfeeding (Reiter, 1980). In fact, both these potentiating factors make the rat more sensitive to the antigonadotrophic effects of daily, afternoon melatonin injection (Blask and Nodelman, 1979; Reiter, 1980; Blask et al., 1981), and ostensibly to the endogenous melatonin signal emitted from the pineal gland as well. Therefore, light deprivation (blinding) in combination with olfactory bulbectomy leads to a decrease in tumor incidence and growth (Chang et al., 1985;

Sanchez Barcelo et al., 1988) as well as an increase in tumor regression (Blask, 1984; Chang et al., 1985). Not only does pinealectomy prevent the oncostatic effects of blinding/olfactory bulbectomy, but tumor growth is even further promoted beyond the level observed in the intact controls (Blask, 1984; Sanchez Barcelo et al., 1988). Interestingly, the combination of blinding with underfeeding is even a more potent inhibitor of breast cancer growth in rats with DMBA-induced tumors in that no tumors develop during a 20-week observation period. In this model system, pinealectomy totally cancels the antitumorigenic effect of blinding/underfeeding (Blask, 1984; Sanchez Barcelo et al., 1988). Interestingly, blinding alone decreases tumor incidence with no change in tumor number. Pinealectomy not only prevents this effect but causes a substantial increase in tumor number over that seen in both the controls and blinded animals (Blask et al., 1988).

Although the exact mechanism(s) by which light deprivation in combination with either olfactory bulbectomy or underfeeding inhibits DMBA tumor growth are unknown, these may involve a pineal gland and perhaps melatonin-induced neuroendocrine suppression of PRL (Leadem and Blask, 1981; Blask, 1984) and E secretion (Vaughan, 1984; Sanchez Barcelo et al., 1988) as well as a decrease in E-receptor number in the tumors themselves (Sanchez Barcelo et al., 1988).

Effects of Melatonin

A study by Hamilton (1969) was the first to document the effects of melatonin on DMBA-induced breast carcinogenesis. Although melatonin had no effect on the percentage of rats with palpable mammary tumors, 80% of tumors in the melatonin group were adenocarcinomas as compared with only 30% in the vehicle-treated controls. These results indicated that melatonin injections begun before tumor initiation actually enhanced mammary oncogenesis. Interestingly, virtually every subsequent study on the influence of melatonin on breast cancer growth has been at odds with the Hamilton study. Regardless of whether animals are maintained in diurnal or constant lighting, daily afternoon injections of melatonin begun on the day of or before tumor initiation with DMBA and continued thereafter decrease the incidence of rats with palpable mammary tumors as compared with vehicle-injected controls (Aubert et al. 1980; Tamarkin et al., 1981; Shah et al., 1984). While Aubert et al. (1980) reported that melatonin injections were still effective in pinealectomized rats, two other groups (Tamarkin et al., 1981; Shah et al., 1984) failed to observe an oncostatic effect of melatonin in animals in which the pineal gland had been removed, suggesting that the presence of the endogenous melatonin signal may be required for the antitumorigenic effect of exogenous melatonin to be expressed. However, if this is indeed the case, it is puzzling that melatonin apparently retains its effectiveness in animals maintained in constant

lighting (Shah et al., 1984), a photoperiodic scheme that functionally pinea-
lectomizes the animal and extinguishes the melatonin signal (Reiter, 1980).
As alluded to earlier in this chapter, melatonin is also effective in inhibiting
melanoma growth in pinealectomized hamsters.

As was the case for transplanted B16 mouse melanoma cells (Narita
and Kudo, 1985), oral melatonin, supplied in the drinking water before
DMBA administration and continued throughout the period of carcinogenesis, effectively suppressed tumor incidence in rats maintained in a
short photoperiod or in constant light (Kothari, 1987). Orally administered
melatonin (400–500 µg/day) is also effective in suppressing tumorigenesis
in rats with NMU-induced tumors (Blask, unpublished results).

It has been proposed by Kothari's group (Shah et al., 1984; Kothari,
1988) that melatonin administration before carcinogen treatment decreases
the susceptibility of the mammary epithelium to neoplastic transformation via neuroendocrine mechanisms, leading to a decrease in the secretion of PRL (particularly the mitogenic 16-kdalton fragment of PRL) and
E and/or by compromising the ability of the parenchymal cells to metabolize
DMBA to its putative reactive intermediates. We have never observed a
clear-cut melatonin-induced suppression in either circulating PRL or E
(Blask et al., 1986; Pelletier et al., 1987; Blask, unpublished results), and
thus we have some reservations about the neuroendocrine hypothesis
(Blask, 1984; Blask and Leadem, 1987). Furthermore, if the increased
susceptibility hypothesis is true, then restricting melatonin treatment to
the tumor initiation period alone (before and during the period of carcinogen administration) should be sufficient to inhibit mammary oncogenesis.
We have tested this hypothesis in both the DMBA and NMU breast cancer
models, and have found that daily afternoon injections of melatonin,
restricted to the tumor initiation phase only, are completely ineffective
in suppressing tumor development and growth (Blask, unpublished results).
However, if melatonin injections encompass both the initiation and promotional phases of oncogenesis, a potent oncostatic effect of melatonin is
observed (Fig. 1) (Pelletier et al., 1987; Blask et al., 1988).

To test the postulate that melatonin inhibits oncogenic processes
associated primarily with tumor promotion, we began treating female rats
with daily afternoon melatonin injections 3 weeks after the administration of DMBA but before the appearance of overt, palpable tumors in
animals that were either fed ad libitum or subjected to the potentiating
factor of moderate underfeeding; the injections were continued throughout
the remainder of the experimental period (Blask et al., 1986). Although
melatonin treatment restricted to the promotional phase suppressed tumorigenesis in both fed and underfed rats, it was clearly much more effective
in the underfed animals, suggesting an increased sensitivity of underfed
animals to melatonin (Blask et al., 1981). Preliminary results in rats with
NMU-induced breast cancers also indicate that melatonin injections

restricted to the promotional phase inhibit mammary oncogenesis (Blask, unpublished observations). Further support for the antipromotion hypothesis is also provided by the study of Aubert et al. (1980), who demonstrated that melatonin injections were effective in causing the regression of established, palpable DMBA-induced mammary tumors. Moreover, previous reports that melatonin injections inhibit the growth of transplantable breast cancers in both mice and rats (Anisimov et al., 1973; Karmali et al., 1978) lend further credence to this postulate.

In Vitro Studies

To further examine the antipromotion hypothesis using a more direct approach, we studied the effects of melatonin on the ability of the human breast cancer cell line MCF-7 to proliferate in culture. For example, we have found that concentrations of melatonin corresponding to the physiological range (10^{-9} M to 10^{-11} M) present in the blood of humans (Vaughan, 1984) cause from 60% to 80% inhibition of cell proliferation after 7 days of continuous incubation; higher or lower concentrations of melatonin had no effect on cell growth. Melatonin's inhibitory effect is reversible, since the logarithmic growth of MCF-7 cells was restored after melatonin-containing medium was replaced with fresh medium lacking this indoleamine. Additionally, both scanning and transmission electron microscopy of these cells reveals several striking morphological changes that correlate with melatonin's inhibition of cell growth. After only 4 days of exposure to melatonin, MCF-7 cells exhibit reduced numbers of surface microvilli (Fig. 2), nuclear swelling, cytoplasmic and ribosomal shedding, disruption of mitochondrial cristae, vesiculation of the smooth endoplasmic reticulum, and increased numbers of autophagic vacuoles (Fig. 3), all characteristics of sublethal but reversible cellular injury (Blask and Hill, 1986a; Hill and Blask, 1988).

In addition to melatonin, we have tested the ability of a number of its precursors, metabolites, or analogues to inhibit MCF-7 cell growth. Neither serotonin, *N*-acetylserotonin, 5-methoxytryptophol, 5-methoxytryptamine, nor 6-hydroxymelatonin have any effect on cell proliferation; however, the melatonin analogue, 6-chloromelatonin, is as effective as melatonin itself in inhibiting cell growth (Blask and Hill, 1986a,b). Therefore, it appears that the in vitro antiproliferative effect of melatonin on MCF-7 cells is a relatively specific characteristic of this molecule. Furthermore, the failure of melatonin's major metabolites, 6-hydroxymelatonin and *N*-acetylserotonin (Young et al., 1985), to inhibit cell proliferation suggests that the oncostatic activity is caused by melatonin itself rather than one of its metabolic products. Moreover, MCF-7 cells apparently do not metabolize melatonin in vitro (Danforth et al., 1983). By contrast, Leone and colleagues (1988) reported that 6-hydroxymelatonin was markedly more potent than melatonin in

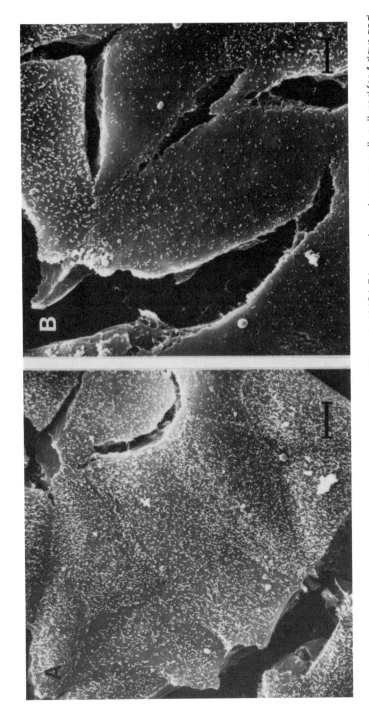

Fig. 2. Scanning electron micrograph of surface morphology of **(A)** control MCA-7 human breast cancer cells cultured for 4 days and **(B)** MCF-7 cells exposed to 10^{-9} M melatonin for 4 days. Bar = 1 μm.
[Reprinted with permission from Hill, S. M., Blask, D. E. 1988. Effects of the pineal hormone melatonin on the proliferation and morphological characteristics of human breast cancer cells (MCF-7) in culture. *Cancer Res.* 48:6121–6126.]

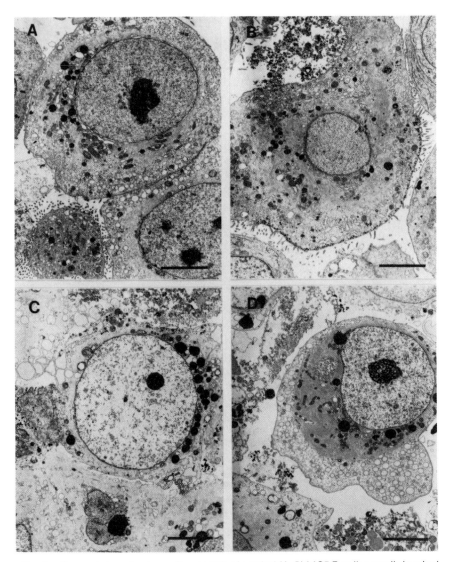

Fig. 3. Transmission electron micrograph of control (**A, B**) MCF-7 cells or cells treated with 10^{-9} M melatonin (**C, D**) in culture for 4 days. Bar=5 μm.

[Reprinted with permission from Hill, S. M., Blask, D. E. 1988. Effects of the pineal hormone melatonin on the proliferation and morphological characteristics of human breast cancer cells (MCF-7) in culture. *Cancer Res. 48:6121–6126.*]

inhibiting the proliferation of a human ovarian carcinoma cell line than melatonin, suggesting that this major metabolite is the active oncostatic form of melatonin for this cell line.

The fact that melatonin's antiproliferative effect is a serum-dependent phenomenon (Hill and Blask, 1986; Blask and Hill, 1986a) initially suggested to us that it might be interacting with some component(s) in the serum such as a hormone(s) and/or a growth factor(s). Because MCF-7 cells are responsive to the mitogenic effects of both E and PRL (Shafie and Brooks, 1977; Osborne, 1985; Biswas and Vonderhaar, 1987) and contain receptors for both hormones, it occurred to us that these may represent at least two constituents of serum which may play an important role in melatonin's ability to directly inhibit breast cancer cell growth. Indeed, not only does melatonin inhibit E-stimulated MCF-7 cell growth, but the loss of its inhibitory effect in serum-free medium is restored by nearly 40% by means of the addition of PRL (Hill and Blask, 1986; Blask and Hill, 1986a). Thus, melatonin's oncostatic effect in vitro may involve an important interaction with these mitogenic hormones in the microenvironment surrounding breast cancer cells.

We became particularly interested in the interaction between melatonin and E, not only on the basis of our own growth data but as a result of the report of Danforth and associates (1983) that melatonin causes a transient increase in the number of nuclear and cytoplasmic E receptors in MCF-7 cells. We decided to test the postulate that melatonin's ability to inhibit breast cancer growth may relate, in part, to the degree of responsiveness of breast cancer cells to E; that is, melatonin's oncostatic effect should decrease as the E responsiveness of breast cancer cells decreases. We confirmed this hypothesis by demonstrating that, in the less E-responsive human breast cancer cell line T47D, melatonin's antiproliferative effect in physiological concentrations was considerably reduced as compared with the more responsive MCF-7 cells. Moreover, the E-unresponsive and E-receptor-negative cell line BT20 was completely immune to the oncostatic effects of any concentration of melatonin (Hill, 1986; Blask et al., 1988).

Another approach to examining the melatonin–E interaction hypothesis involved studying the interaction of melatonin with the nonsteroidal antiestrogen tamoxifen in both serum-containing and a serum-free, chemically defined medium. Tamoxifen, which binds to the E receptor with high affinity, is currently the antiestrogen of choice in the treatment of E-receptor-positive breast cancer (Osborne, 1985). Melatonin at 10^{-9} M not only was more effective than 10^{-9} M tamoxifen (the therapeutic level) in inhibiting the growth of MCF-7 cells in serum-containing medium, but its antiproliferative effect was completely blocked by tamoxifen when the two compounds were coincubated. Interestingly, in serum-free medium, melatonin completely blocks the ability of tamoxifen to inhibit E-stimulated cell growth (Hill, 1986; Blask et al., 1988). Additionally, we found (Hill, 1986;

Blask and Hill, 1988) that melatonin inhibits the release of an E-inducible cathepsin-like 52-kdalton glycoprotein, which is an autocrine growth factor produced by MCF-7 cells (Vignon and Rochefort, 1985). Although indirect, these results further support the melatonin–E interaction hypothesis and suggest that part of the mechanism(s) of melatonin's antipromotion effect may be an interaction directly at the level of the E receptor.

CONCLUSIONS

It would be premature to tacitly assume that the tumor-promoting effects of pinealectomy in experimental animals result from the elimination of melatonin from the circulation, particularly because there are alternate sources of this indole. Perhaps more importantly, pinealectomy is also likely to quench other nonmelatonin compounds (i.e., low molecular weight peptides) that also have potent oncostatic properties (Ebels et al., 1988).

The failure of melatonin to inhibit carcinogen-induced breast cancer growth when its administration is restricted to the initiation phase strongly suggests that its influence on oncogenic mechanisms during the promotional phase is more crucial for its antineoplastic properties to be manifest. Furthermore, the evidence for a neuroendocrine mechanism by which melatonin inhibits breast cancer growth is not as compelling as that for a direct antiproliferative effect on the cancer cells themselves. Coupled with the fact that melatonin can inhibit the action of tumor-promoting mitogenic hormones such as E, this evidence lends even more credence to the antipromotion hypothesis proposed here.

Although the interaction of melatonin with the endocrine system has been emphasized in this discussion of oncogenesis, its relationship to immune function (Angeli et al., 1988) may ultimately turn out to be as important if not more important with respect to its oncostatic role. Obviously, much more work is required to understand the cellular-molecular, as well as the organismal, mechanisms by which melatonin inhibits oncogenic processes. Nevertheless, the ability of melatonin to inhibit in vivo and in vitro growth of two clinically important malignant neoplasms, namely melanoma and breast cancer, is perhaps the best argument yet for the serious consideration of melatonin as a potentially useful biological agent in the diagnosis, the treatment, and possibly the prevention of certain forms of cancer.

ACKNOWLEDGMENT

The author's own work presented in this chapter was supported by PHS Grants 5R23CA-27653 and 1R01CA-42424 from the National Cancer Institute, Bethesda, Maryland.

REFERENCES

Angeli, A., Gatti, G., Sartori, D., Del Ponte, D., Carignola, R. 1988. Effects of exogenous melatonin on human natural killer (NK) cell activity. An approach to the immunomodulatory role of the pineal gland. In: *The Pineal Gland and Cancer*, Gupta, D., Attanasio, A., Reiter, R. J., eds., pp. 145–156. London: Brain Research Promotion.

Anisimov, V. M., Morozov, V. G., Kbavinson, V. K., Dilman, V. M. 1973. Correlations of anti-tumor activity of pineal and hypothalamic extract, melatonin and sygethin in mouse transplantable mammary tumors. *Vopr. Onkol. (Leningr.)* 19:99–101.

Aubert, C., Janiaud, P., Lecalvez, J. 1980. Effect of pinealectomy and melatonin on mammary tumor growth in Sprague-Dawley rats under conditions of lighting. *J. Neural Transm.* 47:121–130.

Aubert, C., Prade, M., Bouhon, C. 1970. Effet de la pinéalectomie sur les tumeurs mélaniques du hamster dore induites par l'administration (per os) d'une seule dose 9-10, dimethyl-1-2 benzanthracene. *C. R. Acad. Sci.* 271:2465–2468.

Bartsch, H., Bartsch, C., Flehmig, B. 1986. Differential effect of melatonin on slow and fast growing passages of a human melanoma cell line. *Neuroendocrinol. Lett.* 8:289–293.

Biswas, R., Vonderhaar, B. K. 1987. Role of serum in the prolactin responsiveness of MCF-7 human breast cancer cells in long-term tissue culture. *Cancer Res.* 47:3509–3514.

Blask, D. E. 1984. The pineal: an oncostatic gland? In: *The Pineal Gland*, Reiter, R. J., ed., pp. 253–284. New York: Raven Press.

Blask, D. E., Hill, S. M. 1986a. Effects of melatonin on cancer: studies on MCF-7 human breast cancer cells in culture. *J. Neural Transm. (Suppl.)* 21:433–449.

Blask, D. E., Hill, S. M. 1986b. Inhibition of human breast cancer and endometrial cancer cell growth in culture by the pineal hormone melatonin and its analogue 6-chloromelatonin. *Biol. Reprod.* 34 (Suppl. 1):176 (Abstr. 254).

Blask, D. E., Hill, S. M. 1988. Melatonin and cancer: basic and clinical aspects. In: *Melatonin: Clinical Perspectives*, Miles, A., Thompson, C., Philbrick, D. R. S., eds. Oxford: Oxford University Press (*in press*).

Blask, D. E., Leadem, C. A. 1987. Neuroendocrine aspects of neoplasia: a review. *Neuroendocrinol. Lett.* 9:1–12.

Blask, D. E., Nodelman, J. L. 1979. Antigonadatrophic and prolactin inhibitory effects of melatonin in anosmic male rats. *Neuroendocrinology* 29:406–412.

Blask, D. E., Hill, S. M., Pelletier, D. B. 1988. Oncostatic signaling by the pineal gland and melatonin in the control of breast cancer. In: *The Pineal Gland and Cancer*, Gupta, D., Attanasio, A., Reiter, R. J., eds., pp. 195–206. London: Brain Research Promotion.

Blask, D. E., Leadem, C. A., Richardson, B. A. 1981. Nutritional status, time of day and pinealectomy: factors influencing the sensitivity of the neuroendocrine-reproductive axis of the rat to melatonin. *Horm. Res. (Basel)* 14:104–112.

Blask, D. E., Hill, S. M., Orstead, K. M., Massa, J. S. 1986. Inhibitory effects of the pineal hormone melatonin and underfeeding during the promotional phase of 7,12-dimethylbenzanthracene (DMBA) -induced mammary tumorigenesis. *J. Neural Transm.* 67:125–138.

Chang, N., Spaulding, T. S., Tseng, M. T. 1985. Inhibitory effects of superior cervical ganglionectomy on dimethylbenz(a)anthracene-induced mammary tumors in the rat. *J. Pineal Res.* 2:331–340.

Danforth, D. N., Tamarkin, L., Lippman, M. 1983. Melatonin increases estrogen receptor hormone binding activity of human breast cancer cells. *Nature (London)* 305:323–325.

Das Gupta, T. K., Terz, J. 1967. Influence of the pineal gland on the growth and spread of melanoma in the hamster. *Cancer Res.* 27:1306–1311.

Ebels, I., Noteborn, H. P. J. M., Bartsch, H., Bartsch, C. 1988. Effects of low molecular weight pineal compounds on neoplastic growth. In: *The Pineal Gland and Cancer*, Gupta, D., Attanasio, A., Reiter, R. J., eds., pp. 261–272. London: Brain Research Promotion.

El-Domieri, A. A. H., Das Gupta, T. K. 1973. Reversal by melatonin of the effect of pinealectomy on tumor growth. *Cancer Res.* 33:2830–2833.

El-Domieri, A. A. H., Das Gupta, T. K. 1976. The influence of pineal ablation and administration of melatonin on growth and spread of hamster melanoma. *J. Surg. Oncol.* 8:197–205.

Georgiou, E. 1929. Uber die Natur und die Pathogenese der Krebstumoren, Radiale Heiling des Crebses bei weissen Mausen. *Z. Krebsforsch.* 38:562–572.

Ghosh, B. C., El-Domieri, A. A. H., Das Gupta, T. K. 1973. Effect of melatonin on hamster melanoma. *Surg. Forum* 24:121–122.

Hamilton, T. 1969. Influence of environmental light and melatonin upon mammary tumor induction. *Br. J. Surg.* 56:764–766.

Hill, S. M. 1986. *Antiproliferative Effect of the Pineal Hormone Melatonin on Human Breast Cancer Cells in Vitro.* Doctoral dissertation, pp. 1–206. Tucson: University of Arizona.

Hill, S. M., Blask, D. E. 1986. Melatonin inhibition of MCF-7 breast cancer cell proliferation: influence of serum factors, prolactin and estradiol. In: *Proceedings of the 68th Annual Meeting of the Endocrinological Society*, p. 246 (Abstr. 863).

Hill, S. M., Blask, D. E. 1988. Effects of the pineal hormone, melatonin, on the proliferation and morphological characteristics of human breast cancer cells (MCF-7) in culture. *Cancer Res.* 48:6121–6126.

Karmali, R. A., Horrobin, D. F., Ghayur, T. 1978. Role of the pineal gland in the aetiology and treatment of breast cancer. *Lancet* 2:1002.

Kothari, L. S. 1987. Influence of chronic melatonin on 9,10-dimethyl-1,2-benzanthracene-induced mammary tumors in female Holtzman rats exposed to continuous light. *Oncology* 44:64–66.

Kothari, L. S. 1988. Effect of melatonin on the mammary gland morphology, DNA synthesis, hormone profiles and incidence of mammary cancer in rats. In: *The Pineal Gland and Cancer*, Gupta, D., Attanasio, A., Reiter, R. J., eds., pp. 207–219. London: Brain Research Promotion.

Kothari, L. S., Shah, P. N., Mhatre, M. C. 1982. Effect of continuous light on the incidence of 9,10-dimethyl-1,2-benzanthracene-induced mammary tumors in female Holtzman rats. *Cancer Lett.* 16.313–317.

Kothari, L. S. Shaf, P. N., Mhatre, M. C. 1984. Pineal ablation in varying photoperiods and the incidence of 9,10-dimethyl-1,2-benzanthracene-induced mammary cancer in rats. *Cancer Lett.* 22:99–102.

Lapin, V. 1976. Pineal gland and malignancy. *Oster. Z-Onkol.* 3:51–60.

Leadem, C. A., Blask, D. E. 1981. Evidence for an inhibitory influence of the pineal on prolactin in the female rat. *Neuroendocrinology* 33:268–275.

Leone, A. M., Silman, R. E., Hill, B. T., Whelan, R. D. H., Shallard, S. A. 1988. Growth inhibitory effects of melatonin and its metabolites against ovarian tumor cell lines in vitro. In: *The Pineal Gland and Cancer*, Gupta, D., Attanasio, A., Reiter, R. J., eds., pp. 273–281. London: Brain Research Promotion.

Meyskens, F. L., Salmon, S. F. 1981. Modulation of clonogenic human melanoma cells by follicle-stimulating hormone, melatonin and nerve growth factor. *Br. J. Cancer* 43:111–115.

Narita, T. 1988. Effect of melatonin on B16 melanoma growth. In: *The Pineal Gland and Cancer*, Gupta, D., Attanasio, A., Reiter, R. J., eds., pp. 345–354. London: Brain Research Promotion.

Narita, T., Kudo, H. 1985. Effect of melatonin on B16 melanoma growth in athymic mice. *Cancer Res.* 45:4175–4177.

Osborne, C. K. 1985. Effects of estrogens and antiestrogens on human breast cancer cell proliferation: in vitro studies in tissue culture and in vivo studies in athymic mice. In: *Hormonally Responsive Tumors*, Hollander, V. P., ed., pp. 93–113. Orlando: Academic Press.

Pelletier, D. B., Hill, S. M., Blask, D. E. 1987. Inhibition of N-methyl-nitrosourea (NMU) -induced mammary carcinogenesis by melatonin. In: *Proceedings of the Annual Meeting of the Society of Neuroscience* 13:1670 (Abstr. 450.4).

Reiter, R. J. 1980. The pineal and its hormones in the control of reproduction in mammals. *Endocrinol. Rev.* 1:109–131.

Sanchez Barcelo, E. J., Cos Corral, S., Mediavilla, M. D. 1988. Influence of pineal gland function on the initiation and growth of hormone dependent breast tumors. Possible mechanisms. In: *The Pineal Gland and Cancer*, Gupta, D., Attanasio, A., Reiter, R. J., eds., pp. 221–232. London: Brain Research Promotion.

Shafie, S., Brooks, S. C. 1977. Effect of prolactin on growth and the estrogen receptor level in human breast cancer cells (MCF-7). *Cancer Res.* 37:792–799.

Shah, P. N., Mhatre, M. C., Kothari, L. S. 1984. Effect of melatonin on mammary carcinogenesis in intact and pinealectomized rats in varying photoperiods. *Cancer Res.* 44:3403–3407.

Stanberry, L. R., Das Gupta, T. K., Beattie, C. W. 1983. Photoperiodic control of melanoma growth in hamster: influence of pinealectomy and melatonin. *Endocrinology* 113:469–475.

Tamarkin, L., Cohen, M., Rosell, D., Reichert, C., Lippman, M., Chabner, B. 1981. Melatonin inhibition and pinealectomy enhancement of 7,12-dimethyl-benz(a)anthracene-induced mammary tumors in the rat. *Cancer Res.* 41:4432–4436.

Tapp, E. 1982. The pineal gland in malignancy. In: *The Pineal Gland, Vol. III*, Reiter, R. J., ed., pp. 171–188. Boca Raton, Florida: CRC Press.

Vaticon, M. D., Fernandez-Galez, C., Esquifino, A., Tejero, A., Aguilar, E. 1980. Effect of constant light on prolactin secretion in adult female rats. *Horm. Res. (Basel)* 12:277–288.

Vaughan, G. M. 1984. Melatonin in humans. In: *Pineal Research Reviews*, Reiter, R. J., ed., pp. 141–201. New York: Alan R. Liss.

Vignon, F., Rochefort, H. 1985. Estrogen-specific proteins released by breast cancer cells in culture and control of cell proliferation. In: *Hormonally Responsive Tumors*, Hollander, V. P., ed., pp. 135–153. Orlando: Academic Press.

Walker, M. J., Chaduri, P. K., Beattie, C. W., Tito, W. A., Das Gupta, T. K. 1978. Neuroendorine and endocrine correlates to hamster melanoma growth *in vitro. Surg. Forum* 29:151–152.

Welsch, C. W. 1985. Host factors affecting the growth of carcinogen-induced rat mammary carcinomas: a review and tribute to Charles Brenton Huggins. *Cancer Res.* 45:3415–3443.

Young, I. M., Leone, A. M., Stovell, F. P., Silman, R. E. 1985. Melatonin is metabolized to N-acetylserotonin and 6-hydroxymelatonin in man. *J. Clin. Endocrinol. Metab.* 60:114–119.

14 Calcium Homeostasis and Oxidative Stress

GARY A. PASCOE
Tetra Tech, Inc.
Bellevue, Washington

CONTENTS

Biological systems exist in a state of continuous change; growth is followed by decay. The health and vitality of a particular biological system depend not only on substantial input of resources and removal of waste, but on defenses against potentially injurious forces. Information emerging from recent research indicates that some of these forces involve oxygen-based radicals that are capable of oxidative reactions with crucial cellular macromolecules. The oxidative state within a cell is a fine balance between pro-oxidant and antioxidant systems. Under optimal cellular conditions, antioxidant systems effectively impart protection against oxidant-induced macromolecular damage of both endogenous and exogenous origin. However, a prolonged imbalance in this equilibrium can lead to a potentially dangerous condition of oxidative stress and resultant cellular malfunctions (Sies, 1985). Lately, this phenomenon has become increasingly

337

recognized for its roles in chemical toxicities, radiation damage, carcinogenesis, and aging (Halliwell, 1981; Floyd, 1982; McBrien and Slater, 1982; Sies, 1985).

Factors regulating the intracellular oxidative balance are (1) dietary input both of antioxidants that are not biosynthesized within mammalian cells and of precursors to those antioxidants that are synthesized; (2) exogenously induced generation of prooxidant species or depletion of endogenous antioxidants; and (3) alterations in normal cell physiology, which regulates the biochemical processes that maintain the balance.

Studies of factors involved in cellular injury have recently focused on the role of the multifunctional electrolyte, Ca^{2+}. A disturbance in Ca^{2+} homeostasis in vitro either by chemical oxidant or by nonchemical means is recognized as leading to oxidative stress-related cell injury. Because Ca^{2+} is involved in a number of cellular processes, it is not surprising that disruption in its regulation may lead to deleterious events in the cell.

The roles of pro- and antioxidants in the cell and current theories about the relationship of cell calcium to oxidation-related alterations in cell function are discussed in this chapter. Two main lines of investigation focus on mechanistic interpretations of the effects of chemical oxidants and physiological alterations in Ca^{2+} homeostasis on oxidative stress-related phenomena. Evidence that oxidative stress may be involved in carcinogenicity is addressed with the understanding that interpretations of these results are mostly controversial and that this is a rapidly changing field of research.

OXIDATIVE STRESS

Oxygen is the supporting element of aerobic life; it can also be a toxic component of the environment. Its reactivity has rendered it particularly functional in biological electron transport/energy-producing systems, but this same reactivity has necessitated the evolution of biochemical mechanisms to remove unusable oxygen and oxygen by-products of metabolism from mammalian cells.

Types and Sources of Reactive Oxygen Species

A number of reactive oxygen species are produced as normal by-products in biological tissues. Those of major concern for reactivity toward biological constituents are listed in Table I. Hydrogen peroxide (H_2O_2) and superoxide anion ($O_2^{\cdot-}$) are the best studied and the predominant species produced in nonperoxisomal enzymatic systems. They are considered to have relatively low reactivity toward cellular macromolecules.

Superoxide anion, formed by a one-electron reduction of molecular oxygen, is produced in numerous enzymatic reactions and electron transport

TABLE I.
Reactive Oxygen Species of Interest in Oxidative Stress[a]

Compound	Remarks
$O_2^{\cdot-}$, superoxide anion	One-electron reduction state, formed in many autooxidation reactions (e.g., flavoproteins; redox cycling)
$HO\cdot_2$, perhydroxy radical	Protonated form of $O_2^{\cdot-}$, more lipid-soluble
H_2O_2, hydrogen peroxide	Two-electron reduction state, formed from $O_2^{\cdot-}$ ($HO\cdot_2$) by dismutation, or directly from O_2
$HO\cdot$ ($OH\cdot$), hydroxyl radical	Three-electron reduction state, formed by Fenton reaction, metal- (iron-) catalyzed Haber–Weiss reaction; highly reactive
$RO\cdot$, alkoxy radical	Oxygen-centered organic (e.g., lipid) radical
$ROO\cdot$, peroxy radical	Formally formed from organic (e.g., lipid) hydroperoxide, ROOH, by hydrogen abstraction
ROOH	Organic hydroperoxide (e.g., lipid-, thymine-OOH)
$\Delta_g O_2$ (also O_2)[b]	Singlet molecular oxygen, first excited state, 22 Kcal/mol above ground state (triplet) $^3O_2$; red (dimol) or infrared (monomol) photoemission
RO (also RO)[b]	Excited carbonyl, blue-green photoemission (e.g., formed via dioxetane as intermediate)

[a] Reprinted with permission from Sies, H., ed. 1985. *Oxidative Stress.* London: Academic Press.
[b] Compounds of radical and nonradical nature are included.

systems (Sies and Cadenas, 1983). Hydrogen peroxide is formed during dismutation of $O_2^{\cdot-}$ and by the two-electron reduction of oxygen by a number of cellular oxidases. For example, the flavoprotein-dependent oxygenases in the oxidative phosphorylation/electron transport system in the mitochondrion, and the cytochrome P-450-dependent monooxygenase system in the endoplasmic reticulum, consume O_2 and produce both H_2O_2 and $O_2^{\cdot-}$ as minor by-products of inefficient O_2 utilization.

Superoxide can also be produced as a by-product of redox cycling of quinones that occur as drugs, environmental chemicals, and dietary substituents. Redox cycling is the shuttling of an electron between a quinone and its semiquinone free radical, via one-electron reduction of the quinone and autoxidation of the radical, that produces $O_2{}^{\cdot-}$ as one of the products. Quinone reduction and subsequent redox cycling can be catalyzed by a variety of flavoenzymes. As an example, the popular analgesic acetaminophen is metabolized by the NADPH-dependent cytochrome P-450 monooxygenase system to a quinone imine intermediate in the liver (Dahlin et al., 1984); this intermediate has been hypothesized to undergo redox cycling and produce oxidative stress-related cellular damage (Gerson et al., 1985). By a different mechanism of activation, the herbicides paraquat and diquat are also converted to quinone-containing compounds that are believed to produce lung, liver, and kidney injury via redox cycling (Sandy et al., 1986).

The hydroxyl radical (HO·) is produced in biological organisms from $O_2{}^{\cdot-}$ and H_2O_2 via the Fe^{2+}-catalyzed Fenton reaction:

$$O_2{}^{\cdot-} + H_2O_2 + H^+ \xrightarrow{Fe^{2+}} O_2 + HO\cdot + H_2O$$

Hydroxyl radical is a highly reactive oxygen species and may be the ultimate initiating agent in cellular oxidative damage from drugs and environmental chemicals as well as endogenous oxidation mechanisms (Sies, 1985). Perhydroxyl radical is the protonated form of $O_2{}^{\cdot-}$, but, although it is much more reactive than $O_2{}^{\cdot-}$, is expected to be formed only at a pH lower than that usually encountered in mammalian systems.

Lipid Peroxidation

Alkyl, alkoxyl, and peroxyl radicals (Table I) arise during the oxidative degradation of membranous lipids of the cell. Because such damage affects cellular membrane integrity, it has become a major focus of investigations into the biochemical and pathological mechanisms of toxic cell injury. Lipid peroxidation may be initiated by oxygen free radicals, usually HO·, or other chemical oxidants, including Fe^{2+}, and is characterized by the formation of oxidized lipid by-products (Kappus, 1985). Although lipid peroxidation has generally been thought to be intimately involved in the process of cell injury, recent evidence suggests that it is frequently a secondary reaction to a toxic insult, propagating simultaneously with necrosis, rather than a precursor to necrosis (Tribble et al., 1987).

Lipid peroxidation in general is initiated by radical-mediated abstraction of a hydrogen atom from the alpha carbon to the unsaturated bond in a polyunsaturated fatty acid, forming the alkyl radical (L·) (Fig. 1). This is followed by electronic rearrangement to form a conjugated diene system and addition of O_2 to the radical, producing the peroxyl radical (LOO·). Abstraction of hydrogen from a neighboring fatty acid reduces the peroxyl radical to a lipid hydroperoxide (LOOH) and results in formation of a second

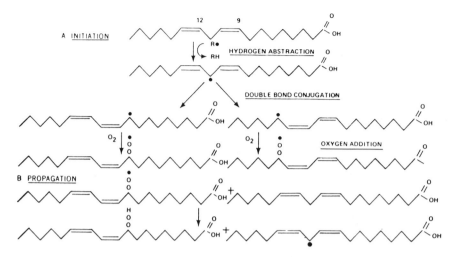

Fig. 1. Sequence of reactions involved in initiation and propagation of free radical-mediated lipid peroxidation.
(Reprinted with permission from McCay, P. B., King, M. M. 1980. Vitamin E: its role as a biologic free radical scavenger. In: *Vitamin E: A Comprehensive Treatise*, Machlin, L. J., ed., pp. 289–317. New York: Marcel Dekker.)

alkyl radical (L·). This second alkyl radical may then propagate the process. Lipid peroxidation can be terminated via radical–radical addition, and by the donation of hydrogen to a lipid peroxyl or alkoxyl radical from a nonlipid source. In biological systems this nonlipid source is usually α-tocopherol (Vitamin E), but may be other cellular macromolecules. This action results in transformation of these molecules to their respective radicals.

In the absence of terminating reactions, continued peroxidative degradation of lipids produces additional toxic by-products. The best characterized of these are malondialdehyde and various alkanes, both of which may be monitored as indirect measurements of lipid peroxidation. Additionally, highly reactive alkenals such as 4-hydroxynonenal are also produced during peroxidative attack on lipids, and may interact with cellular macromolecules to injure the cell (Esterbauer, 1982).

Macromolecular Damage

In addition to damaging the lipid environment of the cell membrane, oxygen radicals and radical by-products of lipid peroxidation may also attack protein and DNA within the cell, inactivating or altering their function. For example, $O_2^{\cdot-}$, HO·, and singlet oxygen have been shown to directly oxidize cysteinyl thiols of protein in vitro, presumably via protein thiyl radical formation (Asada and Kanematsu, 1976; Davies et al., 1987). During extensive lipid peroxidation of model proteins, both lipid peroxyl and oxygen free radicals generate C-, O-, and S-centered radicals in the amino acids tryptophan, tyrosine, and cysteine, respectively (Roubal, 1970; Karel et al., 1975; Schaich and Karel, 1975, 1976; Davies et al., 1987). Additionally, lipid hydroperoxides and aldehyde by-products of peroxidation covalently bind to cysteine sulfhydryl groups of proteins in vitro (Buttkus, 1969; Gardner, 1979). The actual extent of such reactions in vivo is presently unclear, although the skin pigment lipofuscin is believed to result from such protein–lipid radical interactions.

Oxygen free radicals also have been shown to interact with RNA and DNA in vitro, resulting in base alterations and ultimately strand breaks. Most information about interactions of DNA with oxygen radicals has come from radiolysis experiments (Draganic and Draganic, 1971). In x-ray-exposed isolated cell systems, DNA damage results primarily from $HO\cdot$ formed during water radiolysis, but may be caused by secondary radicals arising from oxygen radical interactions with other cell molecules. Hydroxyl radical produced from $O_2 \cdot^-$ via the Fenton reaction in vitro has also been shown to cause DNA strand breakage (Lesko et al., 1980). However, recent evidence indicates that during Fenton reaction-mediated DNA strand breakage, a potentially more stable oxygen radical, the ferryl radical $(Fe^{3+} - HO\cdot)$, is formed as an intermediate that breaks down to release the better characterized $HO\cdot$ radical (Imlay et al., 1988). It has been suggested that this ferryl radical is the actual biologically active radical produced during iron-catalyzed $HO\cdot$ generation (Imlay et al., 1988). The experiments described in Draganic and Draganic (1971) have demonstrated that hydroxyl radicals attack uracil, cytosine, and thymidine, preferentially at the C-5 position. In the presence of oxygen, the resulting radicals are converted to the corresponding peroxyl radicals. These rapidly form their hydroperoxides or decay to a variety of oxidized products (Schulte-Frohlinde and Von Sonntag, 1985).

It has been hypothesized that radicals of lipids may also interact with and damage DNA (Marnett, 1987). The lipid peroxyl radical $(LOO\cdot)$, derived from the lipid hydroperoxide (LOOH), is significantly more stable than $HO\cdot$ and is able to diffuse to cellular loci remote from the site of its generation. It has been suggested that these characteristics may make lipid peroxyl radicals more selective toward DNA and other macromolecules than oxygen-derived radicals (Marnett, 1987).

The important implications of these interactions of oxygen and peroxyl free radical species with the genetic information of the cell are discussed at the end of this chapter.

Cellular Protective Systems

Enzymatic Systems

To protect the cell against oxygen and organic radical damage such as just described, mammalian cells have evolved both enzymatic and nonenzymatic defense systems (Table II). The enzymatic systems consist primarily of three enzymes: superoxide dismutase to catalyze the dismutation of $O_2^{\cdot-}$ to H_2O_2 and H_2O, glutathione (GSH) peroxidase to convert lipid hydroperoxides (LOOH) and H_2O_2 to their respective alcohols, and GSH transferases to catalyze the transfer of GSH to receiving electrophils. The discovery of superoxide dismutase in the late 1960s spearheaded the recent popularity of oxidative stress mechanisms in cytotoxicity (Pryor, 1986). The two mammalian superoxide dismutases contain either Mn or Cu/Zn. GSH peroxidase occurs in two forms, both utilizing GSH as hydrogen donor: one is a Se-containing enzyme active toward H_2O_2 and organic hydroperoxides, and the other form is active only toward organic hydroperoxides. In addition, it has been recently argued that phospholipase A_2 may serve a protective function in the cell against lipid hydroperoxides by releasing them from the membrane to make them available for GSH peroxidase reduction and detoxification (Van Kujik et al., 1987). The GSH transferases, a family of enzymes with overlapping substrates, are mostly found in the cell cytosol with low levels of enzyme activity observable in liver subcellular fractions.

Nonenzymatic Systems

Nonenzymatic cellular defenses consist of a number of chemical antioxidants as well as certain cofactors in enzymatic systems (Table II). The best characterized endogenous antioxidants are the vitamins C and E. Vitamin C (ascorbic acid) is a water-soluble free radical scavenger that acts via the donation of hydrogen atoms to radicals present in the cell cytosol. Vitamin E acts complementary to Vitamin C by scavenging radicals present in the lipid environment of the cell. The name Vitamin E is generic for four major tocopherols that differ by the extent of methylation of the chromanoxyl ring. α-Tocopherol contains a fully substituted ring (Table III) and has the highest in vivo antioxidant activity (Kasparek, 1980). It is also the most common naturally occurring form and is readily absorbed from the diet. Vitamin E is the primary lipid-soluble antioxidant in the cell because it donates phenolic hydrogen to lipid peroxyl and alkoxyl radicals much more effectively than does a second lipid (McCay and King, 1980). This action thereby terminates the propagative process of lipid radical–lipid reactions and generates the α-tocopheryl radical in the process (Fig. 2). The α-tocopheryl radical may be converted back to α-tocopherol via hydrogen donation from Vitamin C (Niki et al., 1982) or via a putative GSH-dependent process presently uncharacterized (McCay et al., 1986),

TABLE II.
Antioxidant Defense in Biological Systems[a]

System	Remarks
Nonenzymatic	
α-Tocopherol	Membrane-bound; receptors? Regeneration
(Vitamin E)	from chromanoxy radical?
Ascorbate	Water-soluble
(Vitamin C)	
Flavenoids	Plant antioxidants (rutin, quercetin, etc.)
Chemical	Food additives, e.g., BHA (butylated hydroxy-anisole), BHT (butylated hydroxytoluene)
β-Carotene,	Singlet oxygen quencher
Vitamin A	
Urate	Singlet oxygen quencher, radical scavenger?
Plasma proteins	Coeruloplasmin (e.g.)
Enzymatic	
Superoxide	CuZn enzyme, Mn enzyme
dismutases	
GSH peroxidases	Selenoenzyme; non-Se enzyme: some GSH *S*-transferases, e.g., isoenzymes B and AA Cytosol and mitochondrial matrix
Catalase	Heme enzyme Predominantly in peroxisomal matrix
Ancillary enzymes	
NADPH-quinone	Two-electron reduction, dicoumarol-sensitive
oxidoreductase	
(DT-diaphorase)	
Epoxide hydrolase	
Conjugation	
enzymes	UDP-glucuronyltransferase Sulfotransferase GSH *S*-transferases
GSSG reductase	
NADPH supply	Glucose-6-phospate dehydrogenase 6-Phosphogluconate dehydrogenase Isocitrate dehydrogenases Malic enzyme Energy-linked transhydrogenase
Transport systems	GSSG export Conjugate export

[a] Reprinted with permission from Sies, H., ed. 1985. *Oxidative Stress*. London: Academic Press.

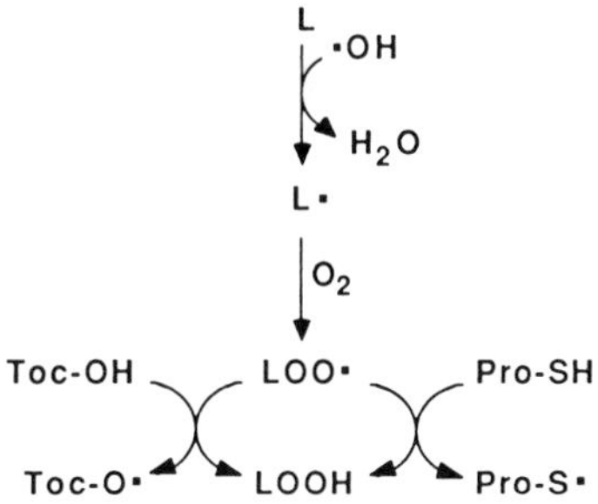

Fig. 2. α-Tocopherol scavenging of lipid peroxyl radical and prevention of radical attacks on protein thiols (ProSH).

TABLE III.
Structure of Natural Tocopherols

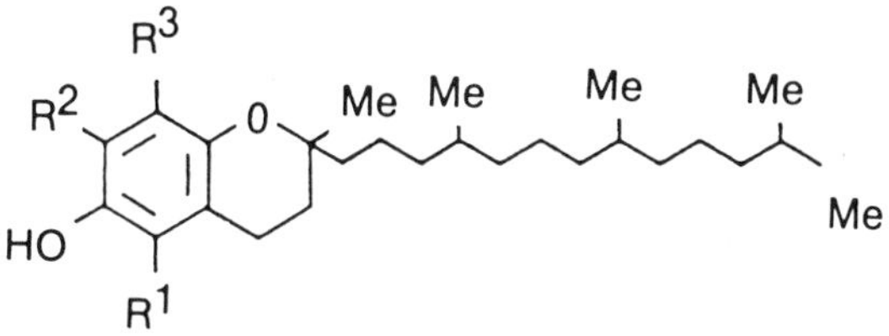

Compound	R^1	R^2	R^3
α-Tocopherol	Me	Me	Me
β-Tocopherol	Me	H	Me
γ-Tocopherol	H	Me	Me
δ-Tocopherol	H	H	Me

or may be reduced to tocopherylquinone and further metabolic products of tocopherol (Gallo-Torres, 1980).

Less well studied nonenzymatic antioxidant defenses are the carotenoids (Vitamin A, β-carotene) and urate, both believed to act as singlet oxygen quenchers in biological systems. In addition, manganese and selenium are frequently considered antioxidants effective against lipid peroxidation in in vitro systems, presumably as cofactors in Mn-superoxide dismutase and Se-GSH peroxidase, respectively.

Glutathione (GSH) is unique in that it acts both nonenzymatically and as the cofactor in GSH peroxidase to remove reactive electrophiles and H_2O_2 from the cell cytoplasm. GSH is an unusual tripeptide of glycine, glutamate, and cysteine in which cysteine and glutamate are joined by

GLUTATHIONE BIOSYNTHESIS

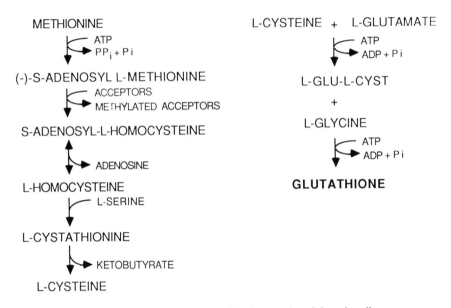

Fig. 3. Glutathione biosynthesis via cystathionine and γ-glutamyl pathways.

a gamma linkage catalyzed by γ-glutamylcysteine synthetase (Fig. 3), the rate-limiting enzyme in GSH biosynthesis. Glycine is added to this dipeptide via GSH synthetase (Beatty and Reed, 1980). The nucleophilic reactivity of GSH lies in the sulfur moiety of cysteine, and nucleophilic scavenging by GSH, whether enzymatic (GSH transferases) or nonenzymatic, involves the formation of thioether linkages at the cysteinyl moiety (Moldéus and Jernström, 1983). GSH also acts as a cellular reductant, losing electrons to form oxidized glutathione (GSSG), which may also arise from glutathionyl radical (GS·)–radical interactions (Ross et al., 1985). To maintain thiol redox balance in the cell, GSH is regenerated from GSSG via GSH reductase, a cytosolic enzyme that utilizes NADPH as cofactor. The relationship between the GSH redox cycle and the enzymatic reduction of H_2O_2 is illustrated in Fig. 4.

GSH serves many functions in the cell in addition to its detoxification reactions. A primary function is to maintain protein thiols (ProSH) in their reduced state, both cytosolic and membrane bound, via thiol-disulfide exchange reactions (Flóhe and Günzler, 1976). This is essential for the activities of many enzymes (Zeigler, 1985), including the calcium translocases responsible for pumping Ca^{2+} across cellular membranes.

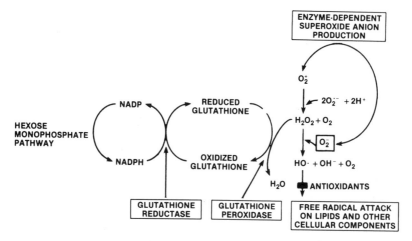

Fig. 4. Scheme of flavoenzyme-mediated H_2O_2 production and enzyme-mediated reduction via the GSH redox cycle.

Regulation of Intracellular GSH

Because of the multitude of functions of GSH, it is not surprising that the regulation of the GSH redox system is complex. The following discussion focuses on rat liver as a model of mammalian GSH regulation. GSH is maintained in the cell cytosol of liver cells at relatively high levels (up to 10 mM) via a highly regulated system of redox cycling between GSH and GSSG, intracellular compartmentation, and interorgan transport (Reed and Fariss, 1984). Cysteine levels are very low in the hepatocyte cytosol (Reed and Orrenius, 1977) and are rate limiting for GSH formation. Most cell types are dependent on extracellular cysteine for GSH synthesis. However, in addition to the small amount of cysteine available from protein breakdown and dietary sources, liver cells are capable of utilizing the cysteinyl sulfur from methionine and cystathionine for GSH biosynthesis (Beatty and Reed, 1980) (see Fig. 3).

Recent reports have provided evidence for the compartmentation of GSH in both mitochondria and nuclei (Meredith and Reed, 1982; Tirminstein and Reed, 1988), and have placed added emphasis on mitochondrial GSH in particular for its role in maintaining cell viability during oxidative changes in liver cells (Olafsdottir et al., 1988). Hepatic GSH is actively transported out of the liver cell to be enzymatically degraded in the kidney (Rankin et al., 1983). These systems thereby regulate not only the cytosolic GSH concentrations, but also those external to the cell.

Interruption of this regulatory system may result in nonphysiological levels of cytosolic GSH. GSH is maintained at relatively high levels in the cytosol to protect against normal physiological production of reactive oxygen species (e.g., H_2O_2) and against dietary sources of electrophilic compounds. It has been demonstrated, in isolated cell systems, that peroxidative degeneration of membranous lipids occurs below certain threshold levels of intracellular GSH (10%–15% of normal) (Anundi et al., 1979). During an increased demand for intracellular GSH as a reductant in certain chemical exposures, intracellular levels may easily become depleted to less than these threshold levels, allowing concentrations of reactive oxygen species to increase in the cell (Reed and Fariss, 1984).

Alterations in physiological levels of other antioxidants also may result in oxidative damage to cellular macromolecules. For example, isolated hepatocytes made deficient in α-tocopherol show elevated lipid peroxidation and macromolecular damage (Hill and Burk, 1982). Alterations in the ability of the cell to regenerate antioxidants, by chemical inhibition of GSH reductase and hence GSH regeneration, for example (Babson et al., 1981), additionally deprive the cell of protective measures against oxidative attack.

In summary, the previously described challenges to the redox balance of the cell may result in oxidative damage to membrane lipids, DNA, and protein, including the loss of critical ProSH groups. Depending on the macromolecular targets of the oxidizing species and on the status of cellular repair mechanisms, one of the toxic endpoints of these challenges, instead of cell death, may be neoplastic transformation of the cell, as described later in this chapter.

CALCIUM HOMEOSTASIS MECHANISMS

A role for Ca^{2+} in the process leading to cell death was proposed decades ago. It was well noted by pathologists that there was a correlation between calcification and tissue necrosis, and early mechanistic studies into this relationship proposed a role for Ca^{2+} in cell death (Gallagher et al., 1956; Reynolds, 1963; Judah et al., 1964; Farber and El-Mofty, 1975). In particular, numerous reports by Trump and colleagues (reviewed in Trump et al., 1981, 1984) have presented evidence that increases in levels of cytosolic Ca^{2+} in excitable tissue following initial cell injury were primary to cell death.

In discerning whether a causal relationship exists between altered Ca^{2+} homeostasis and injurious events in the cell, it is necessary to first understand the mechanisms of Ca^{2+} homeostasis. Intracellular calcium is maintained at a steep concentration gradient from extracellular concentrations in the millimolar range down to the 0.1–0.4 μM range for free Ca^{2+} in the cell cytosol (Thiers and Vallee, 1957; Bresciani and Auricchio, 1962). This gradient, coupled with a large electrochemical gradient, serves as a driving force to transport Ca^{2+} into the cell. Although total cellular calcium ranges from 3 to 30 nmol/mg cell protein, intracellular ionized

Ca^{2+} concentrations are maintained at 0.1% or less of these levels by several mechanisms (Hodgkin and Keynes, 1957). In summary, these mechanisms presently are known to include ATPase-dependent Ca^{2+} translocases in the plasma membrane and endoplasmic reticulum, a Ca^{2+} channel and Na^+/Ca^{2+} exchanger in the plasma membrane, Na^+/Ca^{2+} and Ca^{2+}/H^+ exchangers in the mitochondrion, and a Ca^{2+}-release system in the endoplasmic reticulum (Carafoli, 1987). In addition to these Ca^{2+}-exchange systems, numerous Ca^{2+}-binding proteins, such as calmodulin, have been described for sequestering intracellular Ca^{2+} (Cheung, 1982; Michiel and Wang, 1986). Thus, a complex homeostatic system has evolved to maintain low cytosolic Ca^{2+} by compartmentation in the intracellular matrix and by extrusion to the extracellular milieu (Fig. 5). It is important to note that a major feature of many protein-based regulatory systems is the presence of free SH groups, either at the active site of transport enzymes or proximal enough to affect enzyme activity. For example, the ATPase-dependent translocase of the plasma membrane of rat livers is known to be an SH-containing enzyme, oxidation of which renders the enzyme inactive (Scherer and Deamer, 1986).

Methods used to measure or estimate intracellular free Ca^{2+} levels are listed in Table IV. The ability to measure intracellular Ca^{2+} was greatly increased with the recent development of permeant Ca^{2+}-binding dyes, which were found to offer greatly increased sensitivity and selectivity over

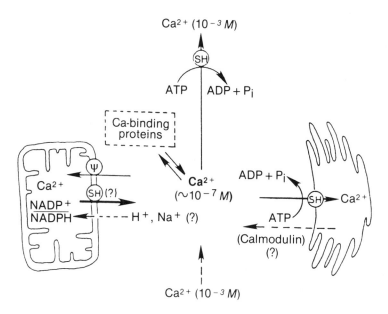

Fig. 5. Thiol-mediated regulation of intracellular Ca^{2+} homeostasis in hepatocytes. (Reprinted with permission from Orrenius, S. 1985. Biochemical mechanisms of toxicity. *Trends Pharmaceut. Sci. FEST (Suppl.)*:1–4.)

TABLE IV.
Measurement in Intact Cells

Microelectrodes

Ca^{2+}-sensitive photoproteins: aequorin

Atomic absorption spectroscopy

Phosphorylase a activity

Metallochromic indicators: murexide, arsenazo III with the uncoupler carbonyl cyanide p-trifluoromethoxyphenyl hydrazone (FCCP) and the ionophore A23187

Fluorescent indicators: "quin 2" analogues

---

earlier methods. The first of these dyes to be developed was quin-2 (Fig. 6). It is loaded into the cell as the hydrophobic acetoxymethyl ester, is released by enzymatic hydrolysis, and fluoresces upon complexing with free Ca^{2+} (Grynkiewicz et al., 1985). Problems of low fluorescent intensity and interference by absorbance from NADPH have been overcome with the development of newer analogues, such as fura-2 (Li et al., 1987). It is anticipated that these newer methods to measure intracellular Ca^{2+} will allow major insights into the relationship between Ca^{2+} homeostasis and toxic cell endpoints.

DISRUPTION OF Ca^{2+} HOMEOSTASIS

Much of the early work that related oxidative stress to disrupted intracellular Ca^{2+} homeostasis involved chemical manipulation of the oxidative

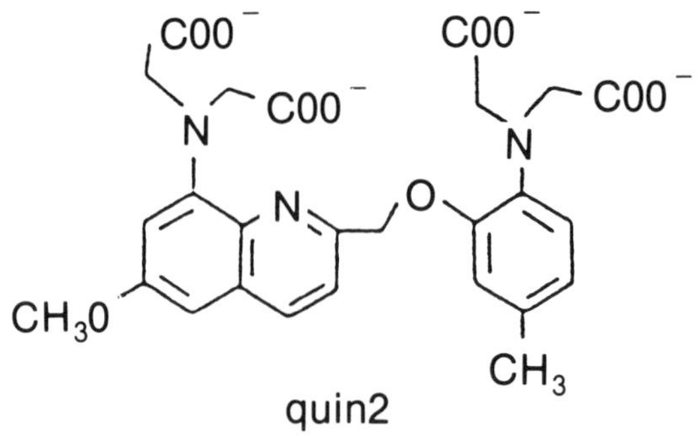

Fig. 6. Structure of Ca^{2+} permeant dye, quin-2.

balance in isolated rat hepatocytes. In a series of experiments by Orrenius and colleagues (Orrenius, 1985), it was demonstrated that certain chemicals that produce oxidative stress in the hepatocyte markedly impair the intracellular compartmentation of Ca^{2+}. Specifically, menadione, a quinone-containing compound, was shown to deplete intracellular GSH and oxidize or arylate ProSH during its redox cycling. Menadione-induced loss of critical ProSH groups of the Ca^{2+} ATPase of hepatocyte plasma membrane impairs the active extrusion of intracellular Ca^{2+} (Nicotera et al., 1985). Additionally, the ability of the endoplasmic reticulum and mitochondrion to take up and retain Ca^{2+} is disturbed by menadione oxidation of membrane ProSH and pyridine nucleotides, respectively (Bellomo et al., 1982; Jones et al., 1983). The loss of the cell's ability to both extrude and store Ca^{2+} results in prolonged elevation of cytosolic Ca^{2+} concentrations. Findings of enhanced activities of Ca^{2+}-dependent membrane-bound phospholipase A_2 and nonlysosomal proteases in the chemical oxidant model (Trump et al., 1976; Smith et al., 1980; Nicotera et al., 1986) suggested that the increased cytosolic Ca^{2+} is directly responsible for enzyme-mediated membrane damage during oxidative insult to isolated cells (Orrenius, 1985). This is illustrated schematically in Fig. 7.

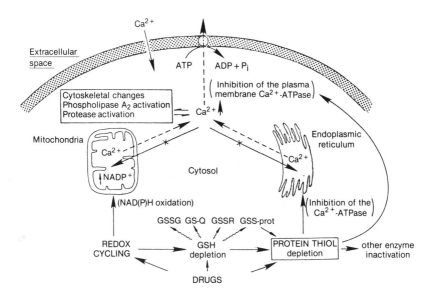

Fig. 7. Scheme proposed by Orrenius (1985) for the cytotoxicity of certain redox cycling quinone drugs. Abbreviations: GS-Q, quinone conjugate with glutathione; GSS-prot, glutathione-protein mixed disulfide.
(Reprinted with permission from Orrenius, S. 1985. Biochemical mechanisms of toxicity. *Trends Pharmaceut. Sci. FEST (Suppl.):*1–4.)

It is evident from these studies that oxidative stress profoundly affects hepatocellular Ca^{2+} homeostasis. However, many facets of the interrelationships between oxidative stress, thiol oxidation, and Ca^{2+} homeostasis remain unclear. The studies mentioned here have relied on an initial chemical-generated oxidative stress with subsequent evaluation of cellular Ca^{2+} status and secondary effects on cell membrane integrity. Reed and colleagues have investigated these relationships using a physiological, rather than a chemical, model of oxidative stress in an isolated hepatocyte system that is referred to as the "Ca^{2+}-omission" model (Pascoe and Reed, 1989). This model consists of isolated hepatocytes incubated in a "complete" medium (Fischer's medium) (Fariss et al., 1985) in the absence of extracellular Ca^{2+} (final extracellular Ca^{2+} levels < 0.8 mM). The omission of extracellular Ca^{2+} depletes cellular Ca^{2+} levels to 10% of those found in Ca^{2+}-adequate cells incubated in 3 mM Ca^{2+} (Pascoe and Reed, 1987). This action rapidly disrupts intracellular Ca^{2+} homeostasis by disturbing Ca^{2+} cycling between mitochondria and the cell cytosol (Thomas and Reed, 1988a) and by causing the rapid release of membrane-bound Ca^{2+} into the cell cytosol (Jewel et al., 1982). As the disturbance in intracellular Ca^+ homeostasis is prolonged, indications of induction of oxidative stress mechanisms in the Ca^{2+}-deficient cell become apparent. The initial indications are a rapid and progressive loss in cellular α-tocopherol content and a concomitant rise in indices of lipid peroxidation (malondialdehyde production) (Pascoe et al., 1987a). The simultaneous loss of α-tocopherol and stimulation of lipid peroxidation most likely indicates that tocopherol consumption results from scavenging of lipid radicals produced secondary to Ca^{2+}-stimulated oxidative processes.

Following these initial indices of oxidative damage to the cellular lipid environment, gradual losses of soluble thiols, ProSH, and cell viability are observed. This scheme of physiologically induced oxidative stress and cell injury is depicted in Fig. 8. The loss in intracellular GSH is attributable both to an enhanced efflux of GSH and its cystathionine pathway precursors and to its oxidation to GSSG (Pascoe et al., 1987a). Although one of the primary roles of GSH in the hepatocyte is to maintain ProSH groups in their reduced state, the loss of ProSH in this model is independent of the oxidative loss of GSH, and correlates only with the levels of cellular α-tocopherol (Reed et al., 1987; Pascoe et al., 1987b). A proposed mechanism for this observation is that prevention of ProSH loss is secondary to α-tocopherol-mediated inhibition of lipid peroxyl radical formation (see Fig. 2). Supplementation of cellular α-tocopherol thereby prevents the loss of ProSH levels and maintains viability in Ca^{2+}-deficient cells. From these observations, we have suggested that the cell damage in the Ca^{2+}-omission model is directly related to the oxidative loss of ProSH, secondary to loss of α-tocopherol (Pascoe and Reed, 1989). The intimate relationship

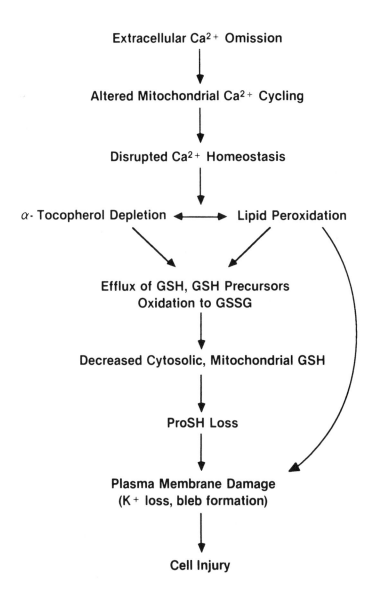

Fig. 8. Ca^{2+}-omission model of oxidative stress in isolated hepatocytes.

between cell viability and ProSH levels in the hepatocyte has been previously established with both chemical oxidant and physiological models of oxidative stress (Orrenius, 1985; Pascoe et al., 1987b).

The correlative inhibition of oxidative stress and cell damage by antioxidants in the Ca^{2+}-omission model suggests that resultant cell damage is related to oxidation of critical cellular constituents. However, recent evidence suggests that the cell damage observed in this model does not simply involve an elevation in cytosolic Ca^{2+} concentration and stimulation of Ca^{2+}-dependent degradative enzymes, in contrast to previous suggestions (Jewel et al., 1982; Pascoe and Reed, 1987). Neither phospholipase A_2 inhibitors (dibucaine, quinacrin) nor protease inhibitors (leupeptin, antipain, trypsin inhibitor), for example, affect the progress of lipid peroxidation or cell injury in the Ca^{2+}-omission model (Thomas and Reed, 1988b). Furthermore, cell injury induced by the addition of the Ca^{2+} ionophore A23187 to this model, which redistributes compartmentalized Ca^{2+}, is not inhibited by antioxidants but is inhibited by Ca^{2+} chelation (Thomas and Reed, 1988b). These studies suggest that cell injury in this model is initiated by a Ca^{2+}-dependent phenomenon that is distinct from, or more complex than, an elevation in cytosolic Ca^{2+} levels. Thus, although the role of oxidative stress in cell injury remains controversial (Tribble et al., 1987), available evidence suggests that during disturbed intracellular Ca^{2+} homeostasis, alterations in cell function are induced by oxidative stress and are responsible for the resultant cell injury. Although the disruption of Ca^{2+} homeostasis apparently induces the enhanced production of active oxygen species, neither the nature of the initiating events nor the underlying mechanism of cell injury is well understood.

OXYGEN RADICALS AND CANCER

A growing body of data now provides suggestive if not compelling evidence that reactive oxygen species produced by radiation and some chemicals including tumor promoters play a role in the process of carcinogenesis. Much of this evidence comes from observations that antioxidants and enzymes that scavenge free radicals or that react with their products serve as powerful anticarcinogens and inhibit the neoplastic process at its various stages. As examples of these phenomena, α-tocopherol and selenium inhibit in vitro carcinogenic transformation induced by x rays and the chemical carcinogens benzo(a)pyrene and tryptophan pyrolysate (Borek et al., 1986); in vivo dietary deficiencies of these antioxidants enhance tumorigenesis of dimethylhydrazine, 7,12-dimethylbenz(a)anthracene, and Adriamycin (Lee and Chen, 1979; Cook and McNamara, 1980; Wang et al., 1982); and α-tocopherol pretreatment of animals prevents the tumorigenicity of methylcholanthrene and dibenzanthracene (Harber and Wissler, 1962; Hultin and Arrhenius, 1965).

Chemical carcinogenesis occurs in two stages: initiation, in which DNA is irreversibly altered, and promotion, in which conditions favoring the expression of carcinogenesis are produced. Most well-studied carcinogens are initiators, but a few are complete carcinogens, acting as both initiator and promoter. In general, the more potent chemical carcinogens are procarcinogens that undergo metabolic transformation to the ultimate carcinogenic species (Floyd, 1982). A number of these transformations are known to produce free radical intermediates that are believed to be one of the proximate forms of the carcinogen. For example, the carcinogens benzo(a)-pyrene, *N*-alkyl-*N*-nitroso compounds, aromatic amines, and acetylamino-fluorene form free radical intermediates during their oxidative metabolism (Nagata et al., 1982). As a further example, there is a strong relationship between the capacity of aminonaphthols and naphthols to form free radicals and their ability to induce cancer in animals (Nagata et al., 1982). These examples of chemical-induced radical damage to the genetic information system of the cell provide strong evidence for the susceptibility of DNA to free radical attack by carcinogenic initiators.

Although less well understood, one mechanism of tumor promotion also appears to involve oxygen radicals. The tumor promoter phorbol-myristate-acetate stimulates the production of $O_2^{\cdot-}$ and H_2O_2 in polymorphonuclear leukocytes, apparently through the action of an NADPH-dependent oxidase (Goldstein et al., 1979). It is speculated that the phorbol esters may enhance tumor formation by this mechanism (Troll and Wiesner, 1985). More direct evidence for a role of free radicals in tumor promotion is that certain free radical-generating compounds (benzoyl peroxide, chloroperbenzoic acid, lauroyl peroxide) are effective skin tumor promoters in mice after initiation with dimethylbenzanthracene (Slaga et al., 1981).

As described earlier, studies on the mechanism of oxygen radical-induced transformation have demonstrated that DNA reacts with oxygen and organic free radicals produced during oxidative stress in vitro (Schulte-Frohlinde and Von Sonntag, 1985), and that it thus becomes adversely altered (Pietronigro et al., 1977). The guanine moiety is the primary target for radical attack on DNA, with resultant production of peroxyl radicals of guanine-DNA. Because antioxidants such as ascorbic acid and α-tocopherol protect against radical attack on cellular macromolecules via scavenging oxygen and peroxyl radicals, they are believed to inhibit formation of radicals of DNA during oxidative stress via donation of hydrogen to the lipid peroxyl radical and perhaps the peroxyl radical of guanine-DNA (Nakayama et al., 1984). The end result is antioxidant-mediated inhibition of free radical-generated carcinogenesis.

Although many of these studies provide circumstantial evidence for a role of free radicals in carcinogenesis, for certain cases, particularly x-ray irradiation, oxygen radicals appear to play an important role. The extent of involvement of oxygen radical mechanisms in nonradiation carcinogenicity remains to be fully determined. Nonetheless, the foregoing examples point to the growing body of work on free radicals, including oxygen radicals, in the mechanism of induction of certain cancers.

REFERENCES

Anundi, I., Högberg, J., Stead, A. H. 1979. Glutathione depletion in isolated hepatocytes: its relation to lipid peroxidation and cell damage. *Acta Pharmacol. Toxicol.* 45:45–51.

Asada, K., Kanematsu, K. 1976. Reactivity of thiols with superoxide radicals. *Agric. Biol. Chem.* 40:1891–1892.

Babson, J. R., Abell, N. S., Reed, D. J. 1981. Protective role of the glutathione redox cycle against Adriamycin-mediated toxicity in isolated hepatocytes. *Biochem. Pharmacol.* 30:2299–2304.

Beatty, P. W., Reed, D. J. 1980. Involvement of the cystathionine pathway in the biosynthesis of glutathione by isolated rat hepatocytes. *Arch. Biochem. Biophys.* 204:80–87.

Bellomo, G., Jewell, S. A., Orrenius, S. 1982. The metabolism of menadione impairs the ability of rat liver mitochondria to take up and retain calcium. *J. Biol. Chem.* 257:11558–11562.

Borek, C., Ong, A., Mason, H., Donahue, L., Biaglow, J. E. 1986. Selenium and vitamin E inhibit radiogenic and chemically induced transformation *in vitro* via different mechanisms. *Proc. Natl. Acad. Sci. USA* 83:1490–1494.

Bresciani, F., Auricchio, F. 1962. Subcellular distribution of some metallic cations in the early stages of liver carcinogenesis. *Cancer Res.* 22:1284–1289.

Buttkus, H. 1969. Reaction of cysteine and methionine with malonaldehyde. *J. Am. Oil Chem. Soc.* 46:88–93.

Carafoli, E. 1987. Intracellular calcium homeostasis. *Annu. Rev. Biochem.* 56:395–433.

Cheung, W. Y. 1982. Calmodulin: an overview. *Fed. Proc.* 41:2253–2257.

Cook, M. G., McNamara, P. 1980. Effect of dietary vitamin E on dimethylhydrazine-induced colonic tumors in mice. *Cancer Res.* 40:1327–1331.

Dahlin, D. C., Miwa, G. T., Lu, A. Y. H., Nelson, S. H. 1984. *N*-Acetyl-*p*-benzoquinone imine: a cytochrome P-450-mediated oxidation product of acetaminophen. *Proc. Natl. Acad. Sci. USA* 81:1327–1331.

Davies, K. J. A., Delsignore, M. E., Lin, S. W. 1987. Protein damage and degradation by oxygen radicals. II. Modification of amino acids. *J. Biol. Chem.* 262:9902–9907.

Draganic, I. G., Draganic, Z. D. 1971. *The Radiation Chemistry of Water.* New York: Academic Press.

Esterbauer, H. 1982. In: *Free Radicals, Lipid Peroxidation, and Cancer,* McBrien, D. C. H., Slater, T. F., eds., pp. 101–128. New York: Academic Press.

Farber, J. L., El-Mofty, S. K. 1975. The biochemical pathology of liver necrosis. *Am. J. Pathol.* 81:237–250.

Fariss, M. W., Brown, M. K., Schmitz, J. A., Reed, D. J. 1985. Mechanism of chemical-induced toxicity. I. Use of a rapid centrifugal technique for the separation of viable and non-viable hepatocytes. *Toxicol. Appl. Pharmacol.* 79:283–295.

Flóhe, L., Günzler, W. A. 1976. Glutathione-dependent enzymatic oxidoreduction reactions. In: *Glutathione: Metabolism and Function*, Kroc Foundation Series, Vol. 6, Arias, I. M., Jakoby, W. B., eds., pp. 17–34. New York: Raven Press.

Floyd, R. A., ed. 1982. *Free Radicals and Cancer*. New York: Marcel Dekker.

Gallagher, C. H., Gupta, D. N., Judah, J. D., Rees, K. R. 1956. Biochemical changes in liver in acute thioacetamide intoxication. *J. Pathol. Bacteriol.* 72:193–201.

Gallo-Torres, H. E. 1980. Transport and metabolism. In: *Vitamin E: A Comprehensive Treatise*, Machlin, L. J., ed., pp. 193–267. New York: Marcel Dekker.

Gardner, H. W. 1979. Lipid hydroperoxide reactivity with proteins and amino acids: a review. *J. Agric. Food Chem.* 27:220–229.

Gerson, R. J., Casini, A., Gilfor, D., Serroni, A., Farber, J. L. 1985. Oxygen-mediated cell injury in the killing of cultured hepatocytes by acetaminophen. *Biochem. Biophys. Res. Commun.* 126:1129–1137.

Goldstein, B. D., Witz, G., Amoruso, M., Troll, W. 1979. Protease inhibitors antagonize the activation of polymorphonuclear leukocyte oxygen consumption. *Biochem. Biophys. Res. Commun.* 88:854–860.

Grynkiewicz, G., Poenie, M., Tsien, R. Y. 1985. A new generation of Ca^{2+} indicators with greatly improved fluorescence properties. *J. Biol. Chem.* 260:3440–3450.

Halliwell, B. 1981. Free radicals, oxygen toxicity, and aging. In: *Age Pigments*, Sohal, R. S., ed., pp. 1–62. Amsterdam: Elsevier/North Holland, Biomedical Press.

Harber, S. L., Wissler, R. W. 1962. Effect of vitamin E on carcinogenicity of methylcholanthrene. *Proc. Soc. Exp. Biol. Med.* 111:774–775.

Hill, K. E., Burk, R. F. 1982. Effect of selenium deficiency and vitamin E deficiency on glutathione metabolism in isolated hepatocytes. *J. Biol. Chem.* 257:10668–10672.

Hodgkin, A. L., Keynes, R. D. 1957. Movements of labelled calcium in squid giant axons. *J. Physiol.* 138:253–281.

Hultin, T., Arrhenius, E. 1965. Effects of carcinogenic amines on amino acid incorporation by liver systems. III. Inhibition by aminofluorene treatment and its dependence on vitamin E. *Cancer Res.* 25:124–131.

Imlay, J. A., Chin, S. M., Linn, S. 1988. Toxic DNA damage by hydrogen peroxide through the Fenton reaction in vivo and in vitro. *Science* 240:640–642.

Jewel, S. A., Bellomo, G., Thor, H., Orrenius, S., Smith, M. T. 1982. Bleb formation in hepatocytes during drug metabolism is caused by disturbances in thiol and calcium ion homeostasis. *Science* 217:1257–1259.

Jones, D. P., Thor, H., Smith, M. T., Jewell, S. A., Orrenius, S. 1983. Inhibition of ATP-dependent microsomal Ca^{2+} sequestration during oxidative stress and its prevention by glutathione. *J. Biol. Chem.* 258:6390–6393.

Judah, J. D., Ahmed, K., McLean, A. E. M. 1964. Possible role of ion shifts in liver injury. In: *Cellular Injury*, de Reuck, A. V. S., Knight, J., eds., pp. 187–208. Boston: Ciba Foundation.

Kappus, H. 1985. Lipid peroxidation: mechanisms, analysis, enzymology, and biological relevance. In: *Oxidative Stress*, Sies, H., ed., pp. 273–310. London: Academic Press.

Karel, M., Schaich, K., Roy, R. B. 1975. Interaction of peroxidizing methyl linoleate with some proteins and amino acids. *J. Agric. Food Chem.* 23:159–163.

Kasparek, S. 1980. Chemistry of tocopherols and tocotrienols. In: *Vitamin E: A Comprehensive Treatise*, Machlin, L. J., ed., pp. 7–65. New York: Marcel Dekker.

Lee, C., Chen, C. 1979. Enhancement of mammary tumorigenesis in rats by vitamin E deficiency (abstract). *Proc. Am. Assoc. Cancer Res. Annu. Meet.* 20:132.

Lesko, S. A., Lorentzen, R. J., Ts'o, P. O. P. 1980. Role of superoxide in deoxyribonucleic acid strand scission. *Biochemistry* 19:3023–3028.

Li, Q., Altschuld, R. A., Stokes, B. T. 1987. Quantitation of intracellular free calcium in single adult cardiocytes by fura-2 fluorescence microscopy: calibration of fura-2 ratios. *Biochem. Biophys. Res. Commun.* 147:120–126.

Marnett, L. J. 1987. Peroxy free radicals: potential mediators of tumor initiation and promotion. *Carcinogenesis* 8:1365–1373.

McBrien, D. C. H., Slater, T. F., eds. 1982. *Free Radicals, Lipid Peroxidation, and Cancer.* New York: Academic Press.

McCay, P. B., King, M. M. 1980. Vitamin E: its role as a biologic free radical scavenger. In: *Vitamin E: A Comprehensive Treatise*, Machlin, L. J., ed., pp. 289–317. New York: Marcel Dekker.

McCay, P. B., Lai, E. K., Powell, S. R., Breuggemann, G. 1986. Vitamin E functions as an electron shuttle for glutathione-dependent "free radical reductase" activity in biological membranes (abstract). *Fed. Proc.* 45:451.

Meredith, M., Reed, D. J. 1982. Status of mitochondrial pool of glutathione in the isolated hepatocyte. *J. Biol. Chem.* 257:3747–3753.

Michiel, D. F., Wang, J. H. 1986. Calcium binding proteins. In: *Intracellular Calcium Regulation*, Bader, H., Gietzen, K., Rosenthal, J., eds., pp. 121–138. Manchester, England: Manchester University Press.

Moldéus, P., Jernström, B. 1983. Interaction of glutathione with reactive intermediates. In: *Functions of Glutathione. Biochemical, Physiological, Toxicological, and Clinical Aspects*, Larsson, S., Orrenius, S., Holmgren, A., Mannervik, B., eds., pp. 99–108. New York: Raven Press.

Nagata, C., Kodama, M., Ioki, Y., Kimura, T. 1982. Free radicals produced from chemical carcinogens and their significance in carcinogenesis. In: *Free Radicals and Cancer*, Floyd, R. A., ed., pp. 1–62. New York: Marcel Dekker.

Nakayama, T., Kodama, M., Nagata, C. 1984. Free radical formation in DNA by lipid peroxidation. *Agric. Biol. Chem.* 48:571–572.

Nicotera, P., Moore, M., Mirabelli, F., Bellomo, G., Orrenius, S. 1985. Inhibition of hepatocyte plasma membrane Ca^{2+}-ATPase activity by menadione metabolism and its restoration by thiols. *FEBS Lett.* 181:149–153.

Nicotera, P., Hartzell, P., Baldi, C., Svensson, S.-Å., Bellomo, G., Orrenius, S. 1986. Cystamine induces toxicity in hepatocytes through the elevation of cytosolic Ca^{2+} and the stimulation of a nonlysosomal proteolytic system. *J. Biol. Chem.* 261:14628–14635.

Niki, E., Tsuchiya, J., Tanimura, R., Kamiya, Y. 1982. Regeneration of vitamin E from α-chromanoxyl radical by glutathione and vitamin C. *Chem. Lett.* 1982:789–792.

Olafsdottir, K., Pascoe, G. A., Reed, D. J. 1988. Mitochondrial glutathione as a critical determinant in calcium ionophore-induced injury to isolated hepatocytes. *Arch. Biochem. Biophys.* 263:226–235.

Orrenius, S. 1985. Biochemical mechanisms of toxicity. *Trends Pharmaceut. Sci. FEST (Suppl.)* :1–4.

Pascoe, G. A., Reed, D. J. 1987. Relationship between cellular calcium and vitamin E metabolism during protection against cell injury. *Arch. Biochem. Biophys.* 253:287–296.

Pascoe, G. A., Reed, D. J. 1989. Cell calcium, vitamin E, and the thiol redox system in cytotoxicity. *Free Rad. Biol. Med.* 6:209–224.

Pascoe, G. A., Fariss, M. W., Olafsdottir, K., Reed, D. J. 1987a. A role of vitamin E in protection against cell injury. Maintenance of intracellular glutathione precursors and biosynthesis. *Eur. J. Biochem.* 166:241–247.

Pascoe, G. A., Olafsdottir, K., Reed, D. J. 1987b. Vitamin E protection against chemical-induced cell injury. I. Maintenance of cellular protein thiols as a cytoprotective mechanism. *Arch. Biochem. Biophys.* 256:150–158.

Pietronigro, D. D., Jones, W. B. G., Kalty, K., Demopoulos, H. B. 1977. Interaction of DNA and liposomes as a model of membrane-mediated DNA damage. *Nature (London)* 267:78–79.

Pryor, W. A. 1986. Oxy-radicals and related species: their formation, lifetimes, and reactions. *Annu. Rev. Physiol.* 48:657–667.

Rankin, B. B., McIntyre, T. M., Curthoys, N. P. 1983. Role of the kidney in the inter-organ metabolism of glutathione. In: *Functions of Glutathione. Biochemical, Physiological, Toxicological, and Clinical Aspects*, Larsson, D. D., Orrenius, S., Holmgren, A., Mannervik, B., eds., pp. 31–38. New York: Raven Press.

Reed, D. J., Fariss, M. W. 1984. Glutathione depletion and susceptibility. *Pharmacol. Rev.* 36:25S–33S.

Reed, D. J., Orrenius, S. 1977. The role of methionine in glutathione biosynthesis by isolated hepatocytes. *Biochem. Biophys. Res. Commun.* 77:1257–1264.

Reed, D. J., Pascoe, G. A., Olafsdottir, K. 1987. Some aspects of cell defense mechanisms of glutathione and vitamin E during cell injury. *Arch. Toxicol. (Suppl.)* 11:34–38.

Reynolds, E. S. 1963. Liver parenchymal cell injury. I. Initial alterations of the cell following poisoning with carbon tetrachloride. *J. Cell Biol.* 19:139–157.

Ross, D., Albano, E., Moldéus, P. 1985. The generation and fate of glutathionyl radicals. In: *Free Radicals in Liver Injury*, Poli, G., Cheeseman, K. H., Dianzani, M. U., Slater, T. F., eds., pp. 17–20. Oxford: IRL Press.

Roubal, W. T. 1970. Trapped radicals in dry lipid-protein systems undergoing oxidation. *J. Am. Oil Chem. Soc.* 47:141–144.

Sandy, M. S., Moldéus, P., Ross, D., Smith, M. T. 1986. Role of redox cycling and lipid peroxidation in bipyridyl herbicide cytotoxicity. *Biochem. Pharmacol.* 35:3095–3101.

Schaich, K. M., Karel, M. 1975. Free radicals in lysozyme reacted with peroxidizing methyl linoleate. *J. Food Sci.* 40:456–459.

Schaich, K. M., Karel, M. 1976. Free radical reactions of peroxidizing lipids with amino acids and proteins: an ESR study. *Lipids* 11:392–400.

Scherer, N. M., Deamer, D. W. 1986. Oxidative stress impairs the function of sarcoplasmic reticulum by oxidation of sulfhydryl groups in the Ca^{2+}-ATPase. *Arch. Biochem. Biophys.* 246:589–601.

Schulte-Frohlinde, D., Von Sonntag, C. 1985. Radiolysis of DNA and model systems in the presence of oxygen. In: *Oxidative Stress*, Sies, H., ed., pp. 11–40. London: Academic Press.

Sies, H. 1985. Oxidative stress: introductory remarks. In: *Oxidative Stress*, Sies, H., ed., pp. 1–8. London: Academic Press.

Sies, H., Cadenas, E. 1983. Biological basis of detoxification of oxygen free radicals. In: *Biological Basis of Disease*, Jakoby, W. B., ed., pp. 181–211. New York: Academic Press.

Slaga, T. J., Klein-Szanto, A. J. P., Triplett, L. L., Yotti, L. P., Trosko, J. E. 1981. Skin tumor promoting activity of benzoyl peroxide, a widely used free radical-generating compound. *Science* 213:1023–1025.

Smith, M. W., Collan, Y., Kahng, M. W., Trump, B. F. 1980. Changes in mitochondrial lipids of rat kidney during ischemia. *Biochim. Biophys. Acta* 618:192–201.

Thiers, R. E., Vallee, B. L. 1957. Distribution of metals in subcellular fractions. *J. Biol. Chem.* 226:911–920.

Thomas, C. E., Reed, D. J. 1988a. Effect of extracellular Ca^{2+} omission on isolated hepatocytes. I. Induction of oxidative stress and cell injury. *J. Pharmacol. Exp. Ther.* 245:493–500.

Thomas, C. E., Reed, D. J. 1988b. Effect of extracellular Ca^{2+} omission on isolated hepatocytes. II. Loss of mitochondrial membrane potential and protection by inhibitors of uniport Ca^{2+} transduction. *J. Pharmacol. Exp. Ther.* 245:501–507.

Tirmenstein, M. A., Reed, D. J. 1988. Characterization of glutathione-dependent inhibition of lipid peroxidation of isolated rat liver nuclei. *Arch. Biochem. Biophys.* 261:1–11.

Tribble, D. L., Aw, T. Y., Jones, D. P. 1987. The pathophysiological significance of lipid peroxidation in oxidative cell injury. *Hepatology* 7:377–387.

Troll, W., Wiesner, R. 1985. The role of oxygen radicals as a possible mechanism of tumor promotion. *Annu. Rev. Pharmacol. Toxicol.* 25:509–528.

Trump, B. F., Berezesky, I. K., Osornio-Vargas, A. R. 1981. Cell death and the disease process. The role of calcium. In: *Cell Death in Biology and Pathology*, Bowen, I. D., Lockshin, R. A., eds., pp. 209–242. New York: Chapman and Hall.

Trump, B. F., Berezesky, I. K., Collan, Y., Kahng, M. W., Mergner, W. J. 1976. Recent studies on the pathophysiology of ischemic cell injury. *Beitr. Pathol.* 158:363–388.

Trump, B. F., Berezesky, I. K., Sato, T., Laiho, K. U., Phelps, P. C., DeClaris, N. 1984. Cell calcium, cell injury, and cell death. *Environ. Health Perspect.* 57:281–287.

Van Kujik, F. J. G. M., Sevanian, A., Handelman, G. J., Dratz, E. A. 1987. A new role for phospholipase A2: protection of membranes from lipid peroxidation damage. *Trends Pharmacol. Sci.* 12:31–34.

Wang, Y.-M., Howell, S. K., Kimball, J. C., Tsai, C. C., Sato, J., Gleiser, C. A. 1982. Alpha-tocopherol as a potential modifier of Daunomycin carcinogenicity in Sprague-Dawley rats. In: *Molecular Interrelations of Nutrition and Cancer*, Arnott, M. S., van Eys, J., Wang, Y.-M., eds., pp. 369–379. New York: Raven Press.

Zeigler, R. 1985. Role of reversible oxidation-reduction of enzyme thiols-disulfides in metabolic regulation. *Annu. Rev. Biochem.* 54:305–329.

15 The Question of Cancer

RICHARD G. STEVENS
BARY W. WILSON
LARRY E. ANDERSON
Pacific Northwest Laboratory
Richland, Washington

CONTENTS

A conceptual framework for the pathogenesis of cancer must be constructed in order to evaluate possible mechanisms of cancer induction. This is not a straightforward task because the amount of observational and experimental data on cancer is vast. We will use the model depicted in Fig. 1 to illustrate our discussion of the ways in which agents such as ELF electric and magnetic fields might affect the risk of cancer.

The first tenet of our discussion is that cancer induction is a stochastic process. That is to say, it is probabilistic in nature; not every smoker gets lung cancer, not every nonsmoker avoids lung cancer. However, smokers are at greatly increased risk, all else being equal. On the crudest level, an epidemiological study can examine the association of an agent and development of cancer. If enough such evidence accumulates, as in the case of smoking and lung cancer, one can conclude that exposure increases risk. This information is useful in and of itself. More useful still would be knowledge of the biological mechanisms that relate exposure to cancer.

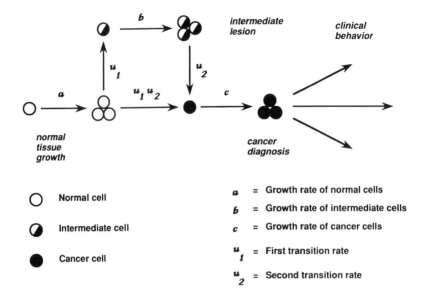

Fig. 1. The model for cancer incorporates growth characteristics of normal cells ("a"), intermediate cells ("b"), and cancer cells ("c"). It also incorporates transition (or "mutation") rates "u_1" and "u_2." Agents that affect growth of cells, transition rates, or clinical behavior will influence the risk of dying of cancer, although the biological mechanisms for these various effects may be very different.

At this time, there are insufficient data to judge whether or not ELF fields influence cancer risk. However, because the exposure is pervasive and is increasing rapidly as societies industrialize, it is important to design studies that can focus on any potential effect that may actually exist.

MODEL FOR CANCER

Cellular processes involved in changing from "normal healthy" to "dead of cancer" are represented in Fig. 1 as a series of stages. The model is based on our distillation of the literature on cancer biology, initiation-promotion, oncogenes, anti-oncogenes, and epidemiology. Much of the rationale for Fig. 1 comes from the work of Knudson (1985) and Moolgavkar and Knudson (1981). Two basic processes are important to the model: damage to DNA (e.g., mutation), and the growth characteristics of cells.

Many investigators believe that a clinically apparent cancer arises from the clonal expansion of a single transformed cell. Experimental evidence for this has been generated from a study of adult leukemia (Fialkow et al., 1987). In addition, a recent acceleration of research into the genetic basis of cancer has occurred (Bishop, 1983a). From a study of cancer-causing viruses (Bishop, 1983b), a class of genes called oncogenes were discovered. Oncogenes arise from proto-oncogenes as a result of initiation and/or over-production of the gene product (Hunter, 1984). Activation may result from such mechanisms as mutation and chromosomal rearrangements. It is possible that two or more different oncogene products are necessary for true malignant conversion of a normal cell (Land et al., 1983). A class of genes called tumor-suppressor genes (anti-oncogenes) was discovered even more recently (Harris, 1986); the lack of these genes or gene products results in cancer (Friend et al., 1986).

In addition to the central role of oncogenes in cancer induction, growth and progression of cells are also important (Slaga, 1983; Shubik, 1984; Vandenbark and Niedel, 1984). If transformation of a normal cell to a malignant cell were the result of two (or more) heritable changes in DNA, then a combination of DNA damage and enhanced growth of normal and intermediate cells would increase the chances of such an occurrence. Once a transformed cell has appeared, the chances that it will grow to kill the host may be affected by many factors, including nutrition and efficiency of immune surveillance.

A two-stage model has been developed to account for a genic etiology of cancer in general (Moolgavkar and Knudson, 1981). The model has been evaluated to some extent with regard to breast cancer (Moolgavkar et al., 1980), and provided the principal basis for Fig. 1. In this model a normal cell divides as part of a normal tissue. However, agents that stimulate cell proliferation increase the likelihood of cancer formation by increasing the number of normal cells at risk of genetic changes that lead to cancer. Thus, normal tissue can give rise directly to a cancer cell, or to an intermediate cell; factors affecting mutation rates in normal cells will affect these probabilities. Intermediate cells can also proliferate. Agents increasing the growth of intermediate cells (such as promoters) increase the number of intermediate cells and hence increase the risk that a cell will undergo an additional genetic event and become malignant. Finally, agents that increase proliferation of cancer cells will increase the chances for formation of a clinically apparent cancer. The virulence of this cancer may be affected by both intrinsic (e.g., immune surveillance) or extrinsic (e.g., nutrition) factors.

INITIATION-PROMOTION

Cancer research at the whole-animal level has been influenced more by the initiation-promotion paradigm than by any other concept. Initiation is believed to be irreversible, whereas promotion has been viewed as reversible. The prototypic experiment in this area used urethane to "initiate" the neoplastic process, and croton oil to "promote" its subsequent manifestation; mouse skin painted with a single dose of urethane followed by repeated applications of croton oil yielded many benign papillomas (Fig. 2A). Cessation of croton oil application resulted in regression of most of the papillomas. A high dose of initiator, however, yielded malignant squamous cell carcinomas (Fig. 2B). Resulting papillomas differed from those arising after exposure to low-dose initiator followed by promoter.

Papillomas from exposure to high-dose initiator arise from many normal cells (Reddy and Fialkow, 1983), whereas papillomas from low-dose initiator arise from one or a few normal cells. In other words, papillomas from high-dose initiator are composed of a much larger number of previously normal cells. This difference in the type of tumor produced by initiator alone compared to that arising from initiator followed by promoter spurred interest in the idea of an "intermediate lesion."

According to the intermediate-lesion hypothesis, two genetic events are necessary for conversion of a normal cell to a malignant cell. Thus, if papillomas represent clones of cells in which only one of the two events has occurred, then an additional application of a small amount of initiator would convert these papillomas to carcinomas with high efficiency (Fig. 2C). This hypothesis was confirmed in experiments by Hennings et al. (1983).

It is now believed that "initiators" are mutagens and capable of inducing complete carcinogenesis if the dose is sufficient (Iversen, 1984). Evidence suggests that the effect of "promoters" results, at least in part, from stimulation of cell proliferation. The promoting factor in croton oil has been shown to be phorbol esters (in particular, 12-*O*-tetradecanoyl-phorbol-13-acetate, TPA). In addition, TPA influences calcium mobilization and

A. Initiation-Promotion

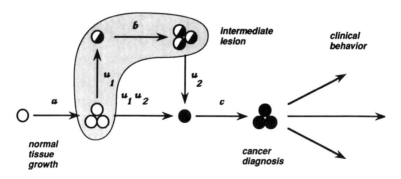

Fig. 2A. The "initiation-promotion" experiment can be used to test agents either for effects on transition rates or for effects on growth of intermediate lesions. This experimental protocol is widely used to screen for putative cancer risk factors.

B. Complete Carcinogenesis

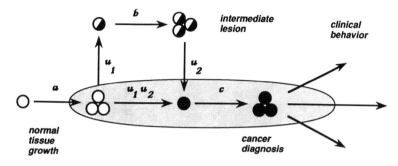

Fig. 2B. An agent that causes the transformation of a normal tissue directly into a cancer tissue is termed a "complete carcinogen." The primary mechanism of action of complete carcinogens is probably by increasing mutation rates, although benign lesions are also often observed in such experiments.

C. Initiation-Promotion-Initiation

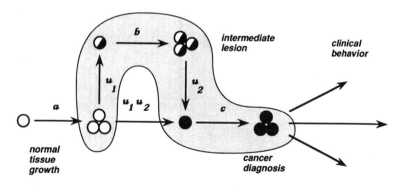

Fig. 2C. The "initiation-promotion-initiation" experiment was designed as a test of the hypothesis that the benign tumors that result from initiation-promotion represent "intermediate lesions" that are heterozygous for an "anti-oncogene." The prediction was made that application of a small amount of "initiator" would transform benign lesions to malignant lesions with high efficiency.

protein kinase activity, providing another possible basis for ELF effects in tumor promotion (see Chapter 10).

POSSIBLE MECHANISMS FOR ELF FIELD EFFECTS

Effects of ELF fields on biological systems may be manifested at the atomic and molecular level, as well as at the levels of cell, organ, and the whole organism. At the atomic level, primary interest now centers around the effects of ELF fields on the motions of biologically important ions such as calcium, magnesium, and lithium (see Chapters 9, 10, 11, and 14). At the molecular level, several workers have studied the effects of ELF fields on isolated DNA, or DNA in cells growing in culture. Most of these studies have not demonstrated effects of ELF fields. However, Goodman et al. (1987) have shown some influence of ELF magnetic fields on the transcription of RNA. If transcription is affected, then the probability of transformation may be increased by enhanced production of oncogene product.

A number of studies have examined the question of ELF effects of cellular proliferation and growth. These studies have not, for the most part, demonstrated field effects in normal cells (Anderson, 1989). In laboratories investigating clonal cells, however, increased growth has been observed to occur with ELF field exposure (Phillips and Winters, 1984). Other work has focused on possible ELF field-induced changes in cell membrane function, including effects on gap-junction formation, and cell-to-cell communication (see Chapter 10).

At the level of the intact animal, a number of laboratories have provided evidence of effects on neuronal and neuroendocrine function. In humans, neuronal and neuroendocrine effects include changes in heartbeat (cardiac interbeat interval) and alterations in urinary excretion of 6-hydroxymelatonin sulfate, a principal metabolite of the pineal hormone melatonin. This work has established the nervous system as a salient mediator of ELF effects in vertebrates, and the results obtained are consistent with the observation of early investigators who described effects of ELF fields in terms of "induced" stress.

A number of ELF field-induced changes in biological systems may be relevant to the model that we have presented regarding the pathogenesis of cancer. The discussion that follows is speculative, and is intended to suggest areas in which additional knowledge may be helpful in determining the role that ELF fields might play in cancer risk.

Growth of normal cells depends, to some extent, on their ability to recognize, and perhaps communicate with, neighboring cells. Thus, ELF field-induced changes that might alter recognition or communication may lead to increased proliferation. In general, this effect may relate to any of the growth rates depicted in the model. These ELF-induced effects may be of particular significance if transformed cells were stimulated to a greater extent than were normal cells. As seen in Chapter 10, changes in cell membrane function can be demonstrated using in vitro models. It remains to be determined if such changes occur in vivo. However, preliminary experiments in our laboratory designed to demonstrate possible ELF magnetic-field effects on memory in rats indicated that calcium efflux across cell membranes may be affected by 60-Hz magnetic fields.

To date, no convincing laboratory evidence has been obtained indicating that ELF fields cause damage to DNA. Therefore, cancer-induction experiments dependent on DNA damage would not seem well advised. However, a recent report offers evidence that ELF exposure can increase micronuclei formation in mouse polychromatic erythrocytes, suggesting possible chromosomal loss (Nahas and Oraby, 1989). Effects on calcium may increase oxidative stress to cells (see Chapter 14), and tumor promoters have been found to increase oxidative stress as well. Thus, experiments designed specifically to identify possible tumor-promoter activity such as initiation-promotion are suggested (Fig. 2A) (see Chapter 10).

Pineal function might be linked to the etiology of cancer in several ways (see Chapters 8 and 13). First, melatonin itself is oncostatic and can inhibit the growth of several cancer cell lines in vitro. Second, melatonin affects the function of the hypothalamic–pituitary–gonadal axis and may thereby affect the availability of hormones required for the growth of hormone-dependent breast, ovarian, and prostate cancers. Changes in melatonin secretion may affect the production of other hormones such as estrogen and prolactin and alter the proliferation of normal, intermediate, or malignant cells.

Experiments using ELF field exposures before administration of a known chemical carcinogen could test for effects on normal cell growth. Effects on growth characteristics of normal stem cells would imply that the exposures affecting risk of cancer would have to occur before the carcinogenic process itself, perhaps 20 or 30 years before the appearance of a cancer in a human. Alternatively, because melatonin has been shown to be directly mitostatic to certain cancer cell lines, an effect of ELF fields on melatonin might mean that exposures in the recent past are important. Such an effect would suggest that a tumor transplantation experiment (Fig. 2D) may provide important information regarding ELF fields and cancer. In such experiments, ELF exposure might enhance the growth of cancers in animals inoculated with malignant cells or tissues by suppressing melatonin production. A more general approach to this question might be a co-carcinogen study (Fig. 2E) in which mortality is examined as a consequence of life-span ELF exposure in conjunction with a known carcinogen.

It is the responsibility of the immune system to recognize and destroy cells that have been transformed. Thus, at some point along the line marked "c" in the diagrams, which depicts growth of a cancerous cell or cell colony, immune surveillance becomes of critical importance in protecting the

D. Transplantation

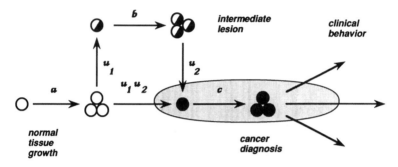

Fig. 2D. In a "transplantation" experiment, cancer cells are implanted into a normal host animal. An agent that affects cancer cell growth will result in early death of the host animal and/or increased yield of cancer cells after a specified period of time.

E. Co-Carcinogen

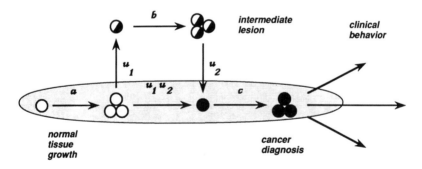

Fig. 2E. A "co-carcinogen" may augment the action of an "initiator" or a "promoter" at any point along the pathway from normal to malignant.

organism. ELF-induced changes in immune system function, although not yet demonstrated directly, might be inferred from results of DMBA-induced breast cancer studies wherein there were more tumors per tumor-bearing animal in ELF electric-field-exposed rats as compared to DMBA-induced controls. As discussed in Chapter 8, the melatonin circadian rhythm, in terms of either amplitude or phase, appears to affect the immune system function. Hence, ELF-induced changes in melatonin rhythms may result in impaired immune function.

In this summarizing chapter, we have attempted to discuss recent findings regarding biological effects of ELF electric and magnetic fields as they relate to a general model of cancer pathogenesis. It is not our intent to argue that ELF exposure increases cancer risk; rather, we wish to suggest areas wherein future experiments may be carried out. Whether or not ELF electric and magnetic fields contribute to increased cancer risk, it is important to conduct scientific studies that will reduce the uncertainty currently associated with the question of cancer.

REFERENCES

Anderson, L. E. 1989. Biological effects of ELF and 60-Hz fields. In: *Biological Effects and Medical Applications of Electromagnetic Fields*. Englewood Cliffs, New Jersey: Prentice Hall (*in press*).

Bishop, J. M. 1983a. Cancer genes come of age. *Cell* 32:1018–1020.

Bishop, J. M. 1983b. Cellular oncogenes and retroviruses. *Annu. Rev. Biochem.* 52:301–354.

Fialkow, P. J., Singer, J. W., Raskind, W. H., et al. 1987. Clonal development, stem-cell differentiation, and clinical remissions in acute nonlymphocytic leukemia. *N. Engl. J. Med.* 317:468–473.

Friend, S. H., Bernards, R., Rogelj, S., et al. 1986. A human DNA segment with properties of the gene that predisposes to retinoblastoma and osteosarcoma. *Nature (London)* 323:643–646.

Goodman, R., Abbott, J., Henderson, A. S. 1987. Transcriptional patterns in the X chromosome of *Sciara coprophila* following exposure to magnetic fields. *Bioelectromagnetics* 8:1–7.

Harris, H. 1986. Malignant tumors generated by recessive mutations. *Nature (London)* 323:582–583.

Hennings, H., Shores, R., Wenk, M. L., et al. 1983. Malignant conversion of mouse skin tumors is increased by tumour initiators and unaffected by tumour promoters. *Nature (London)* 304:67–69.

Hunter, T. 1984. Oncogenes and proto-oncogenes: how do they differ? *J. Natl. Cancer Inst.* 73:773–786.

Iversen, O. H. 1984. Urethane (ethyl carbamate) alone is carcinogenic for mouse skin. *Carcinogenesis* 5:911–915.

Knudson, A. G. 1985. Hereditary cancer, oncogenes, and antioncogenes. *Cancer Res.* 45:1437–1443.

Land, H., Parada, L. F., Weinberg, R. A. 1983. Tumorigenic conversion of primary embryo fibroblasts requires at least two cooperating oncogenes. *Nature (London)* 304:596–606.

Moolgavkar, S. H., Knudson, A. G. 1981. Mutation and cancer: a model for human carcinogenesis. *J. Natl. Cancer Inst.* 66:1037–1052.

Moolgavkar, S. H., Day, N. E., Stevens, R. G. 1980. Two-stage model for carcinogenesis: epidemiology of breast cancer in females. *J. Natl. Cancer Inst.* 65:559–569.

Nahas, S. M., Oraby, H. A. 1989. Micronuclei formation in somatic cells of mice exposed to 50-Hz electric fields. *Environ. Mol. Mutagen.* 13:107–111..

Phillips, J. L., Winters, W. S. 1984. Electromagnetic-field-induced bioeffects in human cells in vitro. In: *Interaction of Biological Systems with Static and ELF Electric and Magnetic Fields*, Anderson, L. E., Kelman, B. J., Weigel, R. J., eds., pp. 279–286. Proceedings of the 23rd Hanford Life Sciences Symposium, Richland, Washington. CONF-830951. Springfield, Virginia: National Technical Information Service.

Reddy, A. L., Fialkow, P. J. 1983. Papillomas induced by initiation-promotion differ from those induced by carcinogen alone. *Nature (London)* 304:69–71.

Shubik, P. 1984. Progression and promotion. *J. Natl. Cancer Inst.* 73:1005–1011.

Slaga, T. J. 1983. Overview of tumor promotion in animals. *Environ. Health Perspect.* 50:3–14.

Vandenbark, G. R., Niedel, J. E. 1984. Phorbol diesters and cellular differentiation. *J. Natl. Cancer Inst.* 73:1013–1019.

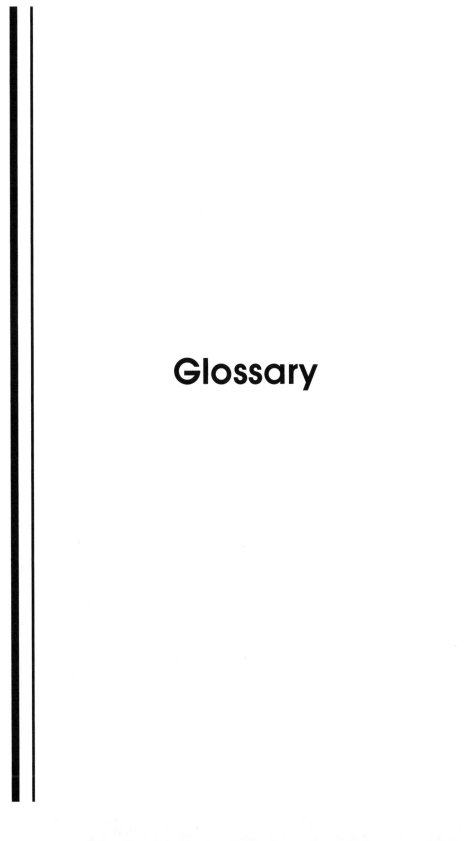

Glossary

Glossary

UNIT ABBREVIATIONS

AC	alternating current
DC	direct current
AM	amplitude-modulated
FM	frequency-modulated
amu	atomic mass unit (u)
Å	angstrom
A	amperes, electrical current
nm	nanometer (1 nm = 10 Å; wavelength)
W	watt, power [1 joule (J) per second]
V	volt, electric potential
C	coulomb, electric charge
Ω	ohm, electrical resistance
f	frequency, cycles per unit time
E	electric field
emf	electromotive force
S/m	Siemens/meter, electrical conductivity
G	gauss, magnetic induction; 1 T (tesla, SI unit) = 0.1 G
Hz	Hertz, cycles per second
B	magnetic field
H	magnetic-field strength
$\mu W/cm^2$	microwatts per centimeter squared, unit of irradiance
M	molar (molal); mol, mole
p	probability
kdalton	molecular weight unit equal to 1000 daltons, nominally = 1000 amu
time	sec, second; min, minute; hr, hour
Vpp/m	volts peak-to-peak per meter in air
V/m	volts root-mean-square (rms) per meter (electric-field strength)

ACRONYMS

AChE	acetylcholinesterase
ACTH	adrenocorticotropin
CNS	central nervous system
cAMP	cyclic adenosine monophosphate
ELF	extremely low frequency
EEG	electroencephalograph
EM	electromagnetic field
GC/MS	gas chromatography/mass spectroscopy
IMP	intramembranous particles
ICR	ion cyclotron resonance
5-MTOL	5-methoxytryptophol
NAT	*N*-acetyltransferase
NADPH	nicotinamide adenine dinucleotide phosphate (reduced)
NIEMR	nonionizing electromagnetic radiation
ODC	ornithine decarboxylase
6-OHMS	6-hydroxymelatonin sulfate
PKC	protein kinase c
PTH	parathyroid hormone
RF	radio frequency
RIA	radioimmune assay
VHF	very high frequency
TPA	tetradecanoyl phorbol acetate

Index

Index

Date Due